Yuri V. Nesterenko   Patrice

Introduction to Algebraic Independence Theory

With contributions from:

F. Amoroso, D. Bertrand, W.D. Brownawell,
G. Diaz, M. Laurent, Yu.V. Nesterenko,
K. Nishioka, P. Philippon, G. Rémond,
D. Roy, M. Waldschmidt

Springer

Editors

Yuri V. Nesterenko
Faculty of Mechanics and Mathematics
Moscow University
119899 Moscow, Russia

E-mail: nest@trans.math.msu.su

Patrice Philippon
Institut de Mathématiques de Jussieu
UMR 7586 du CNRS
4, place Jussieu
75252 Paris Cedex 05, France

E-mail: pph@math.jussieu.fr

Cataloging-in-Publication Data applied for

Die Deutsche Bibliothek - CIP-Einheitsaufnahme

Introduction to algebraic independence theory / Yuri V. Nesterenko ;
Patrice Philippon (ed.). With contributuions from: F. Amoroso .... -
Berlin ; Heidelberg ; New York ; Barcelona ; Hong Kong ; London ;
Milan ; Paris ; Singapore ; Tokyo : Springer, 2001
 (Lecture notes in mathematics ; 1752)
 ISBN 3-540-41496-7

Mathematics Subject Classification (2000): 11J91, 11J85, 11J81, 11J89, 11G05,
12H20, 13F20, 14G40, 14G35, 14L10, 30D15, 33B15, 33E05, 34M35, 34M25

ISSN 0075-8434
ISBN 3-540-41496-7 Springer-Verlag Berlin Heidelberg New York

Springer-Verlag Berlin Heidelberg New York
a member of BertelsmannSpringer Science+Business Media GmbH

© Springer-Verlag Berlin Heidelberg 2001
Printed in Germany

Typesetting: Camera-ready T$_E$X output by the authors
SPIN: 10759897   41/3142-543210 - Printed on acid-free paper

Lecture Notes in Mathematics 1752

Editors:
J.-M. Morel, Cachan
F. Takens, Groningen
B. Teissier, Paris

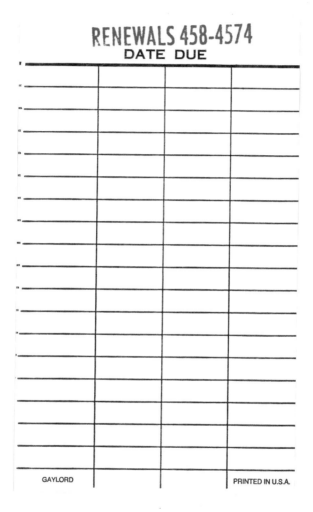

Springer
Berlin
Heidelberg
New York
Barcelona
Hong Kong
London
Milan
Paris
Singapore
Tokyo

Preface

In the last five years there was an essential progress in the development of the theory of transcendental numbers. A new approach to the arithmetic properties of values of modular forms and theta-functions was found. First and up to a recent time there was a unique result of this type established in 1941 by Th. Schneider. It states that the modular function $j(\tau)$ has transcendental values at any algebraic point τ of the complex upper-half plane, distinct from the imaginary quadratic irrationalities. It is well known that the value of the modular function at any imaginary quadratic argument τ with $\text{Im}\tau > 0$, is an algebraic number.

In 1995, K. Barré-Sirieix, G. Diaz, F. Gramain and G. Philibert proved that the value of the function $J(z)$, connected to $j(\tau)$ by the relation $j(\tau) = J(e^{2\pi i\tau})$, at any non-zero algebraic point z of the unit disk is transcendental. In 1969, this assertion was proposed by K. Mahler as a conjecture. The proof rests heavily on the functional modular equations connecting the functions $J(z)$ and $J(z^n)$ over $\mathbf{C}$ for any natural number n. Simultaneously the p-adic analogue of this theorem (conjecture formulated by Yu. Manin, 1971) was proved and it has important consequences in the theory of algebraic numbers.

The solution of the Mahler-Manin's problem was an impulse for further intensive researches. In 1996 Yu. Nesterenko generalized this result, he proved the expected lower bound for the transcendence degree of the field generated over $\mathbf{Q}$ by the values of the Eisenstein series $E_2(\tau)$, $E_4(\tau)$, $E_6(\tau)$ and the exponential function $e^{2\pi i\tau}$. The fact that Eisenstein series satisfy an algebraic system of differential equations is fundamental to the proof. In particular, an effective bound for the multiplicities of zeros for polynomials in the functions involved played an important rôle.

This result of algebraic independence has a number of remarkable consequences. First it implies that the three numbers π, e^π, $\Gamma(\frac{1}{4})$ are algebraically independent (over $\mathbf{Q}$) and, in particular, the two numbers π and e^π are algebraically independent. This last assertion was an old problem in transcendence theory. The theorem implies an even more general assertion : for any natural number D the real numbers π and $e^{\pi\sqrt{D}}$ are algebraically independent. One more consequence is the algebraic independence of the numbers $J(q)$, $J'(q)$ and $J''(q)$ for any algebraic number q satisfying $0 < |q| < 1$. D. Bertrand and independently D. Duverney, Ke. Nishioka, Ku. Nishioka, I. Shiokawa (DNNS) deduced results on algebraic independence of the values of theta-functions at algebraic points and in particular derived the transcendence of the sums $\sum_{n=1}^\infty q^{n^2}$ for any algebraic q satisfying $0 < |q| < 1$. J. Liouville introduced in 1851 this series with $q = \ell^{-1}$ for $\ell \in \mathbf{Z}$, $\ell > 1$ as an example for which his method only proved irrationality. No assertion on transcendence of these sums was known. DNNS have also proved transcendence of the values of the Dedekind eta-function $\eta(q)$ at algebraic points q, $0 < |q| < 1$ and several other interesting corollaries.

P. Philippon introduced a general class of K-functions containing in particular the functions $E_{2k}(\frac{\log z}{2\pi i})$, and he proposed a general approach to study the algebraic independence of the values of these functions. In this way, new results on the algebraic independence of the values of the so-called Mahler's functions at transcendental arguments were obtained. He also found, that the algebraic independence of the numbers π and $e^{\pi\sqrt{D}}$ can be shown using an older measure of algebraic independence of couples of numbers, connected to periods and quasi-periods of elliptic functions with complex multiplications.

It should be stressed, that the above mentioned results were established following a general approach for proofs of algebraic independence of numbers developped in the seventies and eighties. This method uses tools from commutative algebra for establishing zeros and multiplicities estimates (Yu. Nesterenko, W.D. Brownawell, D.W. Masser, G. Wüstholz, P. Philippon) and elimination theory in the form of a criterion for algebraic independence of P. Philippon. It had already been successfully applied before to the study of the values of functions satisfying addition theorems (exponential, elliptic, abelian functions), linear differential equations over $\mathbf{C}(z)$ (the so-called E-functions of Siegel) or special functional equations introduced by K. Mahler (which find applications in the study of dynamical systems).

An instructional conference on algebraic independence was held from September 29 till October 3, 1997 in Luminy (France) co-organized by M. Waldschmidt, R. Tijdeman and Yu. Nesterenko. The aim of this course was to enable graduate students and post-docs to get an acquaintance with the new methods and results in transcendence theory. The lectures delivered on this occasion made the basis for the present book. Parts of this book has been written with the help of the grant INTAS-RFBR, IR97-1904.

This book is ideally divided into four unequal parts.

The first part, consisting of the first four chapters, presents the latest results in transcendence and algebraic independence theory obtained for modular functions and modular forms. It serves as an introduction to and a motivation for reading the three other parts. Chapter 1 establishes the differential equations satisfied by Eisenstein series or modular theta functions as well as Masser's period relations, which play a fundamental rôle in chapter 3 (and 4). It also gives results and conjectures on the values of theta functions and their derivatives. Chapter 2 gives a proof of the transcendance of the values of the modular invariant function J, which has been seminal for the generalisation presented in chapter 3. It also discusses the quantitative aspect of the question with measures of transcendence and conjectures on the values of modular and exponential functions overlapping with those of chapter 1. Chapter 3 gives the proof of the algebraic independence of the values of modular forms and details several remarkable corollaries. It rests on the preliminaries of chapter 1 as well as a fundamental zeros estimate, the proof of which is postponed until chapter 10. Chapter 4 sums up the tools involved in this proof and shows a short-cut to the proof of algebraic independence of π, e^π and $\Gamma(1/4)$ (resp. π, e^π and $\Gamma(1/3)$) using a known measure of algebraic independence of π

and $\Gamma(1/4)$ (resp. $\Gamma(1/3)$) obtained with the help of elliptic functions, rather than modular ones.

The second part consists of chapters 5 to 9, it develops the necessary tools from commutative algebra already presented in chapter 3. Chapter 5 establishes the basic facts from elimination theory which enable one to attach a form to a variety. Its originality is in its multi-homogeneous setting which is new in this context. After proving the elimination theorem and the principality of the eliminating ideal it shows the link with the geometry of multi-projective spaces and studies the so-called resulting form, the use of which is crucial in this multi-homogeneous setting, for example for specialisations. Chapters 6 and 7 introduce, in the homogeneous and multi-homogeneous case respectively, the notions and results from diophantine geometry (height, distance, Bézout theorems, ...) which follow from the study of eliminating and resulting forms. The refined definition of height adopted here is inspired from the calculus in Arakelov theory. In the multi-homogeneous case, the notion of distance from a point to a variety is shown to be more subtle than in the homogeneous case (chapter 7, § 4). Chapter 8 derives from the previous machinery several criteria for algebraic independence which are used in proofs of algebraic independence. In particular, these results include a multi-projective criteria which allows one to take advantage of the different behaviors of numbers in a proof of algebraic independence. One will find another criterion in chapter 13, § 4, involving multiplicities. Finally, we have included (chapter 9) in this part an upper bound for the classical Hilbert function of an homogeneous (or bi-homogeneous) polynomial ideal whose proof rests on a clever use of Bertini's theorem for reducing to the case of dimension 1 which is easy.

The third part (chapters 10 and 11) deals with zeros estimates. Chapter 10 is devoted to the multiplicity estimate used in the proof of the main result of chapter 3, more generally the results here address solutions of algebraic differential equations which satisfy a special property, called D-property. The proof of the general estimate depends on the upper bound for Hilbert functions of ideals given in chapter 9 and then it takes a careful study of the differential system satisfied by the Eisenstein series considered in chapter 3 to show that the D-property is valid for them. Chapter 11 establishes zeros estimate on group varieties which are necessary for the proofs of algebraic independence in chapter 13 and 14. Here one has to control the multiplicities of polynomials at points of a finite type subgroup of the ambiant algebraic group. The degeneracies depend on the distribution of the points with respect to algebraic subgroups and one exhibits a so-called obstructing subgroup which in a way gives the worst degeneracy. The proof, based essentially on a control of degree and multiplicities in intersection by suitable hypersurfaces is presented here in a geometric setting. Yet another multiplicity estimate is to be found in chapter 12.

The fourth part (chapters 12 to 16) is concerned with applications of the tools developed so far. Chapter 12 shows measures of algebraic independence of values of Mahler type functions in one variable (that is satisfying a functional equation for exponentiation of the variable) as well as the zeros estimate for this type of functions. It should be noted that the quality of the measure obtained depends on the precise type of functional equation involved. Chapters 13 and 14 address algebraic independence of two and more numbers respectively in commutative algebraic groups, especially the values of the exponential and elliptic functions. In particular, chapter 13 presents a proof of algebraic independence of two numbers through interpolation determinants which rests on a generalisation of Gelfond's criterion using multiplicities, not covered in chapter 8. Chapter 14 deals with large transcendence degrees of the values of the exponential function. It presents several conjectures in one and several analytic variables and makes some historical comments discussing in particular the so-called technical hypothesis. It gives the proof of the almost best known result towards Gelfond-Schneider's conjecture. Chapter 15 is devoted to metric results, it shows that almost all families of m complex numbers have an order of measure of algebraic independence equal to $m + 1$. The proof uses a kind of "transfer theorem" between measures of algebraic independence and measure of approximation by algebraic numbers. It should be noted that this differs from the approximation properties introduced in chapter 4 only in that here one assumes *a priori* that the numbers are close enough to some variety in order to find good approximations by algebraic numbers. Finally, chapter 16 presents an alternative approach to algebraic independence through effective versions of the Nullstellensatz, which is more direct and enables one to weaken the technical hypothesis. Actually, further developments are mentioned such as Lojasievitch inequalities and algorithmic methods which are relevant for the interpolation problems underlying algebraic independence theory in many ways.

The reader will find thereafter the list of contributors which includes their addresses and the chapters they are responsible for. Chapters authors names are also repeated at the bottom of the first page of each chapter. Statements are numbered afresh in each chapter according to their section of occurrence, and when they are referred to from a different chapter an indication of the chapter number is added. On a different footing, formulas are numbered at a stretch throughout the book. There is a dual system of bibliographical references, one concerns the references that we (rather arbitrarily) declared "in" the subject, the list of which may be found at the end of the book, the other deals with references that we considered "on the rim", and those appear as footnotes on the pages where they show up. However, if such a foot-reference appears twice or more in the same chapter the reader is then prompted to the page of its first occurrence. Finally, a short index is provided after the bibliography, at the very end of the book.

List of Contributors

F. Amoroso (chapter 15)
Département de Mathématiques
Université de Caen
Campus II, BP 5186
F-14032 CAEN Cedex
FRANCE
email : amoroso@math.unicaen.fr

D. Bertrand (chapters 1 & 9)
Institut de Math. de Jussieu
Théorie des Nombres
7 A 33 Chevaleret, case 247
F-75252 PARIS Cedex 05
FRANCE
email : bertrand@math.jussieu.fr

W.D. Brownawell (chapter 16)
Department of Mathematics
Penn State University
405 McAllister bldg
UNIVERSITY PARK, PA 16802
U.S.A
email : wdb@math.psu.edu

G. Diaz (chapter 2)
Équipe de Théorie des Nombres
Université de Saint-Étienne
23, rue du Dr. Paul Michelon
F-42023 SAINT-ETIENNE Cedex
FRANCE

M. Laurent (chapter 13)
Institut de Math. de Luminy
163 av. de Luminy
Case 930
F-13288 MARSEILLE Cedex 9
FRANCE
email : laurent@lmd.univ-mrs.fr

Yu.V. Nesterenko (chapters 3 & 10)
Faculty of Mech. and Math.
Moscow State University
119899 MOSCOW
RUSSIA
email : nest@trans.math.msu.su

Ku. Nishioka (chapter 12)
Department of Mathematics
Keio University
4-1-1 Hiyoshi Kohoku-ku
223 YOKOHAMA
JAPAN
email : nishioka@math.hc.keio.ac.jp

P. Philippon (chapters 4, 6 & 8)
Institut de Math. de Jussieu
Géométrie et Dynamique
8 D 9 Chevaleret, case 247
F-75252 PARIS Cedex 05
FRANCE
email : pph@math.jussieu.fr

G. Rémond (chapters 5 & 7)
Institut de Math. de Jussieu
Théorie des Nombres
Chevaleret, case 247
F-75252 PARIS Cedex 05
FRANCE
email : remond@clipper.ens.fr

D. Roy (chapter 11)
Université d'Ottawa
Département de Mathématiques
585, rue King Edouard
K1N 6N5 OTTAWA, ONTARIO
CANADA
email : droy@mathstat.uottawa.ca

M. Waldschmidt (chapter 14)
Institut de Math. de Jussieu
Théorie des Nombres
7 A 28 Chevaleret, case 247
F-75252 PARIS Cedex 05
FRANCE
email : miw@math.jussieu.fr

Contents

CHAPTER 1

$\Theta(\tau, z)$ and Transcendence

At inter affectus aequationum modularium id maxime
memorabile ac singulare videor animadvertere, quod eidem
omnes aequationi differentiali tertii ordinis satisfaciunt.
C. Jacobi, Fundamenta nova, §32 (1829)

The first two sections of this Chapter are devoted to the *differential* properties of modular forms on which Nesterenko's theorem on the values of Eisenstein series (see Chapter 3, Theorem 1.1 and [**Nes9**]) is based. The emphasis is on purely modular arguments, but we also recall how to establish them via elliptic functions. Similarly, Section 3 describes modular and elliptic proofs of the algebraic relations which connect their *singular values* (i.e. values at CM points), and thanks to which [**Nes9**] becomes a statement on the exponential and the gamma functions.

Cutting the ombilical chord which attaches elliptic functions to their modular offspring has often proved useful in the arithmetic study of the latter. Transcendence theory is no exception to this rule : its first modular results, such as Schneider's theorem on $\{\tau, j(\tau)\}$, were elliptic in nature, but the *cusp* at infinity plays a crucial role in both of its latest advances [**BDGP**], [**Nes9**], whose proofs can therefore not be expressed in purely elliptic terms. In the last section of the paper, we discuss a general conjecture on the values of the Jacobi-Riemann theta function $\theta(\tau, z)$, which covers most known results in this area, and where the elliptic *and* modular variables are treated on an equal footing.

1. Differential rings and modular forms

Let Γ be a congruence subgroup of $SL_2(\mathbf{Z})$ and let k be an even integer. A modular form of weight k relatively to Γ will here be a *meromorphic* function $f(\tau)$ on the upper half-space $\mathfrak{h}$ such that $f(\gamma \cdot \tau) = (c\tau + d)^k f(\tau)$ for any $\gamma = \left(\begin{smallmatrix} a & b \\ c & d \end{smallmatrix}\right)$ in Γ, and which admits a meromorphic continuation at the cusps of Γ; a modular form of weight $k = 0$ is called a *modular* function. We take $q = e^{2\pi i \tau / h}$ (where h is a positive integer) as local parameter at the cusp $i\infty$, and say that f is defined over the field $\mathbf{Q}^{\mathrm{alg}}$ of algebraic numbers if its Fourier expansion

$$f(\tau) = \sum_{n \geq \nu} a_n q^n \qquad (q = \exp(2\pi i \tau / h))$$

at $i\infty$ has algebraic coefficients. Derivatives will be taken with respect to the differential operator

$$D = (1/h)q(d/dq) = (1/2\pi i)d/d\tau .$$

(∗) Chapter's author : Daniel BERTRAND.

The derivative of a modular function is modular of weight 2. But if f is modular of weight $k \neq 0$, then (assuming for notational simplicity that $\tau \to -1/\tau$ occurs in Γ, as will be done throughout the paper):

$$(1) \qquad Df(-1/\tau) = \tau^{k+2} Df(\tau) + k\tau^{k+1} f(\tau)/2\pi i ,$$

and Df is not modular any more. We recover modular forms by differentiating a second time : since $(Df/f)(-1/\tau) = k\tau/2\pi i + \tau^2(Df/f)(\tau)$, we get :

$$
\begin{aligned}
D(Df/f)(-1/\tau) &= k\tau^2/(2\pi i)^2 + (2\tau^3/2\pi i)(Df/f)(\tau) + \tau^4 D(Df/f)(\tau) , \\
(Df/f)^2(-1/\tau) &= k^2\tau^2/(2\pi i)^2 + (2k\tau^3/2\pi i)(Df/f)(\tau) + \tau^4(Df/f)^2(\tau) .
\end{aligned}
$$

Hence $D(Df/f) - (1/k)(Df/f)^2 = D^2f/f - (1+1/k)(Df/f)^2$ is modular of weight 4. The following proposition easily follows from this fact : its first two statements are classical, while the third one is essentially due to Mahler ([**Mah1**]; see [**Nis1**] for a more general version, and for some historical comments).

PROPOSITION 1.1. *Let f be a modular form of weight k, defined over $\mathbf{Q}^{\mathrm{alg}}$. Then*

i) $f^{\#} = kf\, D^2f - (k+1)(Df)^2$ *is modular of weight $2k+4$;*

ii) f, Df, D^2f *and* D^3f *are algebraically dependent over $\mathbf{Q}$;*

iii) *if f is not constant, the functions f, Df and D^2f are algebraically independent over the function field $\mathbf{C}(q)$.*

PROOF. i) has essentially already been verified, and allows us to cook up the modular function $g = f^{\#k}/f^{2k+4}$. If $k \neq 0$ (resp. if $k = 0$) Dg (resp. $(Df)^{\#}$) is a modular form of weight 2 (resp. 8), whose expression as a rational function in f, Df, D^2f and D^3f involves D^3f. Since any three modular forms are algebraically dependent (indeed, the same holds true for any two modular functions, which may be viewed as rational functions on an algebraic curve), we deduce that f satisfies an algebraic differential equation of the third order with constant coefficients, as stated in ii).

iii) We first claim that the modular forms f and $f^{\#}$ are algebraically independent over $\mathbf{C}$. Otherwise, a standard weight argument shows that the modular function $f^{\#k}/f^{2k+4}$ is constant; now :

- if f is a unit at $i\infty$, $f^{\#}$ is not, and must therefore vanish identically; since Df and D^2f have the same order at $i\infty$, we derive a contradiction unless Df vanishes identically;
- if the order ν of f at $i\infty$ is not 0, that of $f^{\#k}$ is equal to $2\nu k$: indeed, the "first" term of its Fourier expansion is given by $\left(k(\frac{\nu}{h})^2 - (k+1)(\frac{\nu}{h})^2\right) a_{\nu}{}^2$, which is not 0. Since $2\nu k \neq (2k+4)\nu$, we again derive a contradiction.

We now prove that if $k \neq 0$ (resp. if $k = 0$), Df/f (resp. $D(Df)/Df$) is transcendental over $\mathbf{C}(f, f^{\#})$. Indeed, this function is h-periodic, and after multiplication by a constant factor, satisfy a relation of the type (cf. (1) :

$$F(-1/\tau) = \tau + \tau^2 F(\tau) .$$

Now, assume that such a function F is algebraic over the ring generated by two algebraically independent modular forms f_1, f_2, and let P be its minimimal polynomial. Expanding $P\left(f_1(-1/\tau), f_2(-1/\tau), F(-1/\tau)\right) = 0$ as a polynomial in τ and looking at its highest degree N in τ, we deduce from the transcendency of the function τ over the field of h-periodic functions that the degrees (i_1, i_2, i_3) of the monomials occuring in P must simultaneously satisfy : $k_1 i_1 + k_2 i_2 + 2i_3 = N$

and $k_1 i_1 + k_2 i_2 + 2i_3 - 1 = N$, unless i_3 always vanishes, in which case P is a relation between f_1 and f_2. We have thus proved that f, Df, $D^2 f$ are algebraically independent over $\mathbf{C}$.

It remains to check that q is transcendental over the differential field $M = \mathbf{C}\left(D^i f;\ i = 0, 1, \dots\right)$ generated by these functions. Assume on the contrary that $[M(q) : M] = d$ is finite. Since M contains modular functions and is stable under D, we deduce from the irreducibility of the modular equations that for infinitely many prime numbers p, the field $M_p = M\left(D^i f(\tau/p);\ i = 0, 1, \dots\right)$ has degree $p+1$ over M, while $q^{1/p}$ has degree d over M_p, and degree p over $M(q)$, so that $dp = [M(q^{1/p}) : M]$ would divide $d(p+1) = \left[M_p\left(q^{1/p}\right) : M\right]$. This final contradiction concludes the proof of Proposition 1.1. □

Copying the specialization argument of [**DNNS1**], we deduce from Theorem 1.1 from Chapter 3 and Proposition 1.1.iii) :

COROLLARY 1.2. *Let f be a non constant meromorphic modular form, defined over $\mathbf{Q}^{\mathrm{alg}}$. For all $\alpha \in \mathfrak{h}$, $\alpha \notin \{\text{poles of } f\}$, such that $e^{2\pi i \alpha} \in \mathbf{Q}^{\mathrm{alg}}$, the numbers $f(\alpha)$, $Df(\alpha)$ and $D^2 f(\alpha)$ are algebraically independent over $\mathbf{Q}$.*

REMARK. i) All the functions occuring in Proposition 1.1.iii) are h-periodic. We may therefore add τ itself (as in [**Nis1**]) to this list of algebraically independent functions. See Conjecture 1.11 in Chapter 3 for a related version of the above corollary.

ii) The automorphic functions considered in [**Mah1**] and [**Nis1**] are relative to general Fuchsian groups, and their proofs of algebraic independence solely appeals to the functional equations. The first step in our argument requires the existence of a cusp, and is closer to Chapter 3, Theorem 3, which is in fact an effective version of Proposition 1.1.iii).

2. Explicit differential equations

The above proof (Proposition 1.1.ii)) of the differential equation for f reduces its actual computation to effective versions of the Riemann-Roch theorem on the modular curve $\mathfrak{h}^*/\Gamma$. The first example below illustrates this *modular* method. We then describe an *elliptic* method, and conclude with the oldest one, which is *cohomological*. For further relations with the Gauss-Manin connection, see [**Kat**][1].

2.1. Eisenstein series. For all integers $k \geq 1$, let M_k be the $\mathbf{C}$-vector space of holomorphic modular forms of weight k for $SL_2(\mathbf{Z})$. As soon as $k > 1$,

$$E_{2k}(\tau) = 1 + (-1)^k \frac{4k}{B_k} \sum_{n \geq 1} \left(\sum_{d|n} d^{2k-1} \right) q^n \qquad (q = e^{2\pi i \tau}),$$

where B_k is the k-th Bernoulli number ($B_1 = 1/6$, $B_2 = 1/30$, $B_3 = 1/42, \dots$), belongs to M_{2k}, and $\Delta = \left(E_4^3 - E_6^2\right)/1728$ to M_{12}. In view of (1), $E_2 = D\Delta/\Delta$ satisfies : $E_2(-1/\tau) = \tau^2 E_2(\tau) + 12\tau/2\pi i$, so that the differential operator

$$D_k = D - kE_2/12$$

---

[1][**Kat**] K. Katz. *p-adic interpolation of real analytic Eisenstein series.* Ann. of Math., 104, (1976), 459–571. See also : Modular functions III, *Lecture Notes in Math.* 350, Springer, (1973), 158–180.

sends M_k into M_{k+2}. Known bases of M_4, M_6, M_8 then yield Ramanujan's relations : $DE_4 - E_2E_4/3 = -E_6/3$, $DE_6 - E_2E_6/2 = -E_4^2/2$, and (in view of Proposition 1.1.i) : $DE_2 - E_2^2/12 = -E_4/12$.

2.2. Modular theta functions. These are the Nullwerte of the elliptic theta functions

$$\theta(\tau, z) := \theta_{0,0}(\tau, z) = 1 + 2 \sum_{n \geq 1} e^{\pi i n^2 \tau} \cos(2\pi n z),$$

$$\theta_{1,0}(\tau, z) = q^{1/4} e^{\pi i z} \theta(\tau, z + \tau/2), \quad \theta_{0,1}(\tau, z) = \theta(\tau, z + 1/2),$$

i.e. in the usual notations : $\theta_2(q) := \theta_{1,0}(\tau, 0)$, $\theta_4(q) := \theta_{0,1}(\tau, 0)$ and

$$\theta_3(q) := \theta_{0,0}(\tau, 0) = 1 + 2 \sum_{n \geq 1} q^{n^2} \quad (\text{with } q = e^{\pi i \tau}) .$$

They are modular forms of weight $1/2$ (with Nebentypus) for the principal congruence subgroup of level 2 (for which $h = 2$).

Denote by $\partial = d/dz$ differentiation with respect to the elliptic variable z. Then $\partial(\partial\theta/\theta)$ is an elliptic function with one pole, of order 2, hence of the form $-\wp(z + c(\tau)) + e(\tau)$, where $\wp$ is the Weierstrass function for the lattice $\mathbf{Z} \oplus \mathbf{Z}\tau$, and $c(\tau) = (1 + \tau)/2$, $e(\tau)$ are independent of z. Now $\wp$ satisfies the differential equation

$$\partial^3 \wp = 12 \wp \cdot \partial \wp ,$$

(a special case of the KdV equation), whose coefficients are independent of τ. Therefore, $-\partial(\partial\theta/\theta) + e(\tau)$ satisfies the same equation. Differentiating again with respect to z and eliminating $\wp$, we get

$$\partial^4 \wp \cdot \partial \wp - \partial^3 \wp \cdot \partial^2 \wp + 12(\partial \wp)^3 = 0 ,$$

and this differential equation is now also satisfied by $-\partial(\partial\theta/\theta)$, thus yielding a non trivial relation with constant coefficients between the $\partial^i \theta$'s for $i = 0, 1, \ldots, 6$. In other words, $\theta(\tau, z)$, *qua* function of z, satisfies an algebraic differential equation of the sixth order with coefficients in $\mathbf{Q}$. Setting $z = 0$ and recalling that θ is an *even* function of z, we get a non trivial relation linking $\partial^6\theta(\tau, 0)$ to $\theta(\tau, 0)$, $\partial^2\theta(\tau, 0)$, $\partial^4\theta(\tau, 0)$. Since only even orders appear here, we may finally appeal to the heat equation :

$$\partial^2 \theta(\tau, z) = 4\pi i (\partial/\partial\tau)\theta(\tau, z) = 2(2\pi i)^2 D\theta(\tau, z)$$

to translate it into a relation linking $D^3(\theta(\tau, 0))$ to $\theta(\tau, 0)$, $D\theta(\tau, 0)$, $D^2\theta(\tau, 0)$, i.e. into a differential equation of the third order for $\theta_3(q)$. Its expression, obtained in a different fashion by Jacobi in 1847, reads as follows (cf. [**Nis1**], §4) :

$$\left(\theta^2 D^3\theta - 15\theta D\theta D^2\theta + 30(D\theta)^3\right)^2 + 32\left(\theta D^2\theta - 3(D\theta)^2\right)^3$$
$$= 4\theta^{10}\left(\theta D^2\theta - 3(D\theta)^2\right)^2 .$$

Of course, $\theta_2(q)$ and $\theta_4(q)$ satisfy the same equation, since the above argument made no use of the actual values of $c(\tau)$, $e(\tau)$.

2.3. Modular functions. (cf. [**Ded**][2], § 5). By way of an example, consider Legendre's function $\lambda(\tau)$. Its inverse function $\tau(\lambda)$ is the quotient of two linearly independent solutions of the Gauss hypergeometric equation :

$$\lambda(1-\lambda)d^2F/d\lambda^2 + (1-2\lambda)dF/d\lambda - (1/4)F = 0 \ ,$$

viz.: $F(\lambda) = F(1/2, 1/2, 1; \lambda)$, which actually is $\theta_3(q)^2$, and $F(1-\lambda)$.

Since $\tau(\lambda)$ is PGL_2-multivalued, its Schwarzian $\{\tau, \lambda\}$ is a single-valued function of λ, hence by regularity a rational function $S(\lambda)$, which can be computed by analyzing the local behaviour of τ near $\lambda = 0, 1, \infty$. Now, the expression of the Schwarzian

$$\{\tau, \lambda\} = \frac{2\frac{d^3\tau}{d\lambda^3} \cdot \frac{d\tau}{d\lambda} - 3\left(\frac{d^2\tau}{d\lambda^2}\right)^2}{2\left(\frac{d\tau}{d\lambda}\right)^2}$$

does not involve τ itself. Thanks to the inverse function theorem, the relation $\{\tau, \lambda\} = S(\lambda)$ therefore becomes a differential equation of order 3 with constant coefficients for $\lambda(\tau)$. Explicitely, one obtains, cf. [**Ded**] (*loc. cit.*), (20) :

$$2\left(\frac{d\lambda}{d\tau}\right)^{-2} \cdot \{\lambda, \tau\} = -\frac{\lambda^2 - \lambda + 1}{\lambda^2(1-\lambda)^2} \ .$$

(NB : in the quotation from our introduction, Jacobi refers not so much to this explicit differential equation as to the fact that for any γ in $GL_2^+(\mathbf{Q})$, the function $\lambda_\gamma(\tau) := \lambda(\gamma \cdot \tau)$ satisfies the same equation as $\lambda(\tau)$.)

3. Singular values

Given two points τ, τ' in the upper half-space $\mathfrak{h}$, the corresponding elliptic curves are isogenous if and only if there exists $\gamma \in GL_2^+(\mathbf{Q})$ such that $\tau' = \gamma \cdot \tau$. Any τ stabilized by a non scalar γ in $GL_2^+(\mathbf{Q})$ is quadratic over $\mathbf{Q}$ and is called a CM (or special) point. Its $SL_2(\mathbf{Z})$-orbit corresponds to an isomorphism class of elliptic curves with complex multiplications by an order in $\mathbf{Q}(\tau)$, and conversely. If N is the determinant of an integral multiple of γ, the modular invariant $j(\tau)$ of this isomorphism class satisfies the modular relation of level N : $\Phi_N(j, j) = 0$. In particular, *any modular function defined over* $\mathbf{Q}^{\mathrm{alg}}$ *assumes algebraic values at non polar CM points*. The following relation ii) on its derivatives seems equally classical, see the last paragraph of [**Sie**][3], but its impact in trancendence theory was noticed only after the work of D. Masser [**Mas1**] on periods of elliptic integrals; we derive it from the more general relation i), which is due to Shimura, who put it in a much broader framework (see [**S1**][4], and (a) below).

PROPOSITION 3.1. *Let w be a CM-point, and let f be a modular form of weight k defined over* $\mathbf{Q}^{\mathrm{alg}}$ *and holomorphic at w.*

i) Assume that $k \neq 0$, and $f(w) \neq 0$; then, the number $Df/f(w)$ is a $\mathbf{Q}^{\mathrm{alg}}$-*linear form in π^{-1} and $f(w)^{2/k}$; in particular, $Df(w)$ and $D^2f(w)$ are algebraic over the field $\mathbf{Q}(\pi, f(w))$;*

[2][**Ded**] R. Dedekind, Schreiben an H. Borchardt über die Theorie der elliptischen Modulfunktionen. *Oeuvres*, t. I, (1877), 174–201.

[3][**Sie**] C.L. Siegel. Bestimmung der elliptischen Modulnfunktion durch eine Transformationsgleichung. *Oeuvres*, t. III, (1964), 366–376.

[4][**S1**] G. Shimura. On some arithmetic properties of modular forms of one and several variables. *Ann. Math.* 102, (1975), 491–515.

ii) Assume that $k = 0$, and $Df(w) \neq 0$; then, the number $d^2f/d\tau^2(w)$ is a $\mathbf{Q}^{\text{alg}}$-linear form in $df/d\tau(w)$ and $(df/d\tau(w))^2$; in particular, $D^2f(w)$ is algebraic over the field $\mathbf{Q}(\pi, Df(w))$.

PROOF. Let $\alpha = \begin{pmatrix} a & b \\ c & d \end{pmatrix}$ be a non scalar element of $GL_2^+(\mathbf{Q})$ stabilizing w. Then, $f_\alpha(\tau) := f(\alpha(\tau))(c\tau + d)^{-k}$ is again modular of weight k, and we may write $f_\alpha(\tau) = f(\tau)h(\tau)$ for some $\mathbf{Q}^{\text{alg}}$-modular function h. Differentiating this relation with respect to D, we get :

$$Df(w)(cw + d)^{-k-2} - (kc/2\pi i)f(w)(cw + d)^{-k-1} = (Df)(w)h(w) + f(w)Dh(w) .$$

Since $h(w) = (cw + d)^{-k} \neq (cw + d)^{-k-2}$, this implies that $Df(w)$ is a linear form in $f(w)/2\pi i$ and $f(w)Dh(w)$ with coefficients in the CM field $\mathbf{Q}(w)$. But Dh has weight 2, so that the $\mathbf{Q}^{\text{alg}}$-modular function $(Dh)^k/f^2$ assumes algebraic values at w. The statement on $D^2f(w)$ follows by considering the modular form $f^\# = kf \, D^2f - (k+1)(Df)^2$ from Proposition 1.1.i), and ii) is a corollary of i), since Df is then a modular form of weight 2.

$\square$

Applying Proposition 3.1.i) to the cusp form Δ and any CM point w, we deduce from Chapter 3 Theorem 1.1 that π, $e^{\pi \, \text{Im}(w)}$ and $\Delta(w)$ are algebraically independent over $\mathbf{Q}$. The latter number is given by the Chowla-Selberg formula, cf. last page of [Wei][5], from which we finally deduce :

COROLLARY 3.2. *For any quadratic imaginary field with discriminant $-D$ and character ε, the numbers π, $e^{\pi\sqrt{D}}$ and $\prod_{a=1,\ldots,D-1} \Gamma(a/D)^{\varepsilon(a)}$ are algebraically independent over $\mathbf{Q}$.*

REMARK. i) The hypothesis $f(w) \neq 0$ in i) cannot be ommitted in general. For instance, ii) does not hold at the elliptic point $w = i$ when f is the modular invariant j. However, the corollary does hold true at the elliptic points $w \equiv i$, $\rho \mod SL_2(\mathbf{Z})$ (with $\rho = \exp(2\pi i/3)$), and in fact :
ii) let E be an elliptic curve defined over $\mathbf{Q}^{\text{alg}}$ and with CM by $\mathbf{Q}(w)$, and let ϖ be a non-zero period of a $\mathbf{Q}^{\text{alg}}$-rational differential of the first kind on E. Up to an algebraic factor, $\Delta(w)$ is given by $(\varpi/2\pi i)^{12}$. Thus, when $w \equiv i$, $\rho \mod SL_2(\mathbf{Z})$, the Chowla-Selberg formula boils down to the elementary remark (using Euler's B-function) that :

- if $w \equiv i$, $E_6(w) = 0$, so that we may take the curve $y^2 = x^3 - x$ for E, and $2\int_1^\infty x^{-1/2}(x^2 - 1)^{-1/2}dx = \Gamma(1/4)^2/(2\pi)^{1/2}$ for ϖ.
- if $w \equiv \rho$, $E_4(w) = 0$, so that we may take the curve $y^2 = x^3 - 1$ for E, and $2\int_1^\infty (x^3 - 1)^{-1/2}dx = \Gamma(1/3)^3/2^{1/3}\pi$ for ϖ.

The proof of Proposition 3.1 given above hides the role played by real-analytic (non holomorphic) functions in these relations. We now explain this, in the *modular*, *elliptic* and *cohomological* points of view already met in §2. Each will give a new (and more explicit) version of the relation

$$E_2(w) = (\text{alg.nb})/\pi + (\text{alg.nb})(\varpi/\pi)^2$$

used in the corollary.

[5][Wei] A. Weil. *Elliptic functions according to Eisenstein and Kronecker.* Ergeb. Math. Grenzgeb. 88, Springer, (1976).

3.1. Near holomorphy and modularity. (cf. [S2][6]). A nearly holomorphic modular form of weight k for a congruence group Γ is a function $f(\tau)$ on the upper half-space $\mathfrak{h}$ such that $f(\gamma \cdot \tau) = (c\tau + d)^k f(\tau)$ for any $\gamma = \left(\begin{smallmatrix} a & b \\ c & d \end{smallmatrix}\right)$ in Γ, and which can be expressed as a polynomial in $(\pi \operatorname{Im}(\tau))^{-1}$, whose coefficients are holomophic functions on $\mathfrak{h} \cup \{\text{cusps}\}$. We say that f is defined over $\mathbf{Q}^{\text{alg}}$ if the Fourier expansions of these coefficients at $i\infty$ have algebraic coefficients. The quotient of two such functions with equal weights will here be called a nearly meromophic $\mathbf{Q}^{\text{alg}}$-modular function. Shimura has shown that just as classical modular functions, *any nearly meromophic modular function defined over $\mathbf{Q}^{\text{alg}}$ assume algebraic values at non-polar CM points* (furthermore, $f(w)$ lies in the maximal abelian extension of $\mathbf{Q}(w)$ if f is defined over $\mathbf{Q}$; cf. [S2] (*loc. cit.*)).

In a vein similar to §2.a, set $\tau = x + iy$, $y = y(\tau) > 0$, and note that

$$[y(-1/\tau)]^{-1} = \tau^2 [y(t)]^{-1} - 2i\tau .$$

Hence, $(kf/4\pi y)(-1/\tau) = \tau^{k+2}(kf/4\pi y)(\tau) + k\tau^{k+1}f(\tau)/2\pi i$ for any modular form f of weight k. Comparing with §1, (1), we see that the differential operator (of Maass type) :

$$\mathcal{D}_k = D - k/4\pi y$$

sends such an f to a nearly meromorphic modular form of weight $k + 2$, and can thus be viewed as a non holomorphic analogue of D_k. Proposition 2 is an immediate corollary of these facts. For instance,

$$E_2^*(\tau) := \mathcal{D}_2\Delta/\Delta = D\Delta/\Delta - 3/\pi y = -3/\pi y(\tau) + E_2(\tau)$$

is a nearly holomorphic modular form of weight 2, so that its value at a CM point w is an algebraic multiple of $\Delta(w)^{1/6}$, i.e. in the notations of Remark 3.ii) :

$$E_2^*(w) = (\text{alg.nb})(\varpi/\pi)^2, \quad \text{and} \quad E_2(w) = 3/\pi \operatorname{Im}(w) + (\text{alg.nb})(\varpi/\pi)^2 .$$

3.2. Real analytic elliptic functions. (cf. [Wei] (*loc. cit.*, p. 6), [Kat] (*loc. cit.*, p. 3)) For any lattice L in $\mathbf{C}$, write $L = (\varpi/2\pi i)(2\pi i\mathbf{Z} \oplus 2\pi i\tau\mathbf{Z})$, with $\operatorname{Im}(\tau) > 0$, and denote by $\sigma(z, L)$, $\zeta(z, L) = \partial\sigma/\sigma$, $\wp(z, L) = -\partial\zeta$ the standard Weierstrass functions attached to L, and by $\eta(z, L)$ be the $\mathbf{R}$-linear form on $\mathbf{C}$ whose values on L are the quasiperiods of ζ, so that in the usual notations :

$$\omega_1 = \varpi, \quad \omega_2 = \tau\varpi, \quad \eta_1 = \eta(\varpi, L), \quad \eta_2 = \eta(\tau\varpi, L).$$

For any $k \geq 2$, $\left[(-1)^k B_k/(2k)!\right] E_{2k}(\tau) = (\varpi/2\pi i)^{2k} G_{2k}(L)$, where :

$$G_{2k}(L) = \sum_{\omega \in L - \{0\}} \omega^{-2k} .$$

In the same way, the nearly holomorphic modular form $E_2^*(\tau)$ is related to Hecke's lattice function (cf. [Kat] (*loc. cit.*, p. 3), 1.3.11, [Ber2])

$$G_2^*(L) = \lim_{s \to 0} \sum_{\omega \in L - \{0\}} \omega^{-2}|\omega|^{-s}$$

by $E_2^*(\tau) = -12(\varpi/2\pi i)^2 G_2^*(L)$, while $E_2(\tau) = -12(\varpi/2\pi i)^2 \eta(\varpi, L)/\varpi$.

---

[6][S2] G. Shimura. Arithmeticity of the special values ... *Sugaku Expositiones* 8, (1995), 17–38; (See also : *Math. Ann.* 278, (1987), 1–28.)

Set : $\zeta^*(z, L) := \zeta(z, L) - \eta(z, L)$. This is an 'almost meromorphic' L-periodic function, whose derivative with respect to $-\partial = -d/dz$ is $\wp(z, L) + G_2^*(L)$, and which satisfies the division formula

$$\sum_{r \in L/L'} \zeta^*(z + r; L') = \zeta^*(z, L)$$

for any sublattice L' of L, cf. [**Wei**] (*loc. cit.*, p. 6), pp. 21, 43.

Differentiating with respect to ∂ and letting z tend to 0, we obtain :

$$\sum_{r \in L/L' - \{0\}} \wp(r, L') = G_2^*(L) - [L : L']G_2^*(L') .$$

In particular, if τ is a CM point w, and if $\delta \in \mathbf{Q}(w)$ is a complex multiplier of the lattice L, so that $L' := \delta L$ is a sublattice of L of index $\delta\bar{\delta}$, we get :

$$(1 - \delta\bar{\delta}/\delta^2)G_2^*(L) = \sum_{r \in L/L' - \{0\}} \wp(r, L') .$$

This shows, in an explicit way, that the algebraic number $(2\pi i/\varpi)^2 E_2^*(w)$ lies in the maximal abelian extension of $\mathbf{Q}(w)$ as soon as L is normalized by the requirement : $G_4(L), \ G_6(L) \in \mathbf{Q}(j(w))$.

3.3. Period relations. (cf. [**Kat**] (*loc. cit.*, p. 3), [**Mas1**]). Under the same requirement, let E be the elliptic curve C/L. Since the action of the CM field $\mathbf{Q}(w)$ on the algebraic de Rham cohomology of E is equivalent to a rational representation, the Hodge filtration splits over $\mathbf{Q}^{\text{alg}}$, and there exists an algebraic number $\beta = \beta(L)$ such $d(\zeta(z, L) - \beta z)$ is an eigenform corresponding to the "complex conjugation" character of $\mathbf{Q}(w)$. Its periods therefore satisfy :

$$\eta_2 - \beta\omega_2 = \bar{w}(\eta_1 - \beta\omega_1) .$$

Combining this with the formula for E_2 given in (b), and with the Legendre relation $\eta_1\omega_2 - \eta_2\omega_1 = 2\pi i$ (i.e. $E_2(-1/\tau) = 12\tau/2\pi i + \tau^2 E_2(\tau)$) yields anew :

$$(w - \bar{w})E_2(w) = -12/2\pi i + (\text{alg.nb})(\varpi/\pi)^2 .$$

(In fact, $\beta(L) = G_2^*(L)$; cf. [**Ber2**]).

4. Transcendence on τ and z

To fix matters, we shall concentrate on the theta function $\theta(\tau, z) = \theta_{0,0}(\tau, z) = \theta_3(\tau, z)$, i.e. :

$$\begin{aligned}
\theta(\tau, z) &= \sum_{n \in Z} e^{\pi i n^2 \tau} e^{2\pi i n z} \\
&= \sum_{n \in Z} q^{n^2} Z^n \quad (\text{with } q = e^{\pi i \tau}, \ Z = e^{2\pi i z}) ,
\end{aligned}$$

which admits the product expansion :

$$\theta(\tau, z) = \prod_{n \geq 1}(1 - q^{2n})(1 + q^{2n-1}Z)(1 + q^{2n-1}/Z) := \Theta(q, Z) ,$$

and on its partial derivatives with respect to

$$D := (1/\pi i)\partial/\partial\tau = q\partial/\partial q, \quad \partial := (1/2\pi i)\partial/\partial z = Z\partial/\partial Z \quad \text{and} \quad ' := \partial/\partial z .$$

Note the changes of notations from the preceding paragraphs about D and ∂. In particular, the heat equation $\theta''(\tau, z) = 4\pi i (\partial/\partial\tau)\theta(\tau, z)$ now reads :

$$\partial^2\theta = D\theta \ .$$

On the other hand, the differential equation of order 6 gotten in §2.b for $\theta(\tau, z)$ as a function of z is homogeneous, and still exhibits rational coefficients when expressed in terms of the new ∂-operator. More precisely :

$$\partial^5(\partial\theta/\theta) \cdot \partial^2(\partial\theta/\theta) - \partial^4(\partial\theta/\theta) \cdot \partial^3(\partial\theta/\theta) - 12\left(\partial^2(\partial\theta/\theta)\right)^3 = 0 \ .$$

In this section, we shall study three equivalent versions of the following "main conjecture" on the Taylor coefficients of $\theta(\tau, z)$.

MAIN CONJECTURE (preliminary version). Let $\alpha \in \mathfrak{h}$, $u \in \mathbf{C}$, $u \notin \mathbf{Q} \oplus \mathbf{Q}\alpha$. Then, at least 6 among the 8 numbers : $e^{\pi i \alpha}$, $e^{2\pi i u}$ and

$$\theta(\alpha, u), \ \theta'(\alpha, u)/\pi, \ \theta''(\alpha, u)/\pi^2, \ \theta'''(\alpha, u)/\pi^3, \ \theta^{\mathrm{iv}}(\alpha, u)/\pi^4, \ \theta^{\mathrm{v}}(\alpha, u)/\pi^5$$

are algebraically independent over $\mathbf{Q}$.

We may equivalently ask : *let κ, U be non-zero and multiplicatively independent complex numbers, with $|\kappa| < 1$. Then at least 6 of the numbers $\{\kappa, U, \partial^i\Theta(\kappa, U), i = 0, \dots, 5\}$ are algebraically independent.* The conjecture can then be translated word for word in the p-adic domains. (Note that when α is not a CM point, the number π does not necessarily occur in the field considered here.)

In view of the differential equation recalled above, the conjecture is equivalent to stating that the field generated over $\mathbf{Q}$ by $e^{\pi i \alpha}$, $e^{2\pi i u}$ and the values of *all* the ∂-derivatives of the function of one variable $\theta_\alpha(z) = \theta(\alpha, z)$ at the point $z = u$ [or equivalently (heat equation), all the (D, ∂)-derivatives of the function of two variables $\theta(\tau, z)$ at the point $(\tau, z) = (\alpha, u)$] has transcendence degree ≥ 6. But it is more interesting (although not surprising) to note that the algebraic closure of the field in question contains all the ∂-derivatives of $\theta_\alpha(z)$ at the point $z = 0$ [or equivalently, all the (∂, D)-derivatives of $\theta(\tau, z)$ at the point $(\tau, z) = (\alpha, 0)$], i.e. eventually the values of all the D-derivatives of the modular theta function $\theta_3(q) = \theta(\tau, 0)$ at $\tau = \alpha$. Indeed, the numbers $D^i\theta_3(e^{\pi i \alpha})$'s $(i \geq 0)$ are algebraic over $\mathbf{Q}\left(E_{2k}(\alpha); k \geq 1\right)$ (cf [**Ber7**], §2); now, reversing the view point on the differential equations of the $\wp$ function, we can express $E_4(\tau)$, $E_6(\tau)$ as polynomials in the ∂-derivatives of the function

$$\begin{aligned} f(z) &= \wp(2\pi i z, 2\pi i (\mathbf{Z} \oplus \mathbf{Z}\tau)) = (2\pi i)^{-2}\wp(z, \mathbf{Z} \oplus \mathbf{Z}\tau) \\ &= -\partial(\partial\theta/\theta)(\tau, z + (1+\tau)/2) + (2\pi i)^{-2}\eta(1, \mathbf{Z} \oplus \mathbf{Z}\tau), \end{aligned}$$

evaluated at any point z, say at $z = u - (1+\tau)/2$, hence in terms of the $\partial^i\theta(\tau, u)$'s and $E_2(\tau)$. Further differentiation with respect to D then implies that the numbers $E_{2k}(\alpha)$'s $(k \geq 1)$ are algebraic over $\mathbf{Q}(\partial^i\theta(\alpha, u); i \geq 0)$.

Variations on this argument eventually lead to the following equivalent reformulation of the conjecture.

MAIN CONJECTURE. Let $\alpha \in \mathfrak{h}$, $u \in \mathbf{C}$, $u \notin \mathbf{Q} \oplus \mathbf{Q}\alpha$. Then, at least 6 among the 8 numbers :

$$e^{\pi i \alpha}, \ \theta(\alpha, 0), \ D\theta(\alpha, 0), \ D^2\theta(\alpha, 0) \text{ and } e^{2\pi i u}, \ \theta(\alpha, u), \ \partial\theta(\alpha, u), \ \partial^2\theta(\alpha, u)$$

are algebraically independent.

Equivalently : *Let κ, U be non-zero and multiplicatively independent complex numbers, with $|\kappa| < 1$. Then at least 6 of the numbers*

$$\kappa, \; \Theta(\kappa, 1), \; D\Theta(\kappa, 1), \; D^2\Theta(\kappa, 1) \text{ and } U, \; \Theta(\kappa, U), \; \partial\Theta(\kappa, U), \; \partial^2\Theta(\kappa, U)$$

are algebraically independent.

Thus, although our conjecture apparently aimed at the elliptic theta function $\theta_\alpha(z)$, it truly concerns the function of two variables $\theta(\tau, z)$. The last formulation on $\Theta(q, Z)$ shows that the role of the cusp $q = 0$ in following Chapters 2-4 (i.e. the most 'modular' aspect of these proofs) could now be played by the point $(q, Z) = (0, 1)$.

We now review what is known on the conjecture.

4.1. On the modular side. Theorem 1.1 from Chapter 3 says that at least 3 of the first four numbers in the new version of the conjecture are algebraically independent. In this respect, note that the fields generated by the first four ones and by the last four ones have the same algebraic closure when we assume (contrary to our hypothesis) that $u \in \mathbf{Q} \oplus \mathbf{Q}\alpha$. Indeed, for any rational numbers a, b, the function $\theta(\tau, a\tau + b)$ is a $\mathbf{Q}^{\text{alg}}$-modular form for some congruence subgroup of $SL_2(\mathbf{Z})$. We may therefore rewrite Theorem 1.1 from Chapter 3 as follows :

PROPOSITION 4.1. *Let $\alpha \in \mathfrak{h}$, $u \in \mathbf{C}$, $u \in \mathbf{Q} \oplus \mathbf{Q}\alpha$. Then, at least 3 of the 4 numbers*

$$e^{\pi i \alpha}, \quad \theta(\alpha, u), \quad \theta''(\alpha, u)/\pi^2, \quad \theta^{\text{iv}}(\alpha, u)/\pi^4$$

are algebraically independent over $\mathbf{Q}$.

(Similarly, Chudnosky's theorem ([**Chu1**], 2.A (4)), which Nesterenko's sharpens, implies that under these hypotheses, $\theta(\alpha, u)$, $\theta''(\alpha, u)/\pi^2$ and $\theta^{\text{iv}}(\alpha, u)/\pi^4$ generate a field of transcendence degre ≥ 2. For a cohomological proof of this result, based on the Gauss-Manin connection, see [**And1**].

4.2. On the elliptic side. Reyssat ([**Rey2**], Corollary 2) proved a general result on the function

$$s_\varpi(z, L) := \sigma(z, L) \exp\left(-(\eta(\varpi, L)/\varpi)z^2/2\right) ,$$

which is close to the study of the last four numbers of the conjecture, but which mix them with the first ones. To allow a comparison, we use the formulae :

$$2\pi i \sigma(z, \mathbf{Z} \oplus \mathbf{Z}\tau) = \sigma(2\pi i z, 2\pi i \mathbf{Z} \oplus 2\pi i \tau \mathbf{Z})$$
$$= q^{1/4}\Delta(\tau)^{-1/8}\theta(\tau, z + (1 + \tau)/2)e^{\left(\eta(1, \mathbf{Z}\oplus\mathbf{Z}\tau)z^2/2 + \pi i z\right)} ,$$

and $\sigma(z, L) = \varpi\sigma(z/\varpi, \mathbf{Z} \oplus \mathbf{Z}\tau)$ for any lattice $L = \varpi(\mathbf{Z} \oplus \mathbf{Z}\tau)$, to write our conjecture yet anew, as follows :

MAIN CONJECTURE (Weierstrass form). *Let $L = \varpi\mathbf{Z} \oplus \alpha\varpi\mathbf{Z}$ be a complex lattice, and let u be a complex number, not in $\mathbf{Q} \cdot L$. Then, at least 6 among the 8 numbers*

$$e^{\pi i \alpha}, \quad (\varpi/2\pi i)^4 G_4(L), \quad (\varpi/2\pi i)^6 G_6(L), \quad (\varpi/2\pi i)^2 \eta(\varpi, L)/\varpi ,$$

$$e^{2\pi i u/\varpi}, \quad (2\pi i/\varpi)s_\varpi(u, L), \quad (\varpi\zeta(u, L) - \eta(\varpi, L)u)/2\pi i, \quad (\varpi/2\pi i)^2 \wp(u, L),$$

are algebraically independent over $\mathbf{Q}$.

The first four numbers generate the same algebraically closed field as those of Proposition 4.1. As for the others, we have :

PROPOSITION 4.2. *Recall the hypotheses of the conjecture, and assume that* $G_4(L)$, $G_6(L)$ *and* $\wp(u, L)$ *are algebraic. Then,*

  i) ([**Chu1**], 2.A, Thm. 4) : $\eta(\varpi, L)/\varpi$ *and* $\zeta(u, L) - \eta(\varpi, L)u/\varpi$ *are algebraically independent over* **Q**;

  ii) ([**Rey2**], Corollary 2) : *at least 2 of the 3 numbers*

$$e^{2\pi i u/\varpi}, \quad s_\varpi(u, L), \quad \zeta(u, L) - \eta(\varpi, L)u/\varpi$$

    are algebraically independent over **Q**;

  iii) ([**Ber4**], Corollary 3) : *if* L *has CM, then* $e^{2\pi i u/\varpi}$ *is transcendental.*

(Moreover, according to [**Rey2**], Corollary 4, the 8 numbers then generate a field of transcendence degre ≥ 3, but this is of course now covered by Nesterenko's theorem.)

REMARK. i) Under the hypotheses of Proposition 4.2, the conjecture states that the 6 numbers $e^{\pi i \alpha}$, $\varpi/2\pi i$, $\eta(\varpi, L)/\varpi$, $\zeta(u, L) - (\eta(\varpi, L)/\varpi)u$, $e^{2\pi i u/\varpi}$ and $s_\varpi(u, L)$ are algebraically independent. All are at present known to be transcendental, *except* $e^{2\pi i u/\varpi}$ in the non-CM case, and $s_\varpi(u, L)$ in all cases.

ii) On the other hand, a transcendence result is known to hold for an analogue of $s_\varpi(z, L)$ involving $E_2^*(\tau)$ (but still belonging to the z-meromorphic world). For any lattice L, set :

$$s^*(z, L) = \sigma(z, L) \exp(-G_2^*(L)z^2/2) .$$

Then, under the hypotheses of Proposition 4.2.iii) (so that by §3.b, $G_2^*(L)$ is algebraic), the number $s^*(u, L)$ is transcendental (cf. [**Ber4**], Corollary 1).

Transcendence theory is not totally devoid of tools to study harmonic functions. We close this survey with a conjecture in this direction, which may be viewed as a non holomorphic analogue of Schneider's theorem on $\{\tau, j(\tau)\}$.

CONJECTURE 4.3. *Let* f *(resp.* g*) be a non-constant meromorphic (resp. a nearly meromorphic and not meromorphic) modular function defined over* **Q**^{alg}, *and let* α *be an element of* $\mathfrak{h}$ *where* f *and* g *are defined. If* $f(\alpha)$ *and* $g(\alpha)$ *are both algebraic, then* α *is a CM point.*

A solution would provide a positive answer to following question, raised by N. Katz ([**Kat**] (*loc. cit.*, p. 3), 4.0.8) : let L be a lattice with algebraic invariants $G_4(L)$, $G_6(L)$. Assume that $G_2^*(L)$ is algebraic. Then, does L necessarily have CM?

CHAPTER 2

Mahler's conjecture and other transcendence results

1. Introduction

The first transcendental result about values of the modular invariant j ("if τ is algebraic but not imaginary quadratic then $j(\tau)$ is transcendental") has been proved in 1937 by Th. Schneider, by means of elliptic functions. To be precise, let $\wp$ be a Weierstrass elliptic function with algebraic invariants; if (ω_1, ω_2) is a basis of periods then ω_1/ω_2 is either quadratic or transcendental. This is transcribed in the above mentioned result on j. There are other translations of this kind; for example :

- We know that ω_1/π is transcendental (Th. Schneider); this gives a modular consequence : if $j(\tau) \notin \{0, 1728\}$ then $j(\tau)$ and $j'(\tau)/\pi$ are not simultaneously algebraic.
- We know that ω_1 is transcendental (Th. Schneider); consequently if τ is quadratic with $j(\tau) \notin \{0, 1728\}$ then $j'(\tau)$ is transcendental.

In the p-adic case, the analogue of the first result and its translation in terms of Eisenstein series (or of the modular invariant) are due to D. Bertrand [**Ber1**].

Along the same lines, there is a lot of algebraic independence results about periods and quasi-periods of elliptic functions (G.V. Chudnovsky [**Chu2**]) having a translation in the modular domain; this has been shown for the first time by D. Bertrand [**Ber2**], who has also given p-adic versions.

So we can say that until 1995, to obtain transcendental results about values of the modular invariant j and its derivatives, it was quite "natural" to work first in the elliptic domain and then to translate results in the modular domain.

It has been shown in 1996 (K. Barré, G. Diaz, F. Gramain and G. Philibert [**BDGP**]) that it is possible to reach transcendental results on values of j without the help of elliptic functions. The results were conjectured in 1969 by K. Mahler (complex domain) and in 1971 by Yu. Manin (p-adic domain).

We shall discuss this idea, which is also the heart of the work of Yu.V. Nesterenko [**Nes8**]-[**Nes9**]. After giving some classical notations and technical results, we present in section 2 a complete proof of Mahler's conjecture. Section 3 is devoted to two results from K. Barré's thesis : a quantitative version of [**BDGP**] and a modular proof of an old result. Section 4 is more speculative, presenting a lot of conjectures on j and the exponential function.

M. Waldschmidt provides in [**Wal9**]-[**Wal10**] a flowering survey of the field, offering a lot of open problems.

(∗) Chapter's author : Guy Diaz.

---

2. A proof of Mahler's conjecture

We give here a complete proof of Mahler's conjecture [**Mah2**] (see [**BDGP**] for the first proof): "$\exp(2i\pi\tau)$ and $j(\tau)$ are not simultaneously algebraic, where $\tau \in \mathbf{C}$ with $\mathrm{Im}\,\tau > 0$." This result is now a corollary of an algebraic independence result of Yu.V. Nesterenko (see [**Nes8**], [**Nes9**]) and P. Philippon (see [**PPh10**], [**PPh11**]). We know for instance that: "If $\exp(2i\pi\tau)$ is algebraic then $j(\tau), j'(\tau)/\pi, j''(\tau)/\pi^2$ are algebraically independent, where $\tau \in \mathbf{C}$ with $\mathrm{Im}\,\tau > 0$".

Let us begin with some classical definitions, notations and results.

2.1. The modular invariant j and the discriminant Δ.
We set

$$\mathcal{H} := \{\tau \in \mathbf{C};\ \mathrm{Im}\,\tau > 0\}, \quad D := \{z \in \mathbf{C};\ |z| < 1\}, \quad D^* := D \setminus \{0\}.$$

We define three functions g_2, g_3, Δ on $\mathcal{H}$:

$$\begin{cases} g_2(\tau) := 60 \sum_{\substack{\omega \in \Lambda_\tau \\ \omega \neq 0}} \omega^{-4}, \text{ where } \Lambda_\tau := \mathbf{Z} + \mathbf{Z}\tau, \\[2ex] g_3(\tau) := 140 \sum_{\substack{\omega \in \Lambda_\tau \\ \omega \neq 0}} \omega^{-6}, \\[2ex] \Delta(\tau) := (2\pi)^{-12}(g_2^3(\tau) - 27g_3^2(\tau)), \text{ the discriminant.} \end{cases}$$

Warning : the coefficient $(2\pi)^{-12}$ is not used in [**Ser**][1].

The function Δ has no zero on $\mathcal{H}$ and *the modular invariant j is the function $j : \mathcal{H} \to \mathbf{C}$ defined by*:

$$j(\tau) := 1728 \frac{g_2^3(\tau)}{g_2^3(\tau) - 27g_3^2(\tau)}.$$

For details on $j, \Delta, g_2, g_3, \ldots$ see [**Ser**] (*loc. cit.*), §VII-2, VII-3, VII-4. For links between elliptic curves and j see [**Sil**][2], [**Lang**][3].

PROPOSITION 2.1. *The modular invariant j*, ([**Ser**] (*loc. cit.*), chap VII, §3.3).

1. *j is an analytic function on $\mathcal{H}$.*
2. *j is invariant under $SL_2(\mathbf{Z})$:*

$$j\left(\frac{a\tau + b}{c\tau + d}\right) = j(\tau) \quad \text{for all} \quad \tau \in \mathcal{H}, \begin{pmatrix} a & b \\ c & d \end{pmatrix} \in SL_2(\mathbf{Z}).$$

In particular $j(\tau + 1) = j(\tau)$, so there is an application $J : D^* \to \mathbf{C}$ such that $j(\tau) = J(e^{2i\pi\tau})$ for all $\tau \in \mathcal{H}$.

3. $J(z) = \dfrac{1}{z} + 744 + \displaystyle\sum_{n=1}^{+\infty} c(n)z^n$ *and $c(n) \in \mathbf{N}$.*

---

[1][**Ser**] J.P. Serre. *Cours d'arithmétique*, Coll. Sup. PUF, seconde édition (1977); English translation : *A course in arithmetic*, Grad. Texts Math. 7, Springer, (1973).

[2][**Sil**] J.H. Silverman. *The arithmetic of elliptic curves*, Grad. Texts Math. 106, Springer, (1986).

[3][**Lang**] S. Lang. *Elliptic functions*, second edition, Grad. Texts Math. 112, Springer, (1987).

There is a "naïve denominator" for J since the function $z \mapsto z.J(z) = 1 + 744z + \sum_{n \geq 1} c(n)z^{n+1}$ is analytical on D. It was used in [**BDGP**] in conjonction with an estimate of Mahler for the $c(n)$'s : $c(n) \leq e^{C_0\sqrt{n}}$. According to a remark of D. Bertrand (see also [**Ber3**]), we can use Δ as a "denominator" for J, and this gives better estimates.

PROPOSITION 2.2. *The discriminant* Δ ([**Ser**] *(loc. cit.,* p. 14)*, Chap VII, §2-2, §4-4).*

1. Δ *is an analytical function on* $\mathcal{H}$.
2. Δ *is modular of weight 12 i.e.*

$$\Delta\left(\frac{a\tau + b}{c\tau + d}\right) = (c\tau + d)^{12}.\Delta(\tau) \quad \text{for all} \quad \tau \in \mathcal{H}, \begin{pmatrix} a & b \\ c & d \end{pmatrix} \in SL_2(\mathbf{Z}) .$$

In particular $\Delta(\tau + 1) = \Delta(\tau)$ *and* Δ *has a Fourier expansion (sometimes called "q-expansion") :*

$$\Delta(\tau) = \sum_{n=1}^{+\infty} \tau(n).z^n \quad \text{where} \quad z := e^{2i\pi\tau} .$$

The coefficients $\tau(n)$ *are all integers. This development defines an analytical function on* D, *which vanishes at* 0 *(and only at* 0*); so* Δ *is said to be a parabolic form of weight 12.*

3. *For all* $\tau \in \mathcal{H}$:

$$\Delta(\tau) = z. \prod_{n=1}^{+\infty} (1 - z^n)^{24} \quad \text{where} \quad z = e^{2i\pi\tau} .$$

PROPOSITION 2.3. *The Eisenstein serie* E_4, *(noted* E_2 *in* [**Ser**] *(loc. cit.,* p. 14)).*

1. $E_4 := 12.(2\pi)^{-4}.g_2$ *is an analytical function on* $\mathcal{H}$.
2. E_4 *is modular of weight 4 i.e.:*

$$E_4\left(\frac{a\tau + b}{c\tau + d}\right) = (c\tau + d)^4.E_4(\tau) \quad \text{for all} \quad \tau \in \mathcal{H}, \begin{pmatrix} a & b \\ c & d \end{pmatrix} \in SL_2(\mathbf{Z}).$$

In particular $E_4(\tau + 1) = E_4(\tau)$ *and* E_4 *has a Fourier expansion; precisely (always with* $z := e^{2i\pi\tau}$ *) :*

$$E_4(\tau) = 1 + 240 \sum_{n \geq 1} \sigma_3(n).z^n, \quad \sigma_3(n) = \sum_{d|n} d^3 .$$

This defines an analytical function on D; *so* E_4 *is said to be* a modular form of weight 4.

3. *We have* $\Delta.j = E_4^3$ *and this is a* modular form of weight 12.

When $f(\tau) = \sum_{n=1}^{+\infty} a_n z^n$ (where $z := e^{2i\pi\tau}, \tau \in \mathcal{H}$) is a *parabolic form of weight* $2k$ $(k \geq 2)$ we have an estimation for the coefficients a_n : $|a_n| \leq c(f).n^k$ for all n (see [**Ser**] *(loc. cit.,* p. 14)*, chap VII, §2-1 and §4-3).*

For example $\Delta^2 j = \Delta.(\Delta.j)$ is a parabolic form of weight 24; so its nth-coefficient has modulus less than $c.n^{12}$ (for $z.J(z)$ the estimate is $e^{c\sqrt{n}}$ in [**Mah3**]).

In the sequel we use only the Fourier expansion of Δ and we call it Δ again:

$$\Delta(z) = \sum_{n \geq 1} \tau(n).z^n = z. \prod_{n=1}^{+\infty} (1 - z^n)^{24}, \; z \in D \; .$$

2.2. Modular polynomials (or modular "equations"). For details see [**Lang**] (*loc. cit.*, p. 14), chap 5, §2.

For all integer $n \geq 2$, there exists a polynomial $\Phi_n \in \mathbf{Z}[X, Y]$ such that:

1. $\Phi_n(j(\tau), j(n\tau)) = 0$ for all $\tau \in \mathcal{H}$, $\Phi_n(J(z), J(z^n)) = 0$ for all $z \in D^*$.
2. The degree in X (resp. Y) of $\Phi_n(X, Y)$ is equal to

$$\psi(n) := n \prod_{p|n} \left(1 + \frac{1}{p} \right) \; .$$

3. The coefficient of $X^{\psi(n)}$ (resp. $Y^{\psi(n)}$) is 1.

We add $\Phi_1(X, Y) := X - Y$.

The first property implies that if $J(q)$ is algebraic, then all the $J(q^n)$, $n \geq 1$, are algebraic. This will be very important in the proof of Mahler's conjecture (§ 2.5).

REMARK. Mahler's method (see [**Mah2**], [**Nis2**]) deals with functions f satisfying a functional equation of the form $P(f(z), f(z^d)) = 0$ where P is a polynomial. Here we have an infinity of functional equations: $\Phi_n(J(z), J(z^n)) = 0$ for all $n \geq 1$.

2.3. Measures of algebraic numbers and polynomials. For more details, we refer to [**Wal8**], III.

If $P \in \mathbf{C}[X_1, \ldots, X_n]$ then $L(P)$ is the sum of the modulus of the coefficients of P. For $n = 1$ write $P(X) = a_0 \prod_{i=1}^{D} (X - \alpha_i)$, the *Mahler measure of P* is the number :

$$M(P) := |a_0| \prod_{i=1}^{D} \max(1, |\alpha_i|) \; .$$

This defines a multiplicative function: $M(P_1.P_2) = M(P_1).M(P_2)$. We also have :

$$M(P) = \exp \left(\int_0^1 \log |P(e^{2i\pi t})| dt \right) \; ,$$

from which follows : $M(P) \leq \sup_{|z| \leq 1} |P(z)|$.

Let α be an algebraic number (over $\mathbf{Q}$), of degree $d(\alpha)$ (*i.e.* $d(\alpha) := [\mathbf{Q}(\alpha) : \mathbf{Q}]$). If $P_\alpha \in \mathbf{Z}[X]$ is the minimal polynomial of α, we define the *Mahler measure of α* by $M(\alpha) := M(P_\alpha)$; and the *(absolute) logarithmic height of α* by $h(\alpha) := \dfrac{\log M(\alpha)}{d(\alpha)}$.

2.4. Technical lemmas.

LEMMA 2.4. Growth of the Taylor coefficients of $\Delta^{2L}.J^k$. *There exists an absolute constant C_1 with the following property. Let L and k be positive integers with $0 \le k \le L$, $1 \le L$; put: $(\Delta^{2L}.J^k)(z) = \sum_{n=1}^{+\infty} c_{Lk}(n).z^n$. Then for all $n \ge 1$, the coefficient $c_{Lk}(n)$ is a rational integer and:*

$$|c_{Lk}(n)| \le C_1^L.n^{12L} \ .$$

PROOF.   1. The Taylor coefficients of $\Delta.J$ and Δ are rational integers; this gives the first point.

2. $\Delta^{2L}.J^k = \Delta^{2L-2k}.(\Delta^2 J)^k$: this is a parabolic form of weight $24L$.

The general result of Hecke ([**Ser**] (*loc. cit.*, p. 14), chap.VII, §4-3) gives only

$$|c_{Lk}(n)| \le C(L,k).n^{12L},$$

with no information on $C(L,k)$. Thus it is necessary to go back to the proof in *loc. cit.* We set, for $\tau \in \mathcal{H}$:

$$f_1(\tau) := \Delta^2(e^{2i\pi\tau}), \ f_2(\tau) := (\Delta^2 J)(e^{2i\pi\tau}), \ f := f_1^{L-k}.f_2^k;$$

f_1, f_2 are parabolic forms of weight 24. The first part of Serre's proof shows that the functions $\tau \mapsto |f_j(\tau)|.(\mathrm{Im}\,\tau)^{12}$, $j = 1, 2$, are bounded on $\mathcal{H}$. So there exists $C \ge 1$ such that :

$$|f_j(\tau)| \le C(\mathrm{Im}\,\tau)^{-12} \quad \text{for} \quad j \in \{1,2\}, \tau \in \mathcal{H}.$$

And so we have $|f(\tau)| \le C^L.(\mathrm{Im}\,\tau)^{-12L}$ for all $\tau \in \mathcal{H}$. We take $y > 0$; for $x \in [0,1]$ the point $z := \exp(2i\pi(x+iy))$ describes a circle $\mathcal{C}y$ centered at 0. The residues formula gives:

$$c_{Lk}(n) = \frac{1}{2i\pi} \int_{\mathcal{C}y} f(x+iy).z^{-n-1}.dz = \int_0^1 f(x+iy).z^{-n}.dx.$$

Therefore :

$$|c_{Lk}(n)| \le C^L.y^{-12L}.\int_0^1 |z|^{-n}dx,$$

but $|z| = e^{-2\pi y}$ and we get : $|c_{Lk}(n)| \le C^L.y^{-12L}.e^{2\pi ny}$. For $y = \dfrac{1}{n}$ this gives : $|c_{Lk}(n)| \le C^L.e^{2\pi}.n^{12L}$, and Lemma 2.4 is proved.   □

LEMMA 2.5. Estimates for modular polynomials. *There exists an absolute constant $C_0 > 0$ such that for all integer $n \ge 2$ we have:*

1. $\deg_X \Phi_n = \deg_Y \Phi_n = \psi(n) \le C_0\, n(\log\log 3n)$;
2. $\log L(\Phi_n) \le C_0 \psi(n)(\log n)$.

See [**BDGP**], lemme 2. The second estimate is a result of P. Cohen [**Coh**]; the older result of K. Mahler [**Mah3**] ($\log L(\Phi_n) << n^{3/2}$) would be sufficient for our purpose!

LEMMA 2.6. Siegel's lemma, (see [**Wal1**] p.32, for example). *Let $(a_{ij})_{\substack{1 \le i \le N \\ 1 \le j \le M}}$ be a matrix with N lines, M columns and with integer coefficients. If $M > N$ there exists a non-zero vector $(x_1, \ldots, x_M) \in \mathbf{Z}^M$ such that:*

$$\sum_{j=1}^{M} a_{ij}.x_j = 0 \quad \text{for} \quad 1 \le i \le N, \quad \text{with} \quad \max_j |x_j| \le \left(\prod_{i=1}^N H_i\right)^{1/(M-N)}$$

where $H_i := \max\left(1; \sum_{j=1}^{M} |a_{ij}|\right)$.

LEMMA 2.7. (See [**BDGP**], lemme 4). *The formal series $J \in \mathbf{C}((X))$ is transcendental over $\mathbf{C}(X)$.*

LEMMA 2.8. (See [**BDGP**], lemme 5). *Let $P \in \mathbf{Z}[X,Y]$ be a polynomial and α, β be algebraic numbers. If $P(\alpha, \beta) = 0$ with $P(\alpha, Y) \neq 0$ then:*

$$\log M(\beta) \leq d(\alpha).(\log L(P) + h(\alpha).\deg_X P).$$

PROOF. We give a short proof. Let $R_\alpha(X) = a_0.\Pi_{j=1}^{d}(X - \alpha_j)$ be the minimal polynomial of α in $\mathbf{Z}[X]$. Put $Q(Y) := a_0^D.\Pi_{j=1}^{d}P(\alpha_j, Y)$ where $D := \deg_X P$. Q is a non zero polynomial (by hypothesis $P(\alpha, Y)$ is not zero and consequently $P(\alpha_j, Y)$ is not zero for all j). We have $Q \in \mathbf{Q}[Y]$ (Q is invariant by permutation on $\{\alpha_1, \ldots, \alpha_d\}$); and on the other hand it is easy to see that coefficients of Q are algebraic integers. Therefore Q is in $\mathbf{Z}[Y]$. Since $P(\alpha, \beta) = 0$ we have $Q(\beta) = 0$, so Q is divisible by the minimal polynomial of β in $\mathbf{Z}[Y]$. This gives:

$$M(\beta) \leq M(Q) = |a_0|^D . \prod_{j=1}^{d} M(P(\alpha_j, Y)).$$

We know that: $M(P(\alpha_j, Y)) \leq \max_{|z|=1} |P(\alpha_j, z)| \leq L(P).\max(1, |\alpha_j|)^D$. Therefore we have: $M(\beta) \leq |a_0|^D . L(P)^d . \prod_{j=1}^{d} \max(1, |\alpha_j|)^D$

i.e. $\quad M(\beta) \leq L(P)^d . M(\alpha)^D.$

Lemma 2.8 is proved. $\hspace{2cm}\square$

CONSEQUENCE. If $J(q)$ is an algebraic number, using $\Phi_n(J(q), J(q^n)) = 0$ we see that $J(q^n)$ is also algebraic. Lemma 2.8 gives an estimate for $M(J(q^n))$; taking into account lemma 2.5 we obtain :

COROLLARY 2.9. *If $J(q)$ is an algebraic number then*

$$\log M(J(q^n)) << n(\log n)(\log \log 3n) \quad \textit{for all} \quad n \geq 2 \ .$$

REMARK. This result is also a consequence of a lemma of D. Bertrand, using a very different approach (Faltings height), [**Ber3**], lemma 2 : *There exists an absolute constant C with the following property. Let $q \in D^*$ be such that $J(q)$ is algebraic; for all integer $n \geq 1$ we have*

$$h(J(q^n)) \leq 2h(J(q)) + 6\log(n+1) + C.$$

LEMMA 2.10. Liouville inequality, ([**Wal8**], §III-2). *Let $P \in \mathbf{Z}[X_1, \ldots, X_m]$ be a polynomial, and $\alpha_1, \ldots, \alpha_m$ be algebraic numbers. If $P(\alpha_1, \ldots, \alpha_m)$ is not zero then it is not "too small" :*

$$\log |P(\alpha_1, \ldots, \alpha_m)| \geq -[\mathbf{Q}(\underline{\alpha}) : \mathbf{Q}].\left(\log L(P) + \sum_{j=1}^{m} h(\alpha_j).\deg_{X_j} P\right) ,$$

where $\mathbf{Q}(\underline{\alpha}) = \mathbf{Q}(\alpha_1, \ldots, \alpha_m)$.

2.5. A proof of Mahler's conjecture.

THEOREM 2.11. (See [**BDGP**]). *Let $q \in D^*$ be an algebraic number. Then $J(q)$ is a transcendental number.*

The same result holds in p-adic and the proof is quite similar. This result is transcribed in terms of theta functions by D. Bertrand in [**Ber7**] (see also the previous chapter).

The theorem can be reformulated in "elliptic" terms: *Let $\wp$ be a Weierstrass's elliptic function with algebraic invariants g_2, g_3. Let $\mathbf{Z}\omega_1 + \mathbf{Z}\omega_2$ be the lattice of periods of $\wp$, with $\omega_2/\omega_1 \in \mathcal{H}$. For all non zero algebraic number α and all determination $\log \alpha$ of its logarithm we have:*

$$\begin{vmatrix} \omega_1 & 2i\pi \\ \omega_2 & \log \alpha \end{vmatrix} \neq 0.$$

Take $q = \exp(2i\pi\omega_2/\omega_1)$ in theorem 2.11. We have no direct proof of this result (there is a mistake in [**Nag**]).

PROOF OF THEOREM 2.11. There are five steps; in steps 1 and 2 we do not need $(q, J(q))$. The parameters N, L_1, L_2 are integers ≥ 1. The "constants" $C_1, C_2, \ldots$ are positive real numbers independent of N, L_1, L_2.

First step. *Construction of an auxiliary function.* We set $\mathcal{L} := \{\lambda = (\lambda_1, \lambda_2) \in \mathbf{Z}^2; 0 \leq \lambda_1 < L_1, 0 \leq \lambda_2 < L_2\}$; we want to construct a non-zero polynomial $A \in \mathbf{Z}[X_1, X_2]$,

$$A(X_1, X_2) = \sum_{\lambda \in \mathcal{L}} a_\lambda . X_1^{\lambda_1} . X_2^{\lambda_2}$$

such that the associated function $F : D \to \mathbf{C}$

$$F(z) := \sum_{\lambda \in \mathcal{L}} a_\lambda . z^{\lambda_1} . (\Delta^{2L_2} . J^{\lambda_2})(z)$$

$$= \sum_{\lambda \in \mathcal{L}} a_\lambda . z^{\lambda_1} . (\Delta(z))^{2L_2 - 2\lambda_2} . (\Delta^2 J(z))^{\lambda_2} ,$$

has a "great zero" in 0, precisely such that: $\mathrm{ord}_0 F \geq N$.

For $z \in D^*$ one has $F(z) = \Delta(z)^{2L_2} . A(z, J(z))$ and F is analytical on D; put : $F(z) = \sum_{n \geq 0} b_n . z^n$. Since $(\Delta^{2L_2} J^{\lambda_2})(z) = \sum_{m \geq 1} c_{L_2\lambda_2}(m) . z^m$, we have

$$F(z) = \sum_{\lambda \in \mathcal{L}} \sum_{m \geq 1} a_\lambda . c_{L_2\lambda_2}(m) . z^{m+\lambda_1}$$

and :

$$b_n = \sum_{\substack{\lambda \in \mathcal{L} \\ \lambda_1 < n}} a_\lambda . c_{L_2\lambda_2}(n - \lambda_1) \in \mathbf{Z} .$$

The condition "$\mathrm{ord}_0 F \geq N$" is also "$b_n = 0$ for $0 \leq n < N$" i.e.:

$(\mathcal{S})$ $\qquad \sum_{\substack{\lambda \in \mathcal{L} \\ \lambda_1 < n}} a_\lambda . c_{L_2\lambda_2}(n - \lambda_1) = 0 \quad$ for $\quad 0 \leq n < N$.

This is a linear system of equations (N equations and $L_1 L_2$ variables, the a_λ, $\lambda \in \mathcal{L}$) with coefficients in $\mathbf{Z}$. In order to use Lemma 2.6 (Siegel's lemma) we ask a first condition

$(\mathcal{C}1)$ $$L_1 L_2 \geq 2N \ .$$

By lemma 2.4 we have the estimate, for each equation in $(\mathcal{S})$:

$$\sum_{\substack{\lambda \in \mathcal{L} \\ \lambda_1 < n}} |c_{L_2 \lambda_2}(n - \lambda_1)| \leq \sum_{\substack{\lambda \in \mathcal{L} \\ \lambda_1 < n}} C_1^{L_2}.(n - \lambda_1)^{12 L_2}$$

$$\leq (L_1 L_2).C_1^{L_2}.N^{12 L_2} \ .$$

By lemma 2.6, there exists a solution $\{a_\lambda; \lambda \in \mathcal{L}\}$ of $(\mathcal{S})$ where the a_λ are integers, not all zero, with

$$|a_\lambda| \leq (L_1 L_2 C_1^{L_2} N^{12 L_2})^{N/(L_1 L_2 - N)} \ .$$

By $(\mathcal{C}1)$ we have $N/(L_1 L_2 - N) \leq 1$, so:

$$|a_\lambda| \leq (L_1 L_2).C_1^{L_2}.N^{12 L_2} \ .$$

In particular, for N sufficiently large, we have:

$$L(A) := \sum_{\lambda \in \mathcal{L}} |a_\lambda| \leq L_1^2.N^{13 L_2}.$$

Since A is not zero the function $z \mapsto A(z, J(z))$ is not the zero function (see Lemma 2.7), set $M := \mathrm{ord}_0 F$. Therefore we have $M \geq N$ and $F(z) = \sum_{n \geq M} b_n.z^n$, $b_M \neq 0$.

Second step. *Upper bound for $|F(z)|$, $z \in D^*$.* From

$$F(z) = z^M . \sum_{\nu \geq 0} b_{M+\nu}.z^\nu$$

we can write

$$G(z) := z^{-M}.F(z) = \sum_{\nu \geq 0} b_{M+\nu}.z^\nu \ .$$

From the expression $b_{M+\nu} = \sum_{\substack{\nu \in \mathcal{L} \\ \lambda_1 < M+\nu}} a_\lambda.c_{L_2 \lambda_2}(M + \nu - \lambda_1)$ we get :

$$|b_{M+\nu}| \leq (L_1 L_2)^2.C_1^{2 L_2}.N^{12 L_2}.(M + \nu)^{12 L_2} \ .$$

For $|z| < 1$ we have:

$$\sum_{\nu \geq 0} (M + \nu)^{12 L_2}.|z|^\nu \leq (2M)^{12 L_2} \left(1 + \sum_{\nu \geq 1} \nu^{12 L_2}.|z|^\nu \right) \ .$$

For all $k \in \mathbf{N}^*$ and all $x \in [0, 1[$ we have (exercise) :

$$1 + \sum_{\nu \geq 1} \nu^k.x^\nu \leq k!/(1 - x)^{k+1} \ .$$

The two inequalities give:

$$\sum_{\nu \geq 0} (M + \nu)^{12 L_2}.|z|^\nu \leq (2M)^{12 L_2}.((12 L_2)!)/(1 - |z|)^{12 L_2 + 1} \ .$$

So for $z \in D$, we have

$$|G(z)| \le \sum_{\nu \ge 0} |b_{M+\nu}|.|z|^{\nu}$$

$$\le (L_1 L_2)^2 . C_1^{2L_2} . N^{12L_2} . \sum_{\nu \ge 0} (M + \nu)^{12L_2} . |z|^{\nu}$$

$$\le (L_1 L_2)^2 . C_1^{2L_2} . N^{12L_2} . (2M)^{12L_2} . ((12L_2)\,!)/(1 - |z|)^{12L_2+1} \ ,$$

and, for L_2 sufficiently large :

$$|G(z)| = \left| \frac{F(z)}{z^M} \right| \le L_1^2 . L_2^{13L_2} . M^{24L_2}/(1 - |z|)^{12L_2+1}.$$

Pause introducing $(q, J(q))$. We suppose now that there exists $q \in D^*$ such that q and $J(q)$ are algebraic numbers. And we want to find a contradiction!

Third step. F *cannot be zero on many points* $q, q^2, q^3 \ldots$ Since F is not the zero function it cannot be zero on all points q^s, $s \in \mathbf{N}^*$. Set $S := \inf\{s \in \mathbf{N}^*; F(q^s) \ne 0\}$; so that : $F(q^s) = 0$ for $1 \le s \le S - 1$ and $F(q^S) \ne 0$. We shall give an upper bound for S (it is a kind of "zero lemma"). On the disk $|z| < r$ (where $r := (1 + |q|)/2$) F vanishes at 0 (with order M) and at $q, q^2, \ldots, q^{S-1}$; we kill these zeros by introducing the function $H : D \to \mathbf{C}$ defined by

$$H(z) := F(z). \frac{1}{z^M} . \prod_{s=1}^{S-1} \frac{r^2 - \overline{q}^s.z}{r(z - q^s)}$$

(we can also take $H(z) := F(z). \frac{1}{z^M} . \prod_{s=1}^{S-1} \frac{1}{z - q^s})$.

The function H is analytical on the disk $|z| < 1$ and the "maximum principle" gives :

$$|H(0)| \le \sup_{|z|=r} |H(z)| =: |H|_r \ .$$

For $|z| = r$, Blaschke's factors $\frac{r^2 - \overline{q}^s z}{r(z - q^s)}$ are ≤ 1 (in modulus) and : $|H(0)| \le |H|_r \le |G|_r$. Using our second step we get :

$$|H(0)| \le L_1^2 . L_2^{13L_2} . M^{24L_2}/(1 - r)^{12L_2+1} \ ,$$

and for L_2 "sufficiently large" : $|H(0)| \le L_1^2 . (L_2 . M)^{24L_2}$. On the other hand :

$$|H(0)| = \left| G(0). \prod_{s=1}^{S-1} \frac{-r}{q^s} \right| = |b_M|. \frac{r^{S-1}}{|q|^{S(S-1)/2}} \ .$$

We know that $b_M \in \mathbf{Z} \setminus \{0\}$ (thus $|b_M| \ge 1$) and

$$|H(0)| \ge \frac{r^{S-1}}{|q|^{S(S-1)/2}} > \frac{|q|^{S-1}}{|q|^{S(S-1)/2}} = \left(\frac{1}{|q|} \right)^{(S-1)(S-2)/2} \ .$$

Combining upper and lower bounds for $|H(0)|$ we get :

$$\frac{1}{2}(S - 1)(S - 2) \log \frac{1}{|q|} < \log |H(0)| \le 2 \log L_1 + 24L_2 \log(L_2 M) \ .$$

If $S \geq 3$ we have $(S-1)(S-2)/2 \geq S^2/9$ and :

$$\left(\frac{1}{9}\log\frac{1}{|q|}\right) S^2 < 2\log L_1 + 24L_2\log(L_2 M) \ .$$

This remains true if $S = 1, 2$ (at least for L_2 "sufficiently large"). So we get :

$$S^2 < C_3(\log L_1 + L_2\log(L_2 M)) \quad \text{where} \quad C_3 = C_3(|q|).$$

Fourth step. *A lower bound for* $|F(q^S)|$. We have

$$F(q^S) = (\Delta(q^S))^{2L_2}.A(q^s, J(q^S))$$

and $F(q^s) \neq 0$. First we give a lower bound for $|A\left(q^S, J(q^S)\right)|$. We have seen (§2.4) that $J(q^S)$ is algebraic, so we can use lemma 2.10 (Liouville inequality) :

$$\log|A(q^S, J(q^S))| \geq -d(q^S).d(J(q^S)). \left(\log L(A) + L_1 h(q^S) + L_2 h(J(q^S))\right) \ .$$

Furthermore $d(q^S) \leq d(q)$, $h(q^S) = Sh(q)$ and we conclude :

$$\log|A(q^S, J(q^S))| \geq -d(q).d(J(q^S)).(\log L(A) + L_1 Sh(q))$$
$$- d(q).L_2.\log M(J(q^S)) \ .$$

Now we use :

- the estimate of $L(A)$ (first step);
- the upper bound of $d(J(q^S))$ coming from lemma 2.5:

$$d(J(q^S)) \leq \psi(S).d(J(q)) \leq C_4.S.\log\log(3S);$$

- the upper bound of $\log M(J(q^S))$ coming from corollary 2.9 :

$$\log M(J(q^S)) \leq C_5.S.\log(2S).\log\log(3S).$$

This gives:

$$\log|A(q^S, J(q^S))| \geq -C_6.S.\log\log(3S).(L_1 S + L_2\log(2NS)) \ .$$

Now, we establish a lower bound for $|\Delta(q^S)|$ by proposition 2.2 :

$$\Delta(q^S) = q^S \prod_{n\geq 1}(1 - q^{nS})^{24} \ ,$$

and the function $z \mapsto \prod_{n\geq 1}(1 - z^n)^{24}$ is analytical on the disk $|z| \leq r$ without zero; so there exists $C_7 > 0$ such that $|\prod_{n\geq 1}(1 - z^n)^{24}| \geq C_7$ for $|z| \leq r$. In particular,

$$|\Delta(q^S)| \geq C_7.|q|^S \ .$$

EXERCISE. For $|z| \leq |q|$ we have $|\Delta(z)| \geq |z|\exp\left(\frac{-|q|}{(1-|q|)^2}\right)$.

Returning to $|F(q^S)|$, all this gives :

$$\log|F(q^S)| \geq -C_8.S.\log\log(3S).(L_1 S + L_2\log(2NS)) \ .$$

Last step. *An upper bound for* $|F(q^S)|$ - *Conclusion.* Using step 2 we have an upper bound for $|F(q^S)|$:

$$|F(q^S)| \leq |q|^{SM}.L_1^2.L_2^{13L_2}.M^{24L_2}/(1-|q|^S)^{12L_2+1} \ .$$

This gives :

$$|F(q^S)| \leq \left(\frac{1}{|q|}\right)^{-SM} . \exp C_9(\log L_1 + L_2\log(L_2M)) \ .$$

Combining the upper and lower bounds for $|F(q^S)|$ we have :

$$- C_8 S \log\log(3S)(L_1S + L_2\log(2NS)) \leq -SM\log\left(\frac{1}{|q|}\right)$$
$$+ C_9(\log L_1 + L_2\log(L_2M)) \ .$$

Therefore :

$$SM \leq C_{10}L_2\log(L_2M) + C_{10}S\log\log(3S)(L_1S + L_2\log(2NS)) \ ,$$

and consequently :

$$M \leq C_{10}(L_2\log(L_2M) + \log\log(3S).(L_1S + L_2\log(2NS))) \ .$$

To reach a contradiction it is sufficient to satisfy the three constraints :

$(\mathcal{C}2)$ $\qquad\qquad M > 3\,C_{10}L_2\log(L_2M) \ ,$

$(\mathcal{C}3)$ $\qquad\qquad M > 3\,C_{10}\log\log(3S)L_2\log(2NS) \ ,$

$(\mathcal{C}4)$ $\qquad\qquad M > 3\,C_{10}\log\log(3S)L_1S \ .$

The problem is now to choose the parameters L_1, L_2, N such that the four constraints $(\mathcal{C}1 - \mathcal{C}2 - \mathcal{C}3 - \mathcal{C}4)$ are satisfied, knowing that $N \leq M$ and $S^2 \leq C_3(\log L_1 + L_2\log(L_2M))$. It is easy to see that the choice $L_1 = L_2 = 2N^{1/2}$ (with N square and sufficiently large) is valid.

So we eventually get a contradiction! The theorem is proved. $\qquad\square$

REMARK. This proof is purely modular (we didnot use elliptic functions or elliptic curves). The main tools are the modular polynomials Φ_n, which reflect the modular properties of J, j; they allow to use the sequence of points $(q^n)_{n\in\mathbf{N}}$. The proof of Yu.V. Nesterenko ([**Nes8**], [**Nes9**]) is quite different : he takes only the point q but with derivatives; and this is possible because he works with the Ramanujan's functions P, Q, R which satisfy a well known differential system.

2.6. An other proof of Mahler's conjecture. P. Philippon presents in [**PPh10**] an "approche méthodique" in order to produce transcendence results and algebraic independence statements, for example he can prove Mahler's conjecture. In the case of $(q, J(q))$, there are two main differences with the proof given here :

- the auxiliary function is replaced by the technics of interpolation determinants of M. Laurent;
- in § 2.5, we made our construction at the point 0 with multiplicity N and we got the result at the point q^S; in [**PPh10**] there is an infinite number of steps, each step being an exchange between the point 0 (with an increasing order of multiplicity) and families $\{q^n \ ; \ 1 \leq n \leq N\}$ with increasing order N.

3. K. Barré's work on modular functions

3.1. Quantitative aspects of Mahler-Manin's conjecture. Theorem 2.11 says that if $q \in D^*$ then q and $J(q)$ are not simultaneously algebraic. So a natural problem is to give a lower bound for $|q - \alpha| + |J(q) - \beta|$ which is explicit in the degrees and heights of the algebraic numbers α, β. This work has been done by K. Barré, in the complex case as well as in the $p-$adic case (see [**Bar2**], [**Bar3**], chap.3). The method of proof is roughly the same as for theorem 2.11 but there are technical difficulties. For example, she needs an upper bound for the order of vanishing at 0 of the auxiliary function ("zero lemma"); here is such a result, improving upon G. Philibert's lemma [**Phi2**].

THEOREM 3.1. ([**Bar3**], chap 2). *Let $P \in \mathbf{C}[X_1, X_2]$ be a non zero polynomial of respective degrees in each variable L_1, L_2 (with $L_2 \geq 1$). The order of vanishing at 0 of the Laurent series $F(T) := P(T, J(T)) \in \mathbf{C}((T))$ satisfies $\mathrm{ord}_0 F \leq 9L_1L_2$.*

In the complex case, the measure of simultaneous approximation for $(q, J(q))$ is the following (where $m(\alpha) := \max(e^e; \log M(\alpha))$ for the algebraic number α).

THEOREM 3.2. ([**Bar2**], théorème 2). *Let $q \in D^*$, there exists a constant $C > 0$ such that for any algebraic numbers α, β, the following inequality holds :*

$$\log(|q - \alpha| + |J(q) - \beta|) \geq -C\, d(\alpha)m(\alpha)^3 d(\beta)^3 (d(\beta) + m(\beta))L_{\alpha,\beta}^3.(\log L_{\alpha,\beta})^4$$

where $L_{\alpha,\beta} := \log(d(\alpha) + m(\alpha) + d(\beta) + m(\beta))$.

There is an explicit formula $C = C(|q|)$ (see [**Bar3**], chap 3). From such an approximation measure we deduce two transcendence measures (for q when $J(q)$ is algebraic and for $J(q)$ when q is algebraic). Here is one example.

THEOREM 3.3. ([**Bar2**], corollaire 1). *Let $q \in D^*$ be an algebraic number. There exists a constant $C' > 0$ such that for any polynomial $P \in \mathbf{Z}[X]$ with degree $D \geq 1$ and height (in the naïve sense) at most H, with $\log \log H \geq e$, the following inequality holds :*

$$\log |P(J(q))| \geq -C'D^3(D + \log H)(\log(D + \log H))^3(\log \log(D + \log H))^4 \ .$$

This gives (only) a transcendence type for $J(q)$ lower than $4 + \varepsilon$, but the measure has a very good dependence in H. Other results, and their p-adic versions, can be found in [**Bar1**]-[**Bar2**]-[**Bar3**]. We note that Yu.V. Nesterenko and P. Philippon have also quantitative results ([**Nes10**], theorem 2 and 4; [**PPh11**], théorème 3). These results are much more general than theorems 3.2-3.3 and are very accurate; but specialised for $(q, J(q))$ they don't cover K. Barré results.

3.2. A modular proof of an old result of D.Bertrand. K. Barré gives a purely modular proof of the following result.

THEOREM 3.4. ([**Bar3**], chap 4). *Let $q \in D^*$ be such that $J(q) \notin \{0, 1728\}$ then $J(q)$ and $q.J'(q)$ are not simultaneously algebraic.*

The same method yields the same result for $p-$adic fields. These are corollaries of D. Bertrand's work on Eisenstein series (see [**Ber1**]) and therefore was known for a long time. But the proofs are very different. D. Bertrand used the role played by Eisenstein series in the theory of elliptic curves; K. Barré doesn't use anything else than J, J' and the modular polynomials (with derivatives, so there are a lot of technical difficulties).

The method can be quantified and gives a measure of simultaneous approximation for $(J(q), q.J'(q))$. This work has been done in the p–adic case ([**Bar3**], chap 4), and so we have a transcendence measure of $J(q)$ when $q.J'(q)$ is algebraic, and a transcendence measure of $q.J'(q)$ when $J(q)$ is algebraic. The proof uses a special zero estimate ([**Bar3**], théorème 4-3), since there is no general result available.

It would be interesting to unify existing zero lemmas of this kind (K. Barré, D. Bertrand, Yu.V. Nesterenko); there is a conjecture and a partial result in [**Ber7**], §4.

4. Conjectures about modular and exponential functions

One of the oldest conjectures of transcendental number theory is the so-called "four exponential conjecture" (noted here (C4E)), which is also the first of the eight problem of Th. Schneider, 1957 (see [**Wal8**], §1.4, for an historical survey on this point).

In 1996, D. Bertrand formulated some conjectures related to the modular function j ([**Ber7**], §5). They are very interesting by themselves; and, moreover, they lead to a special case of the four exponential problem. So, there is a conceivable approach to (C4E) by means of the modular function j!

We present in the sequel four equivalent conjectures which can be seen as a step in this study; they are also of independent interest (see [**Dia3**] for all proofs and some other results).

4.1. Conjectures (C4E), (C4EW), (CDB). Here we state the two strong conjectures and a weaker one.

FOUR EXPONENTIAL CONJECTURE (C4E). *Let x_1, x_2 be two **Q**-linearly independent complex numbers and y_1, y_2 also two **Q**–linearly independent complex numbers. Then at least one of the four numbers $\exp(x_i y_j)$, $(i = 1, 2; j = 1, 2)$, is transcendental over **Q**.*

WEAK FORM (C4EW). *Let α_1, α_2 be positive real algebraic numbers, both different from 1. Then π^2 and the product $(\log \alpha_1)(\log \alpha_2)$ are linearly independent over **Q** (here $\log \alpha_1$, $\log \alpha_2$ are the usual logarithms).*

D.BERTRAND'S CONJECTURE (CDB). ([**Ber7**], conjecture 2-ii). *Let q_1, q_2 be non zero algebraic numbers in D. If q_1, q_2 are multiplicatively independent, then $J(q_1)$, $J(q_2)$ are algebraically independent over **Q**.*

Conjecture (C4EW) is a corollary of (C4E), but also of (CDB) (see [**Ber7**], §5, and theorem 7).

4.2. Four weaker conjectures.

THEOREM 4.1. ([**Dia3**],théorème 1). *The assertions (CI), (CII), (CIII), (CIV) are equivalent.*

- *(CI) Let q_1, q_2 be non zero algebraic numbers in D. If there exists $n \in \mathbf{N} \backslash \{0\}$ such that $\Phi_n(J(q_1), J(q_2)) = 0$ then q_1, q_2 are multiplicatively dependent.*
- *(CII) Let q_1, q_2 be non zero algebraic numbers in D. If $J(q_1) = J(q_2)$ then $q_1 = q_2$.*

- *(CIII) Let $\tau \in \mathcal{H}$; then $\exp(2i\pi\tau)$ and $\exp(-2i\pi/\tau)$ are not simultaneously algebraic.*
- *(CIV) Let α_1, α_2 be non zero algebraic numbers with modulus different from 1. Then π^2 and the product $(\log\alpha_1)(\log\alpha_2)$ are linearly independent over $\mathbf{Q}$ (for any determinations of the logarithms $\log\alpha_1$ and $\log\alpha_2$).*

The link between conjectural statements of § 4.1 and (CI), (CII), (CIII), (CIV) is given by the following result.

THEOREM 4.2. ([**Dia3**], théorème 2).

1. (C4E) *implies each of the assertions* (CI), (CII), (CIII), (CIV).
2. (CDB) *implies each of the assertions* (CI), (CII), (CIII), (CIV).
3. *Each of the assertions* (CI), (CII), (CIII), (CIV) *implies* (C4EW).

So the assertion (CIII) (" $\exp(2i\pi\tau)$ and $\exp(-2i\pi/\tau)$ are not simultaneously algebraic"), which is "purely" exponential, is a corollary of (CDB), which is "purely" modular.

A corollary of (CIII) is worth pointing out: let $\tau \in \mathcal{H}$; if $|\tau| = 1$ then $\exp(2i\pi\tau)$ is a transcendental number! A proof of this assertion would be very interesting in itself.

Toward (CIII) we have the following result.

THEOREM 4.3. ([**Dia3**], proposition 3).
Let $\tau \in \mathcal{H}$; *as soon as one of the following conditions* $(1),\dots,(5)$ *is true then* $\exp(2i\pi\tau)$ *and* $\exp(-2i\pi/\tau)$ *are not simultaneously algebraic:*

1. $2i\pi$ *and* τ *are algebraically dependent over* $\mathbf{Q}$;
2. $\mathrm{Re}(\tau) \in \overline{\mathbf{Q}} \setminus \{0\}$;
3. $\mathrm{Im}(\tau) \in \overline{\mathbf{Q}}$;
4. $\mathrm{Re}(\tau)/|\tau|^2 \in \overline{\mathbf{Q}} \setminus \{0\}$;
5. $\mathrm{Im}(\tau)/|\tau|^2 \in \overline{\mathbf{Q}}$.

A method to solve these conjectures remains to be found! We can hope that the new ideas introduced in [**BDGP**], [**Ber7**], [**Nes9**], [**PPh10**], [**PPh11**] will be useful. But the job is left to be done.

CHAPTER 3

Algebraic independence for values of Ramanujan functions

1. Main theorem and consequences.

Consider the functions

$$P(z) = 1 - 24 \sum_{n=1}^{\infty} \sigma_1(n) z^n, \quad Q(z) = 1 + 240 \sum_{n=1}^{\infty} \sigma_3(n) z^n,$$

$$R(z) = 1 - 504 \sum_{n=1}^{\infty} \sigma_5(n) z^n,$$

where $\sigma_k(n) = \sum_{d|n} d^k$. These functions are connected to Eisenstein series $E_2(\tau)$, $E_4(\tau)$, $E_6(\tau)$ by the identities

$$E_2(\tau) = P(e^{2\pi i \tau}), \quad E_4(\tau) = Q(e^{2\pi i \tau}), \quad E_6(\tau) = R(e^{2\pi i \tau}), \quad \operatorname{Im} \tau > 0.$$

They were specially considered by Ramanujan [**Ram**][1], in 1916, who proved in particular, that these functions satisfy the system of differential equations

$$(2) \qquad DP = \frac{1}{12}(P^2 - Q), \quad DQ = \frac{1}{3}(PQ - R), \quad DR = \frac{1}{2}(PR - Q^2),$$

where $D = z\frac{d}{dz}$.

This Chapter is devoted to the proof of the next theorem.

THEOREM 1.1 ([**Nes9**]). *For any complex $q, 0 < |q| < 1$, the set*

$$q, \quad P(q), \quad Q(q), \quad R(q)$$

contains at least three algebraically independent over **Q** *numbers.*

We point out here several corollaries of the theorem. Take $q = e^{-2\pi}$. Then $P(q) = \frac{3}{\pi}$, $Q(q) = 3\frac{\Gamma(\frac{1}{4})^8}{(2\pi)^6}$, $R(q) = 0$, see Chapter 1, and we derive a Corollary.

COROLLARY 1.2. *The numbers $\pi, e^{\pi}, \Gamma(\frac{1}{4})$ are algebraically independent over rationals.*

Ramanujan functions are related to functions of different classes and this is a reason for many corollaries of Theorem 1.1. We divide the corollaries on three groups in connection with applications to modular, elliptic and theta-functions.

(∗) Chapter's author : Yuri V. NESTERENKO.

---

[1][**Ram**] S. Ramanujan. On certain arithmetical functions, *Trans. Cambridge Phil. Soc.* 22/9, (1916), 159–184; Collected papers, Chelsea Publ. Co., (1962), 136–162.

1.1. Modular functions. We set

$$\Delta(z) = \frac{1}{1728}(Q(z)^3 - R(z)^2), \qquad J(z) = \frac{Q(z)^3}{\Delta(z)} = \frac{1}{z} + 744 + \sum_{n=1}^{\infty} c(n)z^n.$$

The last series gives the Fourier expansion of the modular function $j(\tau) = J(e^{2\pi i \tau})$. Using (2) it is easy to compute, that

$$P = 6\frac{D^2 J}{DJ} - 4\frac{DJ}{J} - 3\frac{DJ}{J - 1728},$$

$$Q = \frac{(DJ)^2}{J(J - 1728)}, \quad R = -\frac{(DJ)^3}{J^2(J - 1728)}.$$

If we take into account that one can have $j(\tau) = 0$, $j(\tau) = 1728$ or $j'(\tau) = 0$ only for points τ that are equivalent to i or $\zeta = e^{2\pi i}$ under the action of modular group, then we obtain the following consequences.

COROLLARY 1.3. *If* $\tau \in \mathbf{C}$, $\mathrm{Im}\,\tau > 0$ *is not equivalent to* i *or* ζ *under the modular group, and we set* $q = e^{2\pi i \tau}$, *then each set*

$$\{q,\ J(q),\ DJ(q),\ D^2 J(q)\}, \qquad \{q,\ j(\tau),\ \pi^{-1} j'(\tau),\ \pi^{-2} j''(\tau)\},$$

contains at least three numbers that are algebraically independent over $\mathbf{Q}$.

Notice that algebraic independence of $DJ(q), D^2 J(q)$ in assumption that $J(q)$ is algebraic and distinct from 0 and 1728 was proved in [**Ber2**].

COROLLARY 1.4. *Suppose that* q *is an algebraic number with* $0 < |q| < 1$. *Then each set*

$$\{P(q), Q(q), R(q)\}, \qquad \{J(q), J'(q), J''(q)\},$$

is algebraically independent over $\mathbf{Q}$. *In particular, all these numbers are transcendental.*

The Corollary 1.3 follows immediately from Theorem 1.1. To prove Corollary 1.4, we note that if $q = e^{2\pi i \tau}$, then by Gelfond-Schneider theorem, τ cannot be an algebraic irrational; hence it is not equivalent to i or ζ under the modular group. The Corollary now follows from Corollary 1.3.

The statement about algebraic independence of second set from Corollary 1.4 was conjectured in 1977 in [**Ber2**].

The transcendence of $J(q)$ for algebraic $q, 0 < |q| < 1$, was proved in 1995 in [**BDGP**]. This assertion was conjectured by Mahler (1969) and Manin (1971, p-adic version). Details of the proof see also in Chapter 2 of this volume.

1.2. Elliptic functions. Let $\wp(z)$ be an elliptic function of Weierstrass with invariants g_2, g_3, periods ω_1, ω_2, corresponding quasi-periods η_1, η_2 and $\tau = \omega_2/\omega_1$, $\mathrm{Im}\,\tau > 0$. Setting $q = e^{2\pi i \tau}$ we have, see [**Lang**][2], Chapter 4,

$$P(q) = 3\frac{\omega_1}{\pi} \cdot \frac{\eta_1}{\pi}, \qquad Q(q) = \frac{3}{4}\left(\frac{\omega_1}{\pi}\right)^4 \cdot g_2, \qquad R(q) = \frac{27}{8}\left(\frac{\omega_1}{\pi}\right)^6 \cdot g_3.$$

These formulae imply that all three numbers $P(q), Q(q), R(q)$ belong to the field $\mathbf{Q}(g_2, g_3, \omega/\pi, \eta/\pi)$ and we derive

---

[2][**Lang**] S. Lang. *Elliptic functions*, second edition, Grad. Texts Math. 112, Springer, (1987).

COROLLARY 1.5. *Let $\wp(z)$ be the Weierstrass $\wp$ –function with algebraic invariants g_2, g_3. Then the numbers*

$$e^{2\pi i(\omega_2/\omega_1)}, \quad \frac{\omega_1}{\pi}, \quad \frac{\eta_1}{\pi}$$

are algebraically independent over **Q**.

COROLLARY 1.6. *Let $\wp(z)$ be the Weierstrass elliptic function with algebraic invariants g_2, g_3 and complex multiplication by the field* **k**. *If ω is any period of $\wp(z)$, η is the corresponding quasi-period, and $\tau \in$* **k**, *$\mathrm{Im}\,\tau \neq 0$, then each of sets*

$$\left\{\pi, \ \omega, \ e^{2\pi i \tau}\right\}, \qquad \left\{\omega, \ \eta, \ e^{2\pi i \tau}\right\}$$

is algebraically independent over **Q**.

In the complex multiplication case the numbers ω_2 and η_2 are algebraic over the field $\mathbf{Q}(\omega_1, \eta_1)$ (see [**Mas1**, Chapter 3]). Using the Legendre relation, we then find that η_1 is algebraic over the field $\mathbf{Q}(\omega_1, \pi)$ and π is algebraic over the field $\mathbf{Q}(\omega_1, \eta_1)$. Since any primitive period ω can be taken as ω_1, this proves the Corollary 1.6.

Notice that algebraic independence of numbers $\frac{\omega_1}{\pi}, \frac{\eta_1}{\pi}$ in conditions of Corollary 1.5, and numbers $\{\pi, \omega\}$, $\{\omega, \eta\}$ in conditions of Corollary 1.6 was proved in [**Chu2**], [**Wal2**]. These results in particular imply algebraic independence of numbers $\{\pi, \Gamma\left(\frac{1}{3}\right)\}$ and $\{\pi, \Gamma\left(\frac{1}{4}\right)\}$, which also are contained in [**Chu2**], [**Wal2**].

For any natural number d there exists a Weierstrass $\wp$ –function with algebraic invariants and with complex multiplication field $\mathbf{Q}(\sqrt{-d})$. Thus with the help of Corollary 1.6 we obtain the following assertion.

COROLLARY 1.7. *For any natural number d the numbers*

$$\pi, \ e^{\pi\sqrt{d}},$$

are algebraically independent over **Q**.

1.3. Theta–functions. We will use the following notations for theta-functions

$$\theta_2(z) = 2z^{1/4} \sum_{n \geq 0} z^{n(n+1)}, \qquad \theta_3(z) = 1 + 2 \sum_{n \geq 1} z^{n^2},$$

$$\theta_4(z) = 1 + 2 \sum_{n \geq 1} (-1)^n z^{n^2}.$$

For $q = e^{2\pi i \tau}, \mathrm{Im}\,\tau > 0$ relations hold, see [**Law**][3],

$$P(q^2) = 4\left(\frac{D\theta_2}{\theta_2} + \frac{D\theta_3}{\theta_3} + \frac{D\theta_4}{\theta_4}\right), \quad Q(q^2) = \frac{1}{2}\left(\theta_2^8 + \theta_3^8 + \theta_4^8\right),$$

$$R(q^2) = \frac{1}{2}(\theta_4^4 - \theta_2^4)(\theta_2^4 + \theta_3^4)(\theta_3^4 + \theta_4^4), \quad D = q\frac{d}{dq}.$$

Here and later we use the notations $\theta_j = \theta_j(q)$. Using formulas, see [**Law**] (*loc. cit.*),

$$\theta_2^4 = 4\left(\frac{D\theta_3}{\theta_3} - \frac{D\theta_4}{\theta_4}\right), \qquad \theta_3^4 = 4\left(\frac{D\theta_2}{\theta_2} - \frac{D\theta_4}{\theta_4}\right),$$

(3)

---

[3][**Law**] D.F. Lawden. *Elliptic functions and applications*, Springer, 1989.

$$\theta_4^4 = 4 \left(\frac{D\theta_2}{\theta_2} - \frac{D\theta_3}{\theta_3} \right),$$

we derive

COROLLARY 1.8 ([**Ber7**]). *For any complex* $q, 0 < |q| < 1$, *the set*

$$q, \quad \frac{D\theta_2}{\theta_2}, \quad \frac{D\theta_3}{\theta_3}, \quad \frac{D\theta_4}{\theta_4}$$

contains at least three algebraically independent numbers.

It is well known that the function

$$\Delta(z) = Q(z)^3 - R(z)^2.$$

satisfies the identities

$$\Delta(z) = z \prod_{n=1}^{\infty} (1 - z^n)^{24}, \quad \Delta(z^2) = 2^{-8} \left(\theta_2(z)\theta_3(z)\theta_4(z) \right)^8.$$

COROLLARY 1.9 ([**DNNS1**]). *Let* $f(z)$ *be one of the functions* $\theta_2(z)$, $\theta_3(z)$, $\theta_4(z)$, $\Delta(z)$. *For any algebraic number* $q, 0 < |q| < 1$, *the numbers*

$$f(q), \quad f'(q), \quad f''(q)$$

are algebraically independent over **Q**. *In particular the number* $f(q)$ *is transcendental.*

The reason for this assertion is the fact that all numbers from corollary 1.7 are algebraic over the field $\mathbf{K} = \mathbf{Q}(f(q), f'(q), f''(q))$ if q is an algebraic number. We present here the proof in the case $f(z) = \theta_3(z)$ (see [**Ber7**]).

PROOF. Using classical relations

$$\theta_2^4 \cdot \theta_4^4 = 8 \left(\frac{D^2\theta_3}{\theta_3} - 3 \left(\frac{D\theta_3}{\theta_3} \right)^2 \right),$$
$$\theta_2^4 + \theta_4^4 = \theta_3^4,$$

we conclude that numbers θ_2, θ_4 are algebraic over the field $\mathbf{K}$. Now relations (3) imply that all the numbers from Corollary 1.8 are algebraic over $\mathbf{K}$. $\qquad\square$

Notice that proof of Corollary 1.9 from [**DNNS1**] do not use the explicit form of the algebraic relations between theta-functions. Another proof of this corollary for theta-functions contained in [**Ber7**] gives results not only for algebraic q.

Applying last Corollary to the function $\theta_3(z)$ we derive that for any algebraic number $\alpha, 0 < |\alpha| < 1$, the numbers

$$\sum_{n\geq 0} \alpha^{n^2}, \quad \sum_{n\geq 1} n^2 \alpha^{n^2}, \quad \sum_{n\geq 1} n^4 \alpha^{n^2}$$

are algebraically independent. In particular we can conclude that the number

(4)
$$\sum_{n\geq 0} \alpha^{n^2}$$

is transcendental. As far back as 1851, in the process of constructing the first examples of transcendental numbers, Liouville presented in [**Lio**] the series (4) for $\alpha = l^{-1}, l \in \mathbf{Z}, l > 1$ as an example for which his method allowed one to prove only the irrationality. Later, in a number of papers, the irrationality of these series at points $\alpha \in \mathbf{Q}$ was proved under certain restrictions imposed on the numerators and denominators of α. No assertions on the transcendence of these sums were known.

COROLLARY 1.10 ([**DNNS1**]). *Let q is algebraic number, $0 < |q| < 1$, and function $f(z)$ is not a constant and is algebraic over the field $\mathcal{K} = \mathbf{Q}(P(z), Q(z), R(z))$. If the number $f(q)$ is defined, then it is transcendental.*

In particular this corollary implies the transcendence of the values of Dedekind eta—function

$$\eta(q^2) = e^{\frac{\pi i \tau}{12}} \cdot \prod_{n \geq 1} \left(1 - q^{2n}\right), \quad , \operatorname{Im} \tau > 0.$$

The reason is the relation $\eta(z)^{24} = \Delta(z)$.

Another example, see [**DNNS2**], is connected to the values of Rogers - Ramanujan continued fraction

$$RR(\alpha) = 1 + \cfrac{\alpha}{1 + \cfrac{\alpha^2}{1 + \cfrac{\alpha^3}{1 + \ddots}}}.$$

It is well known that for any integer $k, k \geq 1$, the functions $\Delta(z^k), \Delta(z^{1/k})$ are algebraic over the field $\mathcal{K}$. If we denote

$$F(q) = \cfrac{q^{1/5}}{1 + \cfrac{q}{1 + \cfrac{q^2}{1 + \cfrac{q^3}{1 + \ddots}}}},$$

then the equality holds

$$\frac{1}{F(q)} - F(q) - 1 = q^{2/5} \frac{\eta(q^{1/5})}{\eta(q^5)}.$$

All these relations imply the transcendence of the number $RR(\alpha)$ for algebraic $\alpha, 0 < |\alpha| < 1$.

One more example, [**DNNS2**], is the transcendence of the sum

$$\sum_{n=0}^{\infty} \frac{1}{F_n^2},$$

where F_n is the Fibonacci sequence.

The article [**DNNS2**] contains many other examples of transcendental series connected to Fibonacci numbers.

In Chapter 1, Corollary 1.2, the assertion containing both Corollaries 1.9 and 1.10 is proved.

Famous Schneider's theorem about transcendence of values of modular function $j(\tau)$ for algebraic not imaginary quadratic $\tau, \operatorname{Im} \tau > 0$, [**Sch2**], does not follow from Theorem 1.1. Both of these assertions can be deduced from the following

CONJECTURE 1.11. *Let be $\tau \in \mathbf{C}, \operatorname{Im} \tau > 0$ and assume that the set*

$$\tau, \quad q = e^{2\pi i \tau}, \quad P(q), \quad Q(q), \quad R(q)$$

contains at most three algebraically independent numbers over **Q**. *Then* τ *is imaginary quadratic and the numbers*

$$q = e^{2\pi i \tau}, \quad P(q), \quad Q(q)$$

are algebraically independent over **Q**.

Of course for imaginary quadratic τ the last assertion of the Conjecture 1.11 follows from Theorem 1.1.

2. How it can be proved?

The proof of the Theorem 1.1 presented in this section is based on the next Theorem 2.1, which is a very special form of P. Philippon's Criteria of algebraic independence, see [**PPh2**] and Chapter 8. For any polynomial A with integer coefficients we use a notation $H(A)$ for the maximum of moduli of its coefficients.

THEOREM 2.1. *Let* $\overline{\omega} = (\omega_1, \dots, \omega_n) \in \mathbf{C}^n$ *and there exists a sequence of polynomials* $A_N \in \mathbf{Z}[x_1, \dots, x_n]$ *such that*

$$\deg A_N \leq \tau(N), \qquad \log H(A_N) \leq \tau(N)$$

and

$$\exp(-\gamma_2 \lambda(N)) \leq |A_N(\overline{\omega})| \leq \exp(-\gamma_1 \lambda(N)),$$

where $\gamma_2 > \gamma_1 > 0$ *are constants,* $\tau(N), \lambda(N)$ *are increasing to* ∞ *real functions satisfying*

$$\lim_{N \to \infty} \frac{\lambda(N+1)}{\lambda(N)} = 1, \qquad \lim_{N \to \infty} \frac{\lambda(N)}{\tau(N)^{k+1}} = \infty$$

for an integer $k \geq 0$. *Then*

$$\operatorname{tr} \deg_{\mathbf{Q}} \mathbf{Q}(\omega_1, \dots, \omega_n) \geq k + 1.$$

PROOF. This Theorem easily follows from [**PPh2**, section 1, last Corollary], if we choose $\delta(N) = \tau(N), U(N) = \frac{\gamma_1}{2}\lambda(N)$ and $\sigma(N) = 3\frac{\gamma_2}{\gamma_1}$. $\qquad\square$

For the proof of Theorem 1.1 we need to construct the sequence of polynomials A_N satisfying the Theorem 2.1 for $\overline{\omega} = (q, P(q), Q(q), R(q))$ and $k = 2$. This construction is realized at the proof of the next lemma.

LEMMA 2.2. *There exists a sequence of polynomials*

$$A_N \in \mathbf{Z}[z, x_1, x_2, x_3], \qquad N \geq N_0,$$

such that

(5) $$\deg A_N \leq \gamma_0 N \ln N, \qquad \ln H(A_N) \leq \gamma_0 N \ln^2 N,$$

(6) $$\exp(-\gamma_2 N^4) \leq |A_N(q, P(q), Q(q), R(q))| \leq \exp(-\gamma_1 N^4),$$

where $\gamma_0, \gamma_1, \gamma_2$ *are constants depending only on* q.

The Theorem 1.1 follows from Theorem 2.1 if we choose $n = 4, k = 2, \tau(N) = \gamma_0 N \ln^2 N, \lambda(N) = N^4, \gamma_1, \gamma_2$ as in the Lemma 2.2 and $\overline{\omega} = (q, P(q), Q(q), R(q))$.

Now the proof of Theorem 1.1 is reduced to the proof of Lemma 2.2. Section 3 of the chapter contains the proof of this Lemma 2.2. It is based on the next Theorem 2.3 about zero bound for polynomials in functions $z, P(z), Q(z), R(z)$ which in some sense gives a bound for the measure of algebraic independence of these functions. Details and the proof of this theorem see in [**Nes9**] and Chapter 10.

THEOREM 2.3. *Let L_1, L_2 be integers, $L_1 \geq 1$, $L_2 \geq 1$. Then for any polynomial $A(z, x_1, x_2, x_3) \in \mathbf{C}[z, x_1, x_2, x_3]$, $A \not\equiv 0$, $\deg_z A \leq L_1$, $\deg_{x_i} A \leq L_2$, the inequality holds*

$$\mathrm{ord}_{z=0} A\big(z, P(z), Q(z), R(z)\big) \leq c L_1 L_2^3,$$

where c is an absolute constant.

It is proved in [**Nes9**] that the bound of Theorem 2.3 is true with $c = 10^{47}$.

In section 4 we include the algebraic tools needed for the proof of Theorem 2.3. The reason for this is that the Section 5 of this Chapter contains another proof of Theorem 1.1, which uses these tools instead of the Criteria of algebraic independence (Theorem 2). As consequence of this other proof is a measure of algebraic independence for numbers from Theorem 1.1 (see Theorem 5.1 in the section 5).

In the case when the point $(Q(q)^3, R(q)^2)$ is proportional to the point with algebraic coordinates the Theorem 1.1 can be reduced with the help of construction from Section 3 (Corollary 3.5) to some sharp estimate for the measure of algebraic independence of two numbers connected with an elliptic function with algebraic invariants, [**PPh10**] and Chapter 6. This estimate was announced in 1980 by G.Chudnovskii (see [**Chu2**]) and proved in 1988 by G.Philibert, [**Phi1**].

3. Construction of the sequence of polynomials

The sequence of polynomials, satisfying the Lemma 2.2, is constructed below with the help of ideas closely related to the method of Siegel-Shidlovskii, [**Shi**], applying to algebraic independence proof of values of E-functions. The next properties of Ramanujan functions are very important for the construction:

1) they have integer coefficients of Taylor expansion at the origine, which grows not very fast;

2) they satisfy the system of algebraic differential equations over $\mathbf{C}(z)$ (Ramanujan);

3) they are algebraic independent over $\mathbf{C}(z)$ (Mahler), see Chapter 1.

A properties similar to 1)-3) are typical for E-functions. In our case the system is not linear, but the coefficients grows slower. The situation was axiomatized by P.Philippon, [**PPh10**], who introduced the class of K-functions.

First of all we state the existence of an auxiliary function $F(z) \neq 0$ depending on integer parameter N, tending to infinity (Lemma 3.1). This function is constructed as a polynomial in $z, P(z), Q(z), R(z)$ which has integer coefficients growing not very fast in dependence on N and it has large order of zero at the origine. This step is based on Siegel's lemma on integer solutions of a linear system of homogeneous algebraic equations and the properties 1), 3) of Ramanujan functions.

Next we prove (Lemma 3.3) that one of numbers $F^{(T)}(q)$ with not very large T can be bounded from below by a function which depends on the order of zero $M = \mathrm{ord}_{z=0} F(z)$. The upper bound for T depends on M and the parameter N. The main tool here is the Interpolation formula, which was used in the same context by Gelfond, [**Gel3**], but specially for exponential function. Taylor coefficients of $F(z)$ at the origine are integers and the lower bound is based on this property.

Since Ramanujan functions satisfy the system of differential equations (2), the number $(12q)^T F^{(T)}(q)$ is a polynomial with integer coefficients in $q, P(q), Q(q)$, $R(q)$. This polynomial is the polynomial A_N from the Lemma 2.2. Coefficients and degrees of this polynomial can be bounded from above in terms of N and M. The upper bound for its value at the point $q, P(q), Q(q), R(q)$ follows from an upper bound for $F^{(T)}(q)$. The last derivative is small since the order of zero M is very large and the proof is based on direct estimates (Lemma 3.2) and Cauchy's integral formula.

All our bounds and estimates are expressed in terms of N and M and the last step in the construction is to state a sharp connection between these two numbers. The lower bound for M in terms of N follows from Lemma 3.1. The upper bound of the same order in N follows from Theorem 2.3. This zero bound is the crucial point of the proof and it is discussed in Chapter 10. In fact this dependence of M in terms of parameter N can be introduced from the beginning of the construction and all entries can be estimated only in terms of parameter N, [**Nes9**].

LEMMA 3.1. *For all sufficiently large integers N there exists a polynomial $A \in$* $\mathbf{Z}[z, x_1, x_2, x_3]$, $A \not\equiv 0$, *such that*

$$\deg_z A \le N, \qquad \deg_{x_i} A \le N, \quad (i = 1, 2, 3), \qquad \log H(A) \le 85N \log N,$$

where $H(A)$ is the maximum of the moduli of the coefficients of A, and the function

$$F(z) = A\big(z, P(z), Q(z), R(z)\big)$$

satisfies the equations

$$F^{(k)}(0) = 0, \qquad k = 0, 1, \dots, \left[\frac{(N+1)^4}{2}\right] - 1.$$

PROOF. It is easy to check that $\sigma_k(n) \le n^{k+1}$. This implies that the functions $P(z), Q(z), R(z)$ can be majorized as follows:

$$P(z) \ll \frac{24 \cdot 2!}{(1-z)^3}, \qquad Q(z) \ll \frac{240 \cdot 4!}{(1-z)^5}, \qquad R(z) \ll \frac{504 \cdot 6!}{(1-z)^7}.$$

In addition, for any natural k, we have

$$z^k \ll \frac{1}{1-z}.$$

We now introduce the set

$$\mathfrak{M} = \big\{\overline{k} = (k_0, k_1, k_2, k_3) \in \mathbf{Z}^4, \ 0 \le k_i \le N, \ i = 0, 1, 2, 3\big\}.$$

It follows from the above bounds that for any vector $\underline{k} \in \mathfrak{M}$, we have

$$z^{k_0} P(z)^{k_1} Q(z)^{k_2} R(z)^{k_3} = \sum_{n=0}^{\infty} d(\overline{k}, n) z^n \ll c_1^{3N} (1-z)^{-16N},$$

where $c_1 = 504 \cdot 6!$ and $d(\overline{k}, n) \in \mathbf{Z}$. From this we find the estimate

(7) $\quad |d(\overline{k}, n)| \le c_1^{3N} (n + 16N)^{16N} \le (nN)^{17N} \ (n \ge 1); \qquad |d(\underline{k}, 0)| \le 1,$

if N is sufficiently large.

Let now

$$A = \sum_{\overline{k} \in \mathfrak{M}} a(\underline{k}) z^{k_0} x_1^{k_1} x_2^{k_2} x_3^{k_3},$$

where the set of integers $a(\underline{k})$ is chosen as the nontrivial solution of the system of linear homogeneous equations

$$\sum_{\underline{k} \in \mathfrak{M}} d(\underline{k}, n) a(\underline{k}) = 0, \qquad n = 0, 1, \dots, \left[\frac{(N+1)^4}{2}\right] - 1.$$

The number of variables $u = (N+1)^4$, and the number of equations $v = \left[\frac{(N+1)^4}{2}\right]$ of this system are connected by the inequality $2v \leq u$. Therefore, by the Siegel's lemma, see Chapter 2, Lemma 2.6] and by (7) this system has a nontrivial integer solution satisfying the inequality

$$\max_{\underline{k} \in \mathfrak{M}} |a(\underline{k})| \leq (N+1)^4 \cdot \left(\frac{(N+1)^4}{2}N\right)^{17N} \leq N^{85N}.$$

Lemma 3.1 is proved. □

Denote

(8) $$M = \operatorname{ord}F(z).$$

Lemma 3.1 implies the inequality

(9) $$M \geq \frac{1}{2}N^4.$$

Also we set

$$r = \min\left\{\frac{1+|q|}{2},\; 2|q|\right\},$$

then $|q| < r < 1$.

LEMMA 3.2. *If the number N is sufficiently large, then for all $z \in \mathbf{C}$, $|z| \leq r$, the following estimate holds:*

$$|F(z)| \leq |z|^M M^{48N}.$$

PROOF. Assume that the function $F(z)$ has the following Taylor series expansion at the origin:

$$F(z) = \sum_{n=M}^{\infty} b_n z^n.$$

Then

$$b_n = \sum_{\underline{k} \in \mathfrak{M}} d(\underline{k}, n)a(\underline{k}) \in \mathbf{Z},$$

and by (7),

$$|b_n| \leq \sum_{\underline{k} \in \mathfrak{M}} (nN)^{17N} N^{85N} \leq n^{17N} N^{103N}, \quad (n \geq M).$$

For $|z| \leq r$ we have

$$
\begin{aligned}
|F(z)| &\leq \sum_{n=M}^{\infty} |b_n| \cdot |z|^n = \sum_{n=0}^{\infty} |b_{n+M}| \cdot |z|^{n+M} \\
&\leq |z|^M N^{103N} \sum_{n=0}^{\infty} (n+M)^{17N} |z|^n \\
&\leq |z|^M N^{103N} (M+1)^{17N} \left(1 + \sum_{n=1}^{\infty} n^{17N} |z|^n\right) \\
&\leq |z|^M N^{103N} (M+1)^{17N} (17N)!(1-r)^{-17N-1} \\
&\leq |z|^M M^{48N},
\end{aligned}
$$

if N is sufficiently large. Lemma 3.2 is proved. □

LEMMA 3.3. *There exists an integer T,*

$$0 \le T \le \gamma N \log M, \quad \gamma = 49 \left(\log \frac{r}{|q|} \right)^{-1},$$

for which the following inequality holds:

$$|F^{(T)}(q)| > \left(\frac{1}{2}|q| \right)^{2M}.$$

PROOF. Let us write

$$L = [\gamma N \log M],$$

and suppose that the following inequalities hold:

(10) $$4^{L+1}|F^{(k)}(q)| \le \left(\frac{1}{2}|q| \right)^M, \quad 0 \le k \le L.$$

On the circle $C : |z| = r$ we have $z\bar{z} = r^2$ and

$$|r^2 - \bar{q}z| = |z| \cdot |\bar{z} - \bar{q}| = r|z - q|.$$

Using these relations and Lemma 1.2, we find that the integral

$$I = \frac{1}{2\pi i} \int_C \frac{F(z)}{z^{M+1}} \cdot \left(\frac{r^2 - \bar{q}z}{r(z-q)} \right)^{L+1} dz$$

can be bounded as follows:

(11) $$|I| \le M^{48N}.$$

Meanwhile,

(12) $$I = \operatorname{Res}_{z=0} G(z) + \operatorname{Res}_{z=q} G(z),$$

where

$$G(z) = \frac{F(z)}{z^{M+1}} \cdot \left(\frac{r^2 - \bar{q}z}{r(z-q)} \right)^{L+1}.$$

The residues at 0 and q are computed to be

$$\operatorname{Res}_{z=0} G(z) = b_M \left(-\frac{r}{q} \right)^{L+1}$$

and

$$\operatorname{Res}_{z=q} G(z) = \frac{1}{L!} \left(\frac{d}{dz} \right)^L \left(F(z) \cdot \frac{(r^2 - \bar{q}z)^{L+1}}{z^{M+1} r^{L+1}} \right) \Bigg|_{z=q}$$

(13) $$= \sum_{k=0}^{L} \frac{F^{(L-k)}(q)}{(L-k)! r^{L+1}} \cdot u_k,$$

where

$$u_k = \frac{1}{k!} \left(\frac{d}{dz} \right)^k \left(\frac{(r^2 - \bar{q}z)^{L+1}}{z^{M+1}} \right) \Bigg|_{z=q} = \frac{1}{2\pi i} \int_{C_1} \frac{(r^2 - \bar{q}z)^{L+1}}{z^{M+1}(z-q)^{k+1}} dz$$

and C_1 is the circle $|z - q| = |q|/2$. Since on C_1 we have

$$|z| \ge |q| - |z - q| \ge \frac{1}{2}|q|,$$

$$|r^2 - \bar{q}z| = |r^2 - q\bar{q} + \bar{q}(q - z)| \le r^2 - |q|^2 + \frac{1}{2}|q|^2 \le r^2 \le 2r|q|$$

and

$$\left|\frac{(r^2 - \bar{q}z)^{L+1}}{(z-q)^{k+1}}\right| \leq (2r|q|)^{L+1}\left(\frac{1}{2}|q|\right)^{-k-1} \leq (4r)^{L+1},$$

it follows that

$$|u_k| \leq (4r)^{L+1}\left(\frac{1}{2}|q|\right)^{-M}, \qquad 0 \leq k \leq L.$$

The last inequality together with (10) and (13) gives us

$$\left|\operatorname{Res}_{z=q}G(z)\right| \leq \sum_{k=0}^{L}\frac{1}{(L-k)!} \leq e.$$

Now since $b_M \in \mathbf{Z}$, from (12) and (11) we obtain

$$\left|\frac{r}{q}\right|^{L+1} \leq \left|b_M\left(-\frac{r}{q}\right)^{L+1}\right| = \left|\operatorname{Res}_{z=0}G(z)\right|$$
$$\leq |I| + \left|\operatorname{Res}_{z=q}G(z)\right|$$
$$\leq M^{48N} + e.$$

But this inequality is false for sufficiently large N. This contradiction shows that there exists an integer $T \leq L$, for which

$$|F^{(T)}(q)| > 4^{-L-1}\left(\frac{1}{2}|q|\right)^{M} > \left(\frac{1}{2}|q|\right)^{2M},$$

if N is sufficiently large. Lemma 3.3 is proved. $\qquad\square$

For any polynomial $E \in \mathbf{C}[z, x_1, x_2, x_3]$, the following identity holds:

$$(14) \qquad \frac{d}{dz}E\big(z, P(z), Q(z), R(z)\big) = z^{-1}DE\big(z, P(z), Q(z), R(z)\big),$$

where

$$(15) \qquad D = z\frac{\partial}{\partial z} + \frac{1}{12}(x_1^2 - x_2)\frac{\partial}{\partial x_1} + \frac{1}{3}(x_1 x_2 - x_3)\frac{\partial}{\partial x_2} + \frac{1}{2}(x_1 x_3 - x_2^2)\frac{\partial}{\partial x_3}$$

is the operator corresponding to the system of differential equations (2).

LEMMA 3.4. *Let $q \in \mathbf{C}$, $0 < |q| < 1$, and N be sufficiently large integer number. Let $A \in \mathbf{Z}[z, x_1, x_2, x_3]$ be the polynomial constructed in Lemma 3.1, and*

$$B(z, x_1, x_2, x_3) = (12z)^T(z^{-1}D)^T A(z, x_1, x_2, x_3),$$

where the integer T was defined in Lemma 3.3. Then for M defined in (8) the following inequalities hold

$$(16) \qquad \deg_z B \leq N, \quad \deg_{x_i} B \leq 2\gamma N \log M, \quad (i = 1, 2, 3),$$

$$(17) \qquad \log H(B) \leq 3\gamma N(\log M)^2,$$

$$(18) \qquad \exp(-\kappa_2 M) \leq |B\big(q, P(q), Q(q), R(q)\big)| \leq \exp(-\kappa_1 M),$$

where

$$\kappa_1 = \frac{1}{2}\log(\frac{1}{r}), \quad \kappa_2 = 3\log(\frac{2}{|q|}).$$

PROOF. By induction, it is easy to prove the identity

$$(19) \qquad (z^{-1}D)^n = z^{-n} \prod_{k=0}^{n-1} (D-k), \qquad n \geq 1.$$

Thus $B \in \mathbf{Z}[z, x_1, x_2, x_3]$. It follows from (14) that

$$B(z, P(z), Q(z), R(z)) = (12z)^T F^{(T)}(z),$$

where the function $F(z)$ was defined in the statement of Lemma 3.2.

Lemma 3.3 and (9) for sufficiently large N, yield the lower bound

$$\left| B(q, P(q), Q(q), R(q)) \right| \geq \left(\frac{1}{2}|q| \right)^{3M}.$$

To obtain the upper bound, we use the formula

$$F^{(T)}(q) = \frac{T!}{2\pi i} \int_{C_2} \frac{F(z)}{(z-q)^{T+1}} \, dz,$$

where C_2 – is the circle $|z - q| = r - |q|$. It follows from the inequality $|z| \leq |z - q| + |q| = r$, (9), Lemma 3.2 and Lemma 3.3 that

$$\begin{aligned} \left| B(q, P(q), Q(q), R(q)) \right| &\leq 12^T \cdot T! \cdot (r - |q|)^{-T} r^M \cdot M^{47N} \\ &\leq \exp(-\kappa_1 M). \end{aligned}$$

This proves the inequalities (18).

To prove (16) and (17), we use the identity (19). Since

$$A \ll N^{85N} (1 + x_1 + x_2 + x_3)^{3N} (1 + z + \cdots + z^N),$$

we easily obtain

$$B(z, x_1, x_2, x_3) \ll N^{85N} M^{2T} (1 + x_1 + x_2 + x_3)^{3N+T} (1 + z + \cdots + z^N).$$

This means that

$$\deg_z B \leq N, \quad \deg_{x_i} B \leq 3N + T \leq 2\gamma N \log M$$

$$H(B) \leq N^{85N} M^{2T} \cdot 4^{3N+T} \cdot (N+1) \leq \exp(3\gamma N (\log M)^2),$$

which proves the estimates (16) and (17). Lemma 3.4 is proved. □

COROLLARY 3.5. *There exists an infinite sequence of integers L and polynomials $B_L \in \mathbf{Z}[z, x_1, x_2, x_3]$ such that*

$$\deg_z B_L \leq 2L, \quad \deg_{x_i} B_L \leq \kappa_3 L \log L, \qquad (i = 1, 2, 3),$$

$$\log H(B_L) \leq \kappa_3 L (\log L)^2,$$

and the following inequalities hold

$$0 < |B_L(q, P(q), Q(q), R(q))| \leq e^{-\kappa_4 L^4},$$

where $\kappa_3 = 96\gamma$, $\kappa_4 = \frac{1}{2}\kappa_1$.

PROOF. Let N be a sufficiently large integer. Define the polynomial $B \in \mathbf{Z}[z, x_1, x_2, x_3]$ as in Lemma 3.4. With M defined in (8) and L the integer such that $L^4 > M \geq (L-1)^4$, in view of (9) we have $N < 2L$ and

$$\deg_z B \leq N < 2L, \quad \deg_{x_i} B \leq 2\gamma N \log M \leq 16\gamma L \log L,$$

$$\log H(B) \leq 3\gamma N (\log M)^2 \leq 96\gamma L (\log L)^2.$$

The inequality (18) implies

$$0 < \left| B\big(q, P(q), Q(q), R(q)\big) \right| \le e^{-\kappa L^4}.$$

Define $B_L = B$. This completes the proof of Corollary 3.5. $\qquad\square$

This Corollary allows to reduce the proof of Corollary 1.5 and in particular the proof of algebraic independence of π and e^π to a sharp estimate of measure of algebraic independence of some numbers connected to an elliptic function with algebraic invariants, see [**PPh10**] and Chapter 4.

There are two differences between Lemma 2.2 and Corollary 3.5: the lower bound for polynomials in Lemma 2.2 is sharper and the sequence of polynomials in Lemma 2.2 is dense. The proof of these properties of constructed sequence is based on the Theorem 2.3.

PROOF OF LEMMA 2.2. The inequality (9) and Theorem 2.3 applied with $L_1 = L_2 = N$ imply

$$\frac{1}{2} N^4 \le M \le c N^4.$$

This allows to express all inequalities of Lemma 3.4 in terms of parameter N and gives the Lemma 2.2. $\qquad\square$

4. Algebraic fundamentals

Here we collect some facts and definitions concerning diophantine properties of ideals in polynomial rings. We refer to [**ZS**][4] for general facts from commutative algebra.

Analogous theory with a few different definitions and assertions is considered in Chapter 6.

Let K be a field of characteristic 0. In section 5 we will use $K = \mathbf{Q}$, in chapter 10 the case $K = \mathbf{C}(z)$ is used.

DEFINITION 4.1. Let $P = \sum a_{\overline{\gamma}} T_1^{\gamma_1} \cdots T_m^{\gamma_m} \in K[T_1, \dots, T_m]$ and $|\ |$ be an absolute value on K. Then we define

$$|\, P\, | = \max_{\overline{\gamma}} |\, a_{\overline{\gamma}}\, |.$$

Let $\mathcal{M}$ be a set such that for any $v \in \mathcal{M}$, there exists an absolute value $|\ |_v$ on the field K, and the set of these absolute values satisfy the following properties:

1. For all $\alpha \in K$, $\alpha \neq 0$, the set $\{v \mid |\, \alpha\, |_v \neq 1\}$ is finite.
2. $\displaystyle\prod_{v \in \mathcal{M}} |\, \alpha\, |_v = 1$ for any $\alpha \in K, \alpha \neq 0$.
3. $\mathcal{M}_\infty = \{v \mid \ |\ |_v$ is archimedean $\}$ is a finite set, denote $\nu = \#\mathcal{M}_\infty$.
4. For every $v \in \mathcal{M}_\infty$, $\alpha \in \mathbf{Q}$, we have $|\, \alpha\, |_v = |\, \alpha\, |_{\mathbf{C}}$ (absolute value in the ordinary sense).

Example 1. Let $K = \mathbf{C}(z)$. Denote $\mathcal{M} = \mathbf{C} \cup \{\infty\}$. For every $v \in \mathcal{M}$, $|\, \alpha\, |_v = e^{-\mathrm{ord}_v \alpha}$. Also $\mathcal{M}_\infty = \emptyset$.

Example 2. Let K be a finite extension of $\mathbf{Q}$ with $\nu = [K : \mathbf{Q}]$. Then $\mathcal{M}$ is the set of all prime ideals of K and all embeddings $\sigma_1, \dots, \sigma_\nu$ of K into $\mathbf{C}$. These embeddings form the set $\mathcal{M}_\infty \subset \mathcal{M}$. We have

$$|\, \alpha\, |_\sigma = |\sigma(\alpha)|_{\mathbf{C}} \quad \text{if } \sigma \in \mathcal{M}_\infty,$$

---

[4][**ZS**] O. Zariski, P. Samuel. *Commutative Algebra*, Vol. I & II. Springer, New-York, (1968).

and

$$|\alpha|_{\mathfrak{m}} = N(\mathfrak{m})^{-\mathrm{ord}_{\mathfrak{m}}\alpha} \ \text{ if } \mathfrak{m} \in \mathcal{M}\backslash\mathcal{M}_\infty.$$

DEFINITION 4.2. For all $P \in K[T_1,\dots,T_m]$, we define height

$$h(P) = \sum_{v \in \mathcal{M}} \log |P|_v.$$

For any $P \not\equiv 0$, since $|P|_v \geq |a|_v$ for some non-zero coeffcient a of P, we have $h(P) \geq \sum_{v \in \mathcal{M}} \log |a|_v = 0$ by the product formula. Also $h(\lambda P) = h(P)$ for $\lambda \in K$ and $\lambda \neq 0$.

Let $I \subset K[\underline{x}] = K[x_0,\dots,x_m]$ be a homogeneous, unmixed[5] (equidimensional) ideal. For an integer r with $1 \leq r \leq m$, let $u_{ij}, 1 \leq i \leq r, 0 \leq j \leq m$ be variables and define

$$L_i = \sum_{j=0}^{m} u_{ij}x_j, \ 1 \leq i \leq r$$

We write $K[U] = K[\underline{u}_1,\dots,\underline{u}_r] = K[u_{10},\dots,u_{rm}]$ where $\underline{u}_i = (u_{i0},\dots,u_{im})$ for $1 \leq i \leq r$ and we set $K[U,\underline{x}] = K[U][x_0,\dots,x_m]$.

DEFINITION 4.3. For any homogeneous ideal $I \subset K[\underline{x}]$, we define $\overline{I}(r)$ as the set of all polynomials $G \in K[U]$ such that

$$Gx_i^M \in (I, L_1,\dots,L_r)$$

for some integer $M > 0$ and all $i, 0 \leq i \leq m$.

In the above definition $(I, L_1,\dots,L_r)$ denotes an ideal generated in $K[U,\underline{x}]$ by the basis polynomials of I and the linear forms $L_1,\dots,L_r$. Note that $\overline{I}(r)$ is an ideal in $K[U]$.

PROPOSITION 4.4. *Let I be a homogeneous, unmixed ideal of the ring $K[\underline{x}]$ with $\dim I = r - 1$. Let $I = I_1 \cap \dots \cap I_s$ be the reduced primary decomposition of I, $\sqrt{I_j} = \mathfrak{p}_j$ and k_j the exponent of I_j for $1 \leq j \leq s$. Then $\overline{I}(r)$ is a principal ideal of $K[U]$ and if $\overline{\mathfrak{p}}_j(r) = (F_j)$, then F_j are irreducible polynomials and the polynomial $F = F_1^{k_1} \cdots F_s^{k_s}$ is a generator of the ideal $\overline{I}(r)$.*

PROOF. See [**Nes2**] and Chapter 6.   □

The generator F of the principal ideal $\overline{I}(r)$ is called an *associated form* of the ideal I. In some cases, this object is also called the *Chow form* or the *Cayley form*

.

Let $I \subset K[\underline{x}]$ be a homogeneous, unmixed ideal and $r = \dim I + 1$. By Proposition 4.4, we have $\overline{I}(r) = (F)$ where F is symmetric in $\underline{u}_i, 1 \leq i \leq r$ and homogeneous.

DEFINITION 4.5. For any homogeneous unmixed ideal $I \subset K[\underline{x}]$ define *degree* and *height* by equalities

$$\deg I = \deg_{\underline{u}_1} F, \qquad h(I) = h(F).$$

---

[5]This means that all components in reduced primary decomposition of I have equal dimensions.

These characteristics of I do not depend on the choise of generator F since for $F_1 = \lambda F, \lambda \in K$, we have $\deg_{\underline{u}_1} F_1 = \deg_{\underline{u}_1} F$ and $h(\lambda F_1) = h(F)$.

Let $|\ |=|\ |_w, w \in \mathcal{M}$, be a certain fixed absolute value on the field K, K_w be a completion of the field K with respect to $|\ |$. The absolute value $|\ |$ can be extended to the algebraic closure of K_w. Let $\mathcal{K}$ be the completion of the algebraic closure of K_w. For any $\overline{\omega} = (\omega_0, \ldots, \omega_m) \in \mathcal{K}^{m+1}$ we set

$$|\overline{\omega}| = \max_{0 \leq j \leq m} |\omega_j|.$$

We now define $|I(\overline{\omega})|$. Consider r skew-symmetric matrices $S^{(1)}, \ldots, S^{(r)}$, $S^{(i)} = \| s_{jk}^{(i)} \|$, where i, j, k vary in the range $1 \leq i \leq r, 0 \leq j \leq m, 0 \leq k \leq m$. We suppose that, except for the skew-symmetry relation $s_{jk}^{(i)} + s_{kj}^{(i)} = 0$, there is no other algebraic relations over $\mathcal{K}$ among the $s_{jk}^{(i)}$. Given a polynomial $E \in K[\underline{u}_1, \ldots, \underline{u}_r]$ and a vector $\overline{\omega} = (\omega_0, \ldots, \omega_m) \in \mathcal{K}^{m+1}$ we let $\varkappa(E)$ denote the polynomial in $s_{jk}^{(i)}, 0 \leq j < k \leq m, 1 \leq i \leq r$, with coefficients in $\mathcal{K}$ that is obtained by substituting the vector $S^{(i)}\overline{\omega}$ in place of the variable $\underline{u}_i = (u_{i0}, \ldots, u_{im})$ in E for $i = 1, \ldots, r$. The map $\varkappa$ is clearly a homomorphism from the ring $\mathcal{K}[\underline{u}_1, \ldots, \underline{u}_r]$ to the polynomial ring $\mathcal{K}[s_{jk}^{(i)}, 0 \leq j < k \leq m, 1 \leq i \leq r]$.

DEFINITION 4.6. For any homogeneous unmixed ideal $I \subset K[\underline{x}]$ define absolute value of I at the point $\overline{\omega}$ by equality

$$|I(\overline{\omega})| = |\varkappa(F)| \cdot |F|^{-1} |\overline{\omega}|^{-r \cdot \deg I}$$

We note the following properties.
1) The above definition does not depend on the choice of generator F of the ideal $\overline{I}(r)$.
2) $|I(\lambda\overline{\omega})| = |I(\overline{\omega})|$ for any $\lambda \in \mathcal{K}, \lambda \neq 0$.

PROPOSITION 4.7. Let I be an unmixed, homogeneous ideal of the ring $K[\underline{x}]$ with $\dim I \geq 0$. Suppose $I = I_1 \cap \ldots \cap I_s$ is the reduced primary decomposition with $\mathfrak{p}_j = \sqrt{I_j}$, and k_j the exponent of I_j. Let $\overline{\omega} \in \mathcal{K}^{m+1}, \overline{\omega} \neq 0$. Then

1) $\displaystyle\sum_{j=1}^{s} k_j \deg \mathfrak{p}_j = \deg I$

2) $\displaystyle\sum_{j=1}^{s} k_j h(\mathfrak{p}_j) \leq h(I) + \nu m^2 \deg I$

3) $\displaystyle\sum_{j=1}^{s} k_j \log |\mathfrak{p}_j(\overline{\omega})| \leq \log |I(\overline{\omega})| + m^3 \deg I.$

Further, if $\mathcal{M}_\infty = \emptyset$, then equality holds in 2). If $|\ |=|\ |_v, v \notin \mathcal{M}_\infty$, then equality holds in 3), and the term $m^3 \deg I$ on the right hand side must be omitted.

PROOF. See [Nes10, Proposition 1.2]. □

Let $P \in K[\underline{x}]$ be a homogeneous polynomial. One can define the normalized value of the polynomial at the point $\overline{\omega}$ as

$$\| P \|_{\overline{\omega}} = |P(\overline{\omega})| \cdot |P|^{-1} |\overline{\omega}|^{-\deg P}.$$

PROPOSITION 4.8. Let $I = (P)$ be a principal ideal of $K[\underline{x}]$ and $\overline{\omega} \in \mathcal{K}^{m+1} \overline{\omega} \neq 0$. Then

1) $\deg I = \deg P$.

2) $h(I) \le h(P) + \nu m^2 \deg P$

3) $\log |I(\overline{\omega})| \le \log \| P \|_{\overline{\omega}} + 2m^2 \deg P$.

If $\mathcal{M}_\infty = \emptyset$, then equality holds in 2). If $| \ | = | \ |_v$, $v \notin \mathcal{M}_\infty$, then the term $2m^2 \deg P$ on the right-hand side of 3) can be omitted.

PROOF. See [**Nes10**, Proposition 1.3]. □

Let $\overline{\varphi} = (\varphi_0, \dots, \varphi_m)$ and $\overline{\psi} = (\psi_0, \dots, \psi_m)$ be two elements of $\mathcal{K}^{m+1}$. We denote the projective distance between these points by

$$\| \overline{\varphi} - \overline{\psi} \| = \max_{0 \le i < j \le m} |\varphi_i \psi_j - \varphi_j \psi_i| \cdot |\overline{\varphi}|^{-1} |\overline{\psi}|^{-1}.$$

Further, let $V(\mathfrak{p})$ denote the variety of zeros of a prime ideal $\mathfrak{p} \subset K[\underline{x}]$ in $\mathbf{P}_\mathcal{K}^m$ and set

(20) $$\rho = \rho(\overline{\omega}) = \min_{\underline{\beta} \in V(\mathfrak{p})} \| \overline{\omega} - \underline{\beta} \|,$$

ρ is the distance from $\overline{\omega}$ to the variety $V(\mathfrak{p})$.

The next two Corollaries easily follow from definitions.

COROLLARY 4.9. *If A is a homogeneous polynomial of the ring $K[\underline{x}]$ contained in $\mathfrak{p}$, then the following inequality holds:*

$$\| A \|_{\overline{\omega}} \le \rho \cdot e^{(2m+1) \deg A}.$$

If the absolute value $| \cdot |$ is nonarchimedean, then the factor $e^{(2m+1) \deg A}$ can be omitted.

PROOF. See [**Nes10**, Corollary 1]. □

COROLLARY 4.10. *The following inequality holds:*

(21) $$| \mathfrak{p}(\overline{\omega}) | \le \rho e^{5m^2 \deg \mathfrak{p}},$$

If the absolute value $| \cdot |$ is nonarchimedean, then the inequality (21) is true without the term $e^{5m^2 \deg \mathfrak{p}}$.

PROOF. See [**Nes10**, Corollary 2]. □

Analogs of the next Proposition are geometric, arithmetic and both metric Bézout theorems from Chapter 6.

PROPOSITION 4.11. *Let $\mathfrak{p} \subset K[\underline{x}]$ be a homogeneous prime ideal with $\dim \mathfrak{p} \ge 0$. Suppose $Q \in K[\underline{x}]$ is a homogeneous polynomial with $Q \notin \mathfrak{p}$. Let $| \ | = | \ |_w$, $w \in \mathcal{M}$. If $r = 1 + \dim \mathfrak{p} \ge 2$, then there exists a homogeneous, unmixed ideal $J \subset K[\underline{x}]$ such that $V(J) = V((\mathfrak{p}, Q))$ with $\dim J = \dim \mathfrak{p} - 1$, and*

1) $\deg J \le \deg \mathfrak{p} \deg Q$

2) $h(J) \le h(\mathfrak{p}) \deg Q + h(Q) \deg \mathfrak{p} + \nu m(r + 1) \deg \mathfrak{p} \deg Q$.

3) For any $\overline{\omega} \in \mathcal{K}^{m+1}$ and ρ as in (15), we have

$$\log |J(\overline{\omega})| \le \log \delta + h(Q) \deg \mathfrak{p} + h(\mathfrak{p}) \deg Q + 11 \nu m^2 \deg \mathfrak{p} \deg Q,$$

where

$$\delta = \begin{cases} \| Q \|_{\overline{\omega}} & \text{if } \rho < \| Q \|_{\overline{\omega}} \\ |\mathfrak{p}(\overline{\omega})| & \text{if } \rho \ge \| Q \|_{\overline{\omega}}. \end{cases}$$

The inequality in 3) above is true in the case $r = 1$ if we take $|J(\overline{\omega})| = 1$.

PROOF. See [**Nes10**, Proposition 1.4]. □

COROLLARY 4.12. *Let* $P \in K[\underline{x}]$, $P \not\equiv 0$ *be a homogeneous polynomial and* $\mathfrak{p}$ *a homogeneous prime ideal in* $K[\underline{x}]$ *with* $\dim \mathfrak{p} \geq 0$, $P \notin \mathfrak{p}$. *Suppose further,*

$$|\mathfrak{p}(\overline{\omega})| \leq e^{-S}, \quad S > 0, \quad \| P \|_{\overline{\omega}} \leq e^{-2m\nu \deg P},$$

and for integer $\eta > 0$,

$$-\eta \log \| P \|_{\overline{\omega}} \geq 2 \, \min(S, \log \frac{1}{\rho}).$$

If $\dim \mathfrak{p} \geq 1$, *then there exists a homogeneous, unmixed ideal* J *with* $\dim J = \dim \mathfrak{p} - 1$, $V(J) = V((\mathfrak{p}, P))$ *and*

1. $\deg J \leq \eta \deg \mathfrak{p} \deg P$
2. $h(J) \leq \eta \Big(h(\mathfrak{p}) \deg P + h(P) \deg \mathfrak{p} + \nu m(r+2) \deg P \deg \mathfrak{p} \Big)$
3.

$$\log |J(\overline{\omega})| \leq -S + \eta \Big(h(\mathfrak{p}) \deg P + h(P) \deg \mathfrak{p} + 12\nu m^2 \deg P \deg \mathfrak{p} \Big).$$

The inequality in 3) above is true in the case $\dim \mathfrak{p} = 0$ *if we take* $|J(\overline{\omega})| = 1$.

PROOF. See [**Nes10**, Corollary 3]. □

The next proposition is comparable with "Closest point property" from Chapter 6.

PROPOSITION 4.13. *Let* $I \subset K[\underline{x}]$ *be a homogeneous unmixed ideal,* $r = 1 + \dim I$. *Then for any* $\overline{\omega} \in \mathcal{K}^{m+1}$, *there exists a zero* $\overline{\beta} \in \mathcal{K}^{m+1}$ *of ideal* I *such that*

$$\deg I \cdot \log \| \overline{\omega} - \overline{\beta} \| \leq \frac{1}{r} \log |I(\overline{\omega})| + \frac{1}{r} h(I) + (\nu + 3)m^3 \deg I.$$

If $\mathcal{M}_{\infty} = \emptyset$, *then the term* $(\nu + 3)m^3 \deg I$ *can be omitted.*

PROOF. See [**Nes10**, Proposition 1.5]. □

5. Another proof of Theorem 1.1.

In this section, we will apply the auxiliary assertions from section 4 in the case $K = \mathbf{Q}$ to the ordinary absolute value $| \cdot |$ on $\mathbf{Q}$. Theorem 1.1 will be deduced from the following assertion.

THEOREM 5.1. *Let* $q \in \mathbf{C}$, $0 < |q| < 1$. *Then for each integer* r, $1 \leq r \leq 3$, *there exists a constant* $\mu_r > 0$ *such that for any homogeneous unmixed ideal* $I \subset \mathbf{Q}[x_0, \ldots, x_4]$, $\dim I = r - 1$, *the following inequality holds:*

$$\log |I(\overline{\omega})| \geq -\mu_r T^{\frac{4}{4-r}} (\log T)^{\frac{8r}{4-r}},$$

where T *is an arbitrary number satisfying the inequality*

$$T \geq \max\{h(I) + \deg I, e\},$$

and $\overline{\omega} = \big(1, q, P(q), Q(q), R(q)\big) \in \mathbf{C}^5$.

We shall show how Theorem 1.1 can be derived from Theorem 5.1. Let $\mathfrak{p}$ be a homogeneous prime ideal generated by all homogeneous polynomials in the ring $\mathbf{Q}[x_0, \ldots, x_4]$ that vanish at $\overline{\omega}$. Then

$$\dim \mathfrak{p} = \operatorname{tr} \deg_{\mathbf{Q}} \mathbf{Q}(q, P(q), Q(q), R(q)).$$

If $\dim \mathfrak{p} \leq 2$, then, by Theorem 5.1, we obtain inequality $|\mathfrak{p}(\overline{\omega})| > 0$, which contradicts to Corollary 4.10. This implies $\dim \mathfrak{p} \geq 3$ and proves Theorem 1.1.

We shall prove the Theorem 5.1 by induction in r. At the proof we will use the notation $t(I) = \deg I + h(I)$ for any homogeneous unmixed ideal of the ring $\mathbf{Q}[x_0, \ldots, x_4]$. The proof is based on Lemma 2.2.

Let $r, 1 \leq r \leq 3$, be the least number for which the assertion of Theorem 5.1 is no longer true. We choose and fix a sufficiently large number λ.

The set of real numbers T for which there exists a prime homogeneous ideal $\mathfrak{p}$ with $\dim \mathfrak{p} = r - 1$,

$$(22) \qquad \log |\mathfrak{p}(\overline{\omega})| < -2\lambda^{12} T^{\frac{4}{4-r}} (\log T)^{\frac{8r}{4-r}}, \quad t(\mathfrak{p}) \leq T,$$

is unbounded. Indeed, otherwise, the inequality

$$\log |\mathfrak{p}(\overline{\omega})| \geq -\gamma t(\mathfrak{p})^{\frac{4}{4-r}} \left(1 + \log t(\mathfrak{p})\right)^{\frac{8r}{4-r}},$$

would hold where with a certain positive constant γ for all homogeneous prime ideals $\mathfrak{p} \subset \mathbf{Q}[X]$, $\dim \mathfrak{p} = r - 1$. Applying now Proposition 4.7 to the arbitrary homogeneous unmixed ideal $I \subset \mathbf{Q}[\underline{x}]$ of dimension $r - 1$, we obtain the inequalities

$$\sum_{j=1}^{s} k_j t(\mathfrak{p}_j) \leq 17 t(I).$$

and

$$\sum_{j=1}^{s} k_j t(\mathfrak{p}_j)^{\frac{4}{4-r}} \left(1 + \log t(\mathfrak{p}_j)\right)^{\frac{8r}{4-r}} \leq c t(I)^{\frac{4}{4-r}} \left(1 + \log t(I)\right)^{\frac{8r}{4-r}},$$

where c is an absolute positive constant. Now the last inequality of Proposition 4.7 yields

$$\begin{aligned} \log |I(\overline{\omega})| &\geq \sum_{j=1}^{s} k_j \log |\mathfrak{p}_j(\overline{\omega})| - 64 \deg I \\ &\geq -\gamma \sum_{j=1}^{s} k_j t(\mathfrak{p}_j)^{\frac{4}{4-r}} \left(1 + \log t(\mathfrak{p}_j)\right)^{\frac{8r}{4-r}} - 64 \deg I \\ &\geq -\mu_r t(I)^{\frac{4}{4-r}} \left(1 + \log t(I)\right)^{\frac{8r}{4-r}}. \end{aligned}$$

But this contradicts the assumption that the assertion of Theorem 4 does not hold for ideals of dimension $r - 1$.

Thus, the set of real numbers T for which there exists a prime homogeneous idea $\mathfrak{p}$ of dimension $r - 1$ satisfying conditions (22) is unbounded. Let now T be sufficiently large, and let $\mathfrak{p}$ be a prime homogeneous ideal of dimension $r-1$ satifying conditions (22).

Let us define the number L by the relation

$$(23) \qquad 2\gamma_2 L^4 = \min\left\{2\lambda^{12} T^{\frac{4}{4-r}} (\log T)^{\frac{3r}{4-r}}, \ \log \frac{1}{\rho}\right\}$$

where ρ is the distance from the point $\overline{\omega}$ to the variety of zeros of the ideal $\mathfrak{p}$, and γ_2 is the constant from Lemma 2.2. It follows from Proposition 4.13 that when T increases, the quantity L also increases up to infinity.

Now we set N as integer part of L and define the polynomial $B = A_N \in \mathbf{Z}[z, x_1, x_2, x_3]$ with the aid of Lemma 2.2. The degree of the polynomial B in the totality of variables does not exceed $n = [\gamma_0 L \log L]$. Let us now define the homogeneous polynomial

$$E = x_0^n B\left(\frac{x_1}{x_0}, \ldots, \frac{x_4}{x_0}\right) \in \mathbf{Z}[x_0, \ldots, x_4].$$

We shall prove that E is not contained in the ideal $\mathfrak{p}$. If the inclusion $E \in \mathfrak{p}$ held, then, by Corollary 4.9, the inequality

$$|E(\overline{\omega})| \leq \rho H(E)|\overline{\omega}|^n e^{9n}.$$

would hold. Since $H(E) = H(B)$ and since $\log H(B) \leq \gamma_0 L \log^2 L$ by Lemma 2.2 and definition of N, it would follow from the obtained inequality in view of (23) that

$$\begin{aligned}
\log|B(\overline{\omega})| &= \log|E(\overline{\omega})| \leq \log\rho + \log H(B) + n(9 + \log|\overline{\omega}|) \\
&\leq -2\gamma_2 L^4 + \gamma_0 L \log^2 L + \gamma_0(9 + \log|\overline{\omega}|)L\log L \\
&< -\gamma_2 N^4.
\end{aligned}$$

However this contradicts Lemma 2.2. Thus $E \notin \mathfrak{p}$.

Let us now apply Corollary 4.12 to the ideal $\mathfrak{p}$ and to the polynomial E. To this end, we set $\eta = 1 + [4\gamma_2/\gamma_1]$, where γ_i are the constants from Lemma 2.2. Since

$$\begin{aligned}
-\eta\log\|E\|_{\overline{\omega}} &\geq -\eta\log|B(\overline{\omega})| \geq \eta\gamma_1 N^4 \\
&> 4\gamma_2 L^4 = 2\min\left\{2\lambda^{12}T^{\frac{4}{4-r}}(\log T)^{\frac{8r}{4-r}}, \ \log\frac{1}{\rho}\right\},
\end{aligned}$$

according to Lemma 2.2, the condition of Corollary 4.12 holds. By this Corollary, we can assert that for $r \geq 2$, there exists a homogeneous unmixed ideal J of the ring $\mathbf{Q}[x_0, \ldots, x_4]$, for which

$$\text{(24)} \qquad\qquad \deg J \leq \eta n \deg\mathfrak{p},$$

$$\text{(25)} \qquad\qquad h(J) \leq \eta(nh(\mathfrak{p}) + h(E)\deg\mathfrak{p} + 24n\deg\mathfrak{p}),$$

$$\text{(26)} \quad \log|J(\overline{\omega})| \leq -2\lambda^{12}T^{\frac{4}{4-r}}(\log T)^{\frac{8r}{4-r}} + \eta(h(E)\deg\mathfrak{p} + nh(\mathfrak{p}) + 192n\deg\mathfrak{p}).$$

Inequality (26) in which we formally set $|J(\overline{\omega})| = 1$ also holds for $r = 1$.

It follows from (23) that

$$\text{(27)} \qquad\qquad L \leq \lambda^4 T^{\frac{1}{4-r}}(\log T)^{\frac{2r}{4-r}},$$

and (22) implies

$$\text{(28)} \qquad\qquad h(\mathfrak{p}) \leq T, \qquad \deg\mathfrak{p} \leq T.$$

Using these inequalities and also Lemma 2.2, we find that

$$\begin{aligned}
\eta(h(E)\deg\mathfrak{p} &+ nh(\mathfrak{p}) + 192n\deg\mathfrak{p}) \\
&\leq \eta(\gamma_0 L\log^2 L \cdot \deg\mathfrak{p} + \gamma_0 L\log L \cdot h(\mathfrak{p}) + 192\cdot\gamma_0 L\log L \cdot \deg\mathfrak{p}) \\
&\leq 2\gamma_0\eta T L(\log L)^2 \leq \lambda^5 T^{\frac{5-r}{4-r}}(\log T)^{\frac{8}{4-r}}.
\end{aligned}$$

The obtained inequality, together with (26), yields

$$\text{(29)} \qquad\qquad \log|J(\overline{\omega})| \leq -\lambda^{12}T^{\frac{4}{4-r}}(\log T)^{\frac{8r}{4-r}}.$$

In the case $r = 1$, inequality (29) does not hold; therefore, in what follows we assume that $r \geq 2$.

It follows from inequalities (24), (25), (27), (28) and (5) that

$$t(J) \leq \lambda^5 T^{\frac{5-r}{4-r}}(\log T)^{\frac{8}{4-r}}.$$

and

$$t(J)^{\frac{4}{5-r}}(\log t(J))^{\frac{8r-8}{5-r}} \leq \lambda^{11}T^{\frac{4}{4-r}}(\log T)^{\frac{3r}{4-r}}.$$

Since $\dim J = r - 2$, according to the definition of the number r, the inequality

$$\log|J(\overline{\omega})| \geq -\mu_{r-1}t(J)^{\frac{4}{5-r}}(\log t(J))^{\frac{8r-8}{5-r}} \geq -\mu_{r-1}\lambda^{11}T^{\frac{4}{4-r}}(\log T)^{\frac{8r}{4-r}}.$$

must hold for the ideal J. If we choose $\lambda > \mu_{r-1}$, then this inequality contradicts inequality (29). The obtained contradiction completes the proof of Theorem 5.1.

It is possible to deduce an estimate for measure of algebraic independence for some numbers from Theorem 5.1.

COROLLARY 5.2. *Let $q \in \mathbf{C}$, $0 < |q| < 1$, and let $\xi_1, \xi_2, \xi_3 \in \mathbf{C}$ be such that all numbers $q, P(q), Q(q), R(q)$ are algebraic over the field $\mathbf{Q}(\xi_1, \xi_2, \xi_3)$. Then there exists a constant $\mu > 0$ which depends only on on the numbers q, ξ_i and is such that for any polynomial $A \in \mathbf{Z}[x_1, x_2, x_3]$, $A \neq 0$, the following inequality holds:*

$$(30) \qquad \log |A(\xi_1, \xi_2, \xi_3)| \geq -\mu T^4 (\log T)^{24},$$

where T is an arbitrary number satisfying the inequality

$$T \geq \max\big(H(A) + \deg A, e\big),$$

In particular, inequality (30) holds for each of the collections of numbers $\{\pi, e^\pi, \Gamma(\frac{1}{4})\}$ and $\{\pi, e^{\pi\sqrt{3}}, \Gamma(\frac{1}{3})\}$.

Note that the exponent 4 can not be decreased in (30). Nevertheless the estimate can be improved in its dependense in logarithmic term by more delicate choise of parameters (see [**PPh10, Nes10**]). If the number q is algebraic there exists a sharp improvement of the estimate in dependese in $H(A)$, [**Nes10**].

PROOF. Denote by $\mathfrak{p}$ a homogeneous prime ideal generated by all homogeneous polynomials of the ring $\mathbf{Q}[x_0, \ldots, x_4]$ which vanish at the point $\overline{w} = (1, q, P(q), Q(q), R(q))$. The values $h(\mathfrak{p})$, $\deg \mathfrak{p}$ can be uniquely determined from $\overline{w}$. According to Theorem 1.1 and to the conditions of the Corollary, we have $\dim \mathfrak{p} = 3$.

For any polynomial $B \in \mathbf{Z}[x_1, x_2, x_3, x_4]$, $B(\overline{w}) \neq 0$, we denote by

$$C(x_0, x_1, x_2, x_3, x_4) = x_0^{\deg B} B\Big(\frac{x_1}{x_0}, \frac{x_2}{x_0}, \frac{x_3}{x_0}, \frac{x_4}{x_0}\Big),$$

a homogeneous polynomial which is not contained in $\mathfrak{p}$. Then $C(\overline{w}) = B(\overline{w})$. We denote by J the homogeneous unmixed ideal in the ring $\mathbf{Q}[\underline{x}] = \mathbf{Q}[x_0, \ldots, x_4]$ which is constructed for $\mathfrak{p}$ and C in accordance with Proposition 4. Then $\dim J = 2$,

$$\deg J \leq c_1 \deg C, \qquad h(J) \leq c_1\big(h(C) + \deg C\big).$$

In addition

$$\log |J(\overline{w})| \leq \log \|C\|_{\overline{w}} + c_1\big(h(C) + \deg C\big).$$

Applying Theorem 5.1 with $T = c_2 t(B)$ to the ideal J for a sufficiently large constant c_2 and taking into account that $\deg C = \deg B$, $h(C) = h(B) \leq \log H(B)$, $r = 3$, $\rho = 0$, we obtain the inequality

$$(31) \qquad \log\big|B\big(q, P(q), Q(q), R(q)\big)\big| \geq -c_3 t(B)^4 \log^{24} t(B).$$

It follows from Theorem 1.1 and the hypothesis of the Corollary that all numbers ξ_1, ξ_2, ξ_3 are algebraic over the field $\mathbf{Q}\big(q, P(q), Q(q), R(q)\big)$. If the polynomial $E \in \mathbf{Q}[x_1, \ldots, x_4]$ is such that with the number $d = E\big(q, P(q), Q(q), R(q)\big)$ all numbers $d\omega_i$ are algebraic integers over the ring $\mathbf{Z}[q, P(q), Q(q), R(q)]$, then

$$\mathrm{Norm}\big(d^{\deg A} A(\omega_1, \omega_2, \omega_3)\big) = B\big(q, P(q), Q(q), R(q)\big),$$

where $B(x_1, x_2, x_3, x_4)$ is a polynomial with integer coefficients. Taking into account that

$$\deg B \leq c_4 \deg A, \qquad \log H(B) \leq c_4 \log H(A),$$

we easily obtain the estimate of Corollary 5.2 with the aid of (31).   □

CHAPTER 4

Some remarks on proofs of algebraic independence

A fascinating aspect of numbers is that a given number may appear in many different contexts, to this respect numbers such as e, π, ... are certainly remarkable. Let's recall once more time how the couple of numbers $(\pi, \Gamma(1/4))$, for example, is related to elliptic and modular functions.

1. Connection with elliptic functions

Let $\wp$ be a *Weierstrass elliptic function* associated to a lattice $\Lambda \subset \mathbf{C}$

$$\wp(z) = \frac{1}{z^2} + \sum_{\lambda \in \Lambda \setminus \{0\}} \left(\frac{1}{(z - \lambda)^2} - \frac{1}{\lambda^2} \right) .$$

This function satisfies a differential equation

$$(\wp')^2 = 4\wp^3 - g_2(\Lambda)\wp - g_3(\Lambda)$$

whose coefficients are the second and third *invariants of the lattice* : $g_2(\Lambda) = 60. \sum_{\lambda \in \Lambda \setminus \{0\}} \lambda^{-4}$ and $g_3(\Lambda) = 140. \sum_{\lambda \in \Lambda \setminus \{0\}} \lambda^{-6}$. Furthermore, the function $\wp(z)$ is periodic with respect to Λ and its primitive $-\zeta(z)$ is therefore quasi-periodic. That is for any $\omega \in \Lambda$ there exists η such that $\zeta(z + \omega) = \zeta(z) + \eta$.

Let $\omega \in \Lambda \setminus \{0\}$ and η the associated quasi-period, and suppose $g_2(\Lambda), g_3(\Lambda)$ are algebraic numbers, an important result of G.V. Chudnovsky [**Chu2**] states that the two numbers $\frac{\pi}{\omega}, \frac{\eta}{\omega}$ are algebraically independent (over $\mathbf{Q}$).

In case the lattice has *complex multiplications* (*i.e.* there exists $t \in \mathbf{C} \setminus \mathbf{Z}$ such that $t\Lambda \subset \Lambda$) then Λ may be written $\Lambda = (\mathbf{Z} + \mathbf{Z}\tau).\omega_1$ where τ is a complex imaginary quadratic number.

In general, write $\Lambda = \mathbf{Z}\omega_1 + \mathbf{Z}\omega_2$ and η_1, η_2 the quasi-periods associated to ω_1 and ω_2 respectively. Let us recall *Legendre's relation*

$$\omega_2\eta_1 - \omega_1\eta_2 = 2i\pi$$

and, if Λ has complex multiplication, *Masser's relation*

$$\eta_2 - \bar{\tau}\eta_1 \in \omega_1.\mathbf{Q}(\tau, g_2(\Lambda), g_3(\Lambda))$$

where $\tau = \omega_2/\omega_1$. These relations imply that in the case of complex multiplication, when the lattice is normalized so that it has algebraic invariants, the five numbers $\pi, \omega_1, \omega_2, \eta_1, \eta_2$ are algebraic over the field generated by $\frac{\pi}{\omega_1}$ and $\frac{\eta_1}{\omega_1}$. Therefore we have a complete answer to the problem of finding all the algebraic relations among the periods and quasi-periods of an elliptic curve with complex multiplication, this is not the case for elliptic curves with no complex multiplication ... not to speak of the case of abelian varieties.

(∗) Chapter's author : Patrice PHILIPPON.

Following an approach of G.V. Chudnovsky, G. Philibert [**Phi1**] has established, for any lattice with algebraic invariants, a *measure of algebraic independence* of the two numbers $\frac{\pi}{\omega_1}$, $\frac{\eta_1}{\omega_1}$:

THEOREM 1.1. *With the above notations the following property holds. For any $\epsilon > 0$ there exists a real $c = c(\epsilon, \Lambda) > 0$ such that for all $S \in \mathbf{Z}[X, Y] \setminus \{0\}$ one has*

$$\left| S\left(\frac{\pi}{\omega_1}, \frac{\eta_1}{\omega_1} \right) \right| > \exp\left(-c.t(S)^{3+\epsilon} \right) \ .$$

Now, where are the numbers π and $\Gamma(1/4)$? Answer : in the example given by the lattice $\mathbf{Z} + \mathbf{Z}i$. More exactly, if we put $\Lambda = (\mathbf{Z} + \mathbf{Z}i).\omega_1$ with $\omega_1 = \frac{\Gamma(1/4)^2}{2\sqrt{2\pi}}$ a computation gives $g_2(\Lambda) = 4$, $g_3(\Lambda) = 0$ and $\eta_1 = \frac{\pi}{\omega_1} = \frac{(2\pi)^{3/2}}{\Gamma(1/4)^2}$. The result of Philibert implies in this case that for any $\epsilon > 0$ there exists a real $c(\epsilon) > 0$ such that for all $S \in \mathbf{Z}[X, Y] \setminus \{0\}$ one has

$$|S(\pi, \Gamma(1/4))| > \exp\left(-c(\epsilon).t(S)^{3+\epsilon} \right) \ .$$

2. Connection with modular series

We have introduced in § 1 the numbers $\frac{\pi}{\omega_1}$ and $\frac{\eta_1}{\omega_1}$ through periods and quasi-periods of elliptic functions, but we can view these same numbers as values of functions of Λ. From this point of view, instead of considering Λ so that $g_2(\Lambda)$ and $g_3(\Lambda)$ are algebraic numbers it is more natural to take Λ of the form $\Lambda = \mathbf{Z} + \mathbf{Z}\tau$ and consider η_1/ω_1, $g_2(\Lambda)$ and $g_3(\Lambda)$ as functions of τ or $q := e^{2i\pi\tau}$.

With this normalization $(\Lambda = \mathbf{Z} + \mathbf{Z}\tau)$ we have

$$(\eta_1/\omega_1)(\Lambda) = 2\zeta(2).E_2(q)$$
$$g_2(\Lambda) = 120\zeta(4).E_4(q)$$
$$g_3(\Lambda) = 280\zeta(6).E_6(q)$$

where $E_2 = P$, $E_4 = Q$ and $E_6 = R$ are the *Ramanujan* (or *Eisenstein*) *functions* (and ζ the Riemann zeta function). Recall from chapter 1, § 2.1 that for $k \in \mathbf{N}^*$ one defines :

$$E_{2k}(z) = 1 + \gamma_k. \sum_{i \geq 1} \sigma_{2k-1}(i).z^i \ ,$$

where $\sigma_\ell(i) = \sum_{h|i} h^\ell$ and $\gamma_1 = -24$, $\gamma_2 = 240$, $\gamma_3 = -504$, ...

The functions $E_2 = P$, $E_4 = Q$, $E_6 = R$ being homogeneous of weights 2, 4 and 6 respectively, we can write for $\Omega \in \mathbf{C}$

$$(\eta_1/\omega_1)(\Lambda.\Omega) = \frac{(\eta_1/\omega_1)(\Lambda)}{\Omega^2}$$
$$g_2(\Lambda.\Omega) = \frac{g_2(\Lambda)}{\Omega^4}$$
$$g_3(\Lambda.\Omega) = \frac{g_3(\Lambda)}{\Omega^6} \ .$$

EXAMPLE. (see also chapter 3) For $\Lambda = \mathbf{Z} + \mathbf{Z}i$ one has $q = e^{-2\pi}$, $g_2(\Lambda) = \frac{\Gamma(1/4)^8}{16\pi^2}$, $g_3(\Lambda) = 0$ and $\frac{\eta_1}{\omega_1}(\Lambda) = \frac{8\pi^2}{\Gamma(1/4)^4}$, from which we deduce

$$P(e^{-2\pi}) = \frac{3}{\pi} \ , \quad Q(e^{-2\pi}) = \frac{3\Gamma(1/4)^8}{64\pi^6} \ , \quad R(e^{-2\pi}) = 0 \ ,$$

and

$$\mathbf{Q}(P(e^{-2\pi}), Q(e^{-2\pi}), R(e^{-2\pi})) = \mathbf{Q}(\pi, \Gamma(1/4)^8) \ .$$

LESSON. A key point in the proof of algebraic independence of e^π, π and $\Gamma(1/4)$ is to consider them in connection with values of the functions z, $P(z)$, $Q(z)$, $R(z)$ rather values of the functions $e^{z/\Omega}$, $\wp(z)$, $\zeta(z)$, for example.

3. Another proof of algebraic independence of π, e^π and $\Gamma(1/4)$

Let us first recall the result of K. Barré-Sirieix, G. Diaz, F. Gramain and G. Philibert [**BDGP**], proved in chapter 2.

THEOREM 3.1. *Let* $J = 1728.\frac{Q^3}{Q^3 - P^2}$ *be the normalized modular function of weight 0 and* $q \in \mathbf{C}$, $0 < |q| < 1$, *then at least one of the numbers* q *or* $J(q)$ *is transcendental over* $\mathbf{Q}$.

In order to establish this theorem, its authors have devised a new transcendence method that one can view as a natural extension of Mahler's method. More precisely, they construct an auxiliary function with a high order at the origin and extrapolate successively on points close to the origin and on derivatives at the origin, the process can be continued indefinitely as in [**PPh10**], see chapter 2, § 2.5 and 2.6. Applying this transcendence method to the functions $z, P(z), Q(z), R(z)$, one gets :

LEMMA 3.2. *Let* $q \in \mathbf{C}$, $0 < |q| < 1$ *and* $D \in \mathbf{N}$ *big enough, there exists a real* $c = c(q) > 0$, *an integer* $T \geq D^4/2$ *and a polynomial* $F \in \mathbf{Z}[X_1, X_2, X_3, X_4]$ *of size* $\leq c.D(\log T)^2$ *satisfying*

$$0 < |F(q, P(q), Q(q), R(q))| < \exp\left(-c^{-1}.T\right) \ .$$

REMARK. It is worth noting that the proof of the above lemma doesn't appeal to any *zeros* or *multiplicity estimate*.

Take $q = e^{-2\pi}$ and suppose that $x = (q, P(q), Q(q), R(q)) \in \mathbf{C}^4$ belongs to an algebraic subset of dimension 2 defined over $\mathbf{Q}$. This means that the four numbers can be expressed algebraically from two of them and, since $R(q) = 0$ and $(\pi, \Gamma(1/4))$ is a couple of algebraically independent numbers, q must be algebraic over $\mathbf{Q}(P(q), Q(q))$. Let $G \in \mathbf{Z}[X_1, X_2, X_3]$ be a corresponding irreducible relation of algebraic dependence, then G does not divide F, $G(q, P(q), Q(q)) = 0$ and we may eliminate X_1 between F and G. We obtain a polynomial $S \in \mathbf{Z}[X, Y]$ of size $\leq c_1.D.(\log T)^2$, defined by

$$S(\pi, \Gamma(1/4)) = (\pi\Gamma(1/4))^{8.\,\deg(\mathrm{Res}_{X_1}(F,G))}.\mathrm{Res}_{X_1}(F, G)(P(q), Q(q), R(q)) \ ,$$

satisfying $0 < |S(\pi, \Gamma(1/4))| < \exp(-c_2.T)$. But this contradicts Philibert's measure (theorem 1.1) since $T \geq D^4/2$ and $t(S) \leq c_3.D(\log T)^2 \leq c_3.(2T)^{1/4}(\log T)^2$. It shows that the point x doesnot lies on an algebraic subset of $\mathbf{C}^4$ defined over $\mathbf{Q}$ of dimension ≤ 2 or equivalently at least three of the numbers $e^{-2\pi}$, $P(e^{-2\pi}) = \frac{3}{\pi}$, $Q(e^{-2\pi}) = \frac{3\Gamma(1/4)^8}{64\pi^6}$ and $R(e^{-2\pi}) = 0$ are algebraically independent.

4. Approximation properties

In the above proof we did not appeal to a criteria for algebraic independence (CIA) to conclude, instead we used a measure of algebraic independence which is in some sense a property external to the transcendence construction (proved with the help of Weierstrass rather than Ramanujan functions). A new idea which has emerged recently in algebraic independence theory (but which is not yet completely matured) is that a part of the information used in the machinery of CIA should be independent of the transcendence construction and even could be refined using other relevant properties than the ones involved in the transcendence construction. As it stands, a part of the proof of CIA is encompassed in the following (*see* chapters 6 and 7 for explanation of the terminology used here)

APPROXIMATION PROPERTY 1. *Let* $n \in \mathbf{N}^*$, *there exists a real* $c(n) \geq 1$ *such that for* $x \in \mathbf{P}_n(\mathbf{C})$, $d \in \{0, \ldots, n\}$ *and reals* $\Delta \geq c(n)$, $H \geq 1$, *it exists an algebraic subset* Z *of* $\mathbf{P}_n(\mathbf{C})$ *defined over* $\mathbf{Q}$, *equidimensional of dimension* d, *satisfying* $\deg(Z) \leq (c(n)\Delta)^{n-d}$, $h(Z) \leq c(n)^{n-d} . H \Delta^{n-d-1}$ *and*

$$\log \mathrm{Dist}(x, Z) \leq -c(n)^{d-n} . (h(Z)\Delta + \deg(Z)(H + \log \Delta)) \Delta^d .$$

This statement is presently a conjecture verified only when d belongs to the restricted set of values $\{n-3, n-2, n-1, n\}$. For $n = 1, 2$ the following assertion is also proved.

APPROXIMATION PROPERTY 2. *Let* $n \in \mathbf{N}^*$, *there exists a real* $c'(n) \geq 1$ *such that for* $x \in \mathbf{P}_n(\mathbf{C})$ *and reals* $\Delta \geq c'(n)$, $H \geq c'(n) \log(\Delta + 1)$, *it exists* $\alpha \in \mathbf{P}_n(\bar{\mathbf{Q}})$ *satisfying* $d(\alpha) := [\mathbf{Q}(\alpha) : \mathbf{Q}] \leq (c'(n)\Delta)^n$, $d(\alpha)h(\alpha) \leq c'(n)^n . H \Delta^{n-1}$ *and*

$$\log \mathrm{Dist}(x, \alpha) \leq -c'(n).d(\alpha). (h(\alpha)\Delta + H) .$$

On an other hand the transcendence construction applied to Ramanujan functions (lemma 3.2), plus the multiplicity estimate of chapter 10 leads to

PROPOSITION 4.1. *Let* $q \in \mathbf{C}$, $0 < |q| < 1$, $x_1, \ldots, x_n$ *a transcendence basis of the field* $\mathbf{Q}(q, P(q), Q(q), R(q))$ *and* $x = (1 : x_1 : \cdots : x_n) \in \mathbf{P}_n(\mathbf{C})$, *there exists a real* $c = c(q) > 0$ *such that for all* $\alpha \in \mathbf{P}_n(\bar{\mathbf{Q}})$ *one has*

$$\log \mathrm{Dist}(x, \alpha) \geq -c.(t(\alpha)d(\alpha))^{4/3} . (\log(t(\alpha)d(\alpha)))^{8/3}$$

where $t(\alpha) = h(\alpha) + \log d(\alpha)$.

PROOF. Indeed, the multiplicity estimate ensures $T < c_4 D^4$ in lemma 3.2. Taking $D = [c_5(t(\alpha)d(\alpha))^{1/3} . (\log(t(\alpha)d(\alpha)))^{2/3}]$, with c_5 large enough, in this lemma and substituing α to x in the value $F(q, P(q), Q(q), R(q))$ one gets a non zero algebraic number of size $\leq c_6.t(\alpha).D(\log D)^2$ and absolute value

$$\leq \max \left(c_7^D . \mathrm{Dist}(x, \alpha); e^{-c_8 D^4} \right) .$$

The proposition follows directly from the size inequality.   □

REMARK. A better control of the size of the polynomial F in lemma 3.2 allows to show the stronger inequality :

$$\log \mathrm{Dist}(x, \alpha) \geq -c.(t(\alpha)d(\alpha))^{4/3} . \log(t(\alpha)d(\alpha))$$

in proposition 4.1.

Now, taking $H = \Delta = (t(\alpha)d(\alpha))^{1/n}$ in AP2 gives a contradiction with the proposition if we assume $n < 3$, which shows again that at least three of the numbers $q, P(q), Q(q), R(q)$ are algebraically independent. This application demonstrates how approximation properties can be substituted to CIA, we will show in chapter 8 how they are actually embodied in the proof of CIA.

CHAPTER 5

Élimination multihomogène

1. Introduction

Ce chapitre et le chapitre 7 entendent donner des outils de géométrie diophantienne dans les espaces multiprojectifs c'est-à-dire du type

$$\mathbf{P} = \mathbf{P}_K^{n_1} \times \cdots \times \mathbf{P}_K^{n_q}$$

où $n_1, \ldots, n_q$ sont des entiers naturels et K un corps de nombres. Pour la plupart nos résultats fournissent les généralisations naturelles d'énoncés connus dans le cas projectif (correspondant à $q = 1$) et dus principalement à P. Philippon (voir [**PPh2**, **PPh7**] et [**Jad**]). On pourra comparer avec les approches de Y.V. Nesterenko [**Nes2**] et W.D. Brownawell [**Bro6**], voir aussi le chapitre 16.

Notre démarche, comme celle des articles cités, se base sur l'étude des formes de Chow des sous-schémas fermés de l'espace envisagé, ici **P**. La première partie (le présent chapitre) consiste donc en une théorie de l'élimination multihomogène; on travaille avec l'anneau multigradué suivant, associé à **P** :

$$K[X] = K[X_0^{(1)}, \ldots, X_{n_1}^{(1)}, \ldots, X_0^{(q)}, \ldots, X_{n_q}^{(q)}].$$

On attache alors à un idéal multihomogène I de $K[X]$ des idéaux éliminants $\mathfrak{E}_d(I)$ (voir définitions plus bas) puis, lorsque ces idéaux sont principaux, des formes éliminantes. Bien que ce paragraphe (comme le suivant) ne fasse appel qu'à des résultats de base en algèbre commutative, la démonstration du théorème de principalité 2.13 (qui généralise [**PPh2**, I.5]) s'avère plus délicate que dans le cas homogène. En effet la formulation des "conditions de hauteurs" qui interviennent est moins aisée car elles doivent prendre en compte les dégénérescences possibles sur chaque sous-ensemble de facteurs de **P**. L'obtention de notre théorème nécessite des résultats nouveaux sur le polynôme de Hilbert-Samuel dans le cas multihomogène.

Le deuxième paragraphe introduit les formes résultantes (ou de Chow) des idéaux multihomogènes. Dans le cas homogène on pouvait essentiellement se dispenser de cette étude en se ramenant à des idéaux premiers pour lesquels formes éliminantes et résultantes coïncident. Ceci tombe en défaut pour les idéaux (mêmes premiers) multihomogènes et l'on doit donc utiliser les formes de Chow. L'un des résultats cruciaux est la proposition 3.5 de séparation des variables (bien sûr sans équivalent si $q = 1$).

Les résultats présentés ici sont extraits de la thèse [**Rem**, chap. III].

(∗) Chapter's author : Gaël RÉMOND.

2. Formes éliminantes des idéaux multihomogènes

Dans toute la suite $q \geq 1$ sera un entier fixé et on appellera anneau multigradué un anneau B admettant une décomposition (comme groupe abélien) de la forme $B = \bigoplus_{n \in \mathbf{Z}^q} B_n$ de telle sorte que $B_n \cdot B_{n'} \subset B_{n+n'}$ pour tout $(n, n') \in \mathbf{Z}^q \times \mathbf{Z}^q$. Un module multigradué M sur B est un B-module muni d'une décomposition $M = \bigoplus_{n \in \mathbf{Z}^d} M_n$ avec $B_n \cdot M_{n'} \subset M_{n+n'}$ pour tout $(n, n') \in \mathbf{Z}^q \times \mathbf{Z}^q$. Un sous-module d'un module M (en particulier un idéal de B) est dit multihomogène s'il est engendré par ses éléments multihomogènes, ces derniers étant naturellement les éléments de $\bigcup_{n \in \mathbf{Z}^q} M_n$.

2.1. Idéaux éliminants. Désormais $n_1, \ldots, n_q$ sont des éléments de $\mathbf{N}$ fixés (on note $n = n_1 + \cdots + n_q$) et $X_0^{(1)}, \ldots, X_{n_1}^{(1)}, \ldots, X_0^{(q)}, \ldots, X_{n_q}^{(q)}$ des indéterminées. On désignera éventuellement leur collection par $X^{(1)}, \ldots, X^{(q)}$ voire par X.

Soit A un anneau nœthérien. On introduit l'anneau

$$B = A[X_0^{(1)}, \ldots, X_{n_1}^{(1)}, \ldots, X_0^{(q)}, \ldots, X_{n_q}^{(q)}]$$

que l'on considérera comme multigradué par $\deg(X_i^{(j)}) = \varepsilon_j$ pour tous i, j où ε_j est $(0, \ldots, 1, \ldots, 0)$ le j-ème élément de la base canonique de $\mathbf{Z}^q$ (on écrira aussi $\varepsilon = \varepsilon_1 + \cdots + \varepsilon_q = (1, \ldots, 1) \in \mathbf{Z}^q$). Pour $k \in \mathbf{N}^q$ on note $\mathcal{M}_k$ l'ensemble des monômes unitaires de multidegré k c'est-à-dire

$$\left\{ X_0^{(1)^{\alpha_{1,0}}} \cdots X_{n_1}^{(1)^{\alpha_{1,n_1}}} \cdots X_0^{(q)^{\alpha_{q,0}}} \cdots X_{n_q}^{(q)^{\alpha_{q,n_q}}} \,\Big|\, \alpha_{i,0} + \cdots + \alpha_{i,n_i} = k_i, \; \forall i \right\}.$$

Lorsque $\mathfrak{m} = X_0^{(1)^{\alpha_{1,0}}} \cdots X_{n_1}^{(1)^{\alpha_{1,n_1}}} \cdots X_0^{(q)^{\alpha_{q,0}}} \cdots X_{n_q}^{(q)^{\alpha_{q,n_q}}}$ est un tel monôme, on utilisera le symbole $\binom{k}{\mathfrak{m}}$ pour le coefficient multinomial

$$\frac{k_1! \ldots k_q!}{\alpha_{1,0}! \ldots \alpha_{1,n_1}! \ldots \alpha_{q,0}! \ldots \alpha_{q,n_q}!}.$$

Pour $r \geq 0$ et $d = (d_1, \ldots, d_r) \in (\mathbf{N}^q)^r$ on note $A[d]$ l'anneau des polynômes à coefficients dans A en les indéterminées $u_{\mathfrak{m}}^{(l)}$ où $1 \leq l \leq r$ et $\mathfrak{m} \in \mathcal{M}_{d_l}$. On désignera à l'occasion la collection des $u_{\mathfrak{m}}^{(l)}$ pour $\mathfrak{m} \in \mathcal{M}_{d_l}$ par $u^{(l)}$. On pose aussi $B[d] = A[d] \underset{A}{\otimes} B$ et on définit, pour $1 \leq l \leq r$,

$$U_l = \sum_{\mathfrak{m} \in \mathcal{M}_{d_l}} u_{\mathfrak{m}}^{(l)} \mathfrak{m} \in B[d].$$

Lorsque I est un idéal de B on désigne par $I[d]$ l'idéal de $B[d]$ engendré par I et les éléments $U_1, \ldots, U_r$.

DÉFINITION 2.1. Soit I un idéal de B. On définit l'idéal caractéristique d'indice d de I par

$$\mathfrak{U}_d(I) = \{ f \in B[d] \mid \exists k \in \mathbf{N}^q \; f\mathcal{M}_k \subset I[d] \}$$

et l'idéal éliminant d'indice d de I par

$$\mathfrak{E}_d(I) = \mathfrak{U}_d(I) \cap A[d].$$

Le théorème suivant, dit de l'élimination multihomogène, donne une interprétation de l'idéal éliminant en termes d'équations polynomiales.

THÉORÈME 2.2. *Soit* $\rho \colon A[d] \to K$ *un morphisme d'anneaux dans un corps* K. *Alors, pour tout idéal multihomogène* I *de* B, *les conditions suivantes sont équivalentes :*

1. $\rho(\mathfrak{E}_d(I)) = 0$;
2. *il existe une extension de corps* L/K *et un zéro non trivial de* $\rho(I[d])$ *dans* $L^{n_1+1} \times \cdots \times L^{n_q+1}$, *c'est-à-dire* $z \in (L^{n_1+1} \setminus 0) \times \cdots \times (L^{n_q+1} \setminus 0)$ *tel que pour tout* $P \in I[d]$ *on ait* $\rho(P)(z) = 0$;
3. *il existe une extension finie de corps* L/K *et un zéro non trivial de* $\rho(I[d])$ *dans* $L^{n_1+1} \times \cdots \times L^{n_q+1}$.

DÉMONSTRATION. L'implication $3 \implies 2$ est triviale. Montrons $2 \implies 1$. Posons $z = (x_0^{(1)}, \ldots, x_{n_1}^{(1)}, \ldots, x_0^{(q)}, \ldots, x_{n_q}^{(q)})$ et soit donc $f \in \mathfrak{E}_d(I)$, choisissons $j_1, \ldots, j_q$ tels que $x_{j_i}^{(i)} \neq 0$. Par définition il existe $k \in \mathbf{N}^q$ tel que $X_{j_1}^{(1)\,k_1} \ldots X_{j_q}^{(q)\,k_q} f \in I[d]$ d'où $X_{j_1}^{(1)\,k_1} \ldots X_{j_q}^{(q)\,k_q} \rho(f) \in \rho(I[d])$. L'hypothèse 2 donne $x_{j_1}^{(1)\,k_1} \ldots x_{j_q}^{(q)\,k_q} \rho(f) = 0$ puis $\rho(f) = 0$ ce qui montre 1. Établissons à présent $1 \implies 3$. Montrons pour commencer que $(*)$ si $k \in \mathbf{N}^q$ alors $\mathcal{M}_k \not\subset \rho(I[d]).K[X]$. Si k ne vérifiait pas cela, on aurait, en notant K_0 le corps des fractions de $\rho(A[d])$ vu comme sous-corps de K,

$$\mathcal{M}_k \subset \rho(I[d]).K[X] \cap K_0[X] = \rho(I[d]).K_0[X] = \rho(I[d]).K_0.$$

Comme $\mathcal{M}_k$ est fini, on peut chasser les dénominateurs et on trouve qu'il existe $a \in A[d]$ tel que $\rho(a) \neq 0$ et $\rho(a)\mathcal{M}_k \subset \rho(I[d])$. Par conséquent pour tout $\mathfrak{m} \in \mathcal{M}_k$ on peut trouver $Q_{\mathfrak{m}} \in I[d]$ tel que $\rho(a)\mathfrak{m} = \rho(Q_{\mathfrak{m}})$; on peut supposer que $Q_{\mathfrak{m}}$ est multihomogène de multidegré k car ρ et $I[d]$ sont multihomogènes. On écrit alors $Q_{\mathfrak{m}} = \sum_{\mathfrak{m}' \in \mathcal{M}_k} \alpha_{\mathfrak{m},\mathfrak{m}'} \mathfrak{m}'$ où $\alpha_{\mathfrak{m},\mathfrak{m}'} \in A[d]$. Notons aussi $\Delta = \det(\alpha_{\mathfrak{m},\mathfrak{m}'})_{\mathfrak{m},\mathfrak{m}'}$. Par définition on a $\Delta.\mathcal{M}_k \subset (Q_{\mathfrak{m}})_{\mathfrak{m} \in \mathcal{M}_k}$ donc $\Delta.\mathcal{M}_k \subset I[d]$; ainsi $\Delta \in \mathfrak{E}_d(I)$ et donc, d'après l'hypothèse 1, $\rho(\Delta) = 0$. Par ailleurs la relation $\rho(a)\mathfrak{m} = \sum_{\mathfrak{m}' \in \mathcal{M}_k} \rho(\alpha_{\mathfrak{m},\mathfrak{m}'})\mathfrak{m}'$ entraîne que $\rho(\Delta) = \rho(a)^{\mathrm{Card}(\mathcal{M}_k)}$ et donc $\rho(a) = 0$ ce qui est une contradiction. On a donc $(*)$ qui implique l'existence de $j_1, \ldots, j_q$ tels que pour tout $k \in \mathbf{N}^q$ on a $X_{j_1}^{(1)\,k_1} \ldots X_{j_q}^{(q)\,k_q} \not\in \rho(I[d]).K[X]$. Nous travaillons ensuite dans l'anneau $C = K[X]/\rho(I[d]).K[X]$ qui est multigradué. L'élément $1 - X_{j_1}^{(1)} \ldots X_{j_q}^{(q)}$ n'est pas inversible dans C : en effet si c en était un inverse on aurait pour tout N

$$c = 1 + X_{j_1}^{(1)} \ldots X_{j_q}^{(q)} + \cdots + X_{j_1}^{(1)\,N} \ldots X_{j_q}^{(q)\,N} + c X_{j_1}^{(1)\,N+1} \ldots X_{j_q}^{(q)\,N+1};$$

en décomposant c en éléments multihomogènes et en choisissant N assez grand cela donnerait

$$c = 1 + X_{j_1}^{(1)} \ldots X_{j_q}^{(q)} + \cdots + X_{j_1}^{(1)\,N} \ldots X_{j_q}^{(q)\,N}$$

et

$$c X_{j_1}^{(1)\,N+1} \ldots X_{j_q}^{(q)\,N+1} = 0$$

donc, c étant inversible, on aurait $X_{j_1}^{(1)\,N+1} \ldots X_{j_q}^{(q)\,N+1} = 0$ dans C ce qui est contraire à l'hypothèse sur $j_1, \ldots, j_q$. Pour conclure il existe donc un idéal maximal de C contenant $1 - X_{j_1}^{(1)} \ldots X_{j_q}^{(q)}$. On note L le quotient de C par cet idéal maximal et $x_0^{(1)}, \ldots, x_{n_q}^{(q)}$ les images dans L de $X_0^{(1)}, \ldots, X_{n_q}^{(q)}$. On vérifie immédiatement que L/K est une extension finie de corps et que $z = (x_0^{(1)}, \ldots, x_{n_1}^{(1)}, \ldots, x_0^{(q)}, \ldots, x_{n_q}^{(q)})$

est un zéro non trivial au sens de l'énoncé car d'une part $x_{j_1}^{(1)} \ldots x_{j_q}^{(q)} = 1$ et d'autre part $\rho(P)(z)$ est l'image de $\rho(P)$ par $K[X] \to L$ et donc nul si $P \in I[d]$. $\qquad\square$

Nous établissons les quelques propriétés algébriques des idéaux éliminants qui nous serviront par la suite. Pour chaque $k \in \mathbf{N}^q$ on choisit un ordre total $\preceq$ sur $\mathcal{M}_k$. Introduisons alors $\mathfrak{S}B[d]$ l'anneau des polynômes à coefficients dans B en les indéterminées $s_{\mathfrak{m},\mathfrak{m}'}^{(l)}$ pour $1 \leq l \leq r$ et $\mathfrak{m} \prec \mathfrak{m}'$ dans $\mathcal{M}_{d_l}$. Pour simplifier les notations nous définissons des éléments $s_{\mathfrak{m},\mathfrak{m}'}^{(l)}$ pour $\mathfrak{m} \succeq \mathfrak{m}'$ dans $\mathcal{M}_{d_l}$ en posant $s_{\mathfrak{m},\mathfrak{m}}^{(l)} = 0$ pour tout $\mathfrak{m}$ et $s_{\mathfrak{m},\mathfrak{m}'}^{(l)} = -s_{\mathfrak{m}',\mathfrak{m}}^{(l)}$ pour tous $\mathfrak{m} \succ \mathfrak{m}'$. On considère alors le morphisme de B-algèbres

$$
\begin{array}{ccc}
B[d] & \xrightarrow{\;\partial\;} & \mathfrak{S}B[d] \\
u_{\mathfrak{m}}^{(l)} & \longmapsto & \displaystyle\sum_{\mathfrak{m}' \in \mathcal{M}_{d_l}} s_{\mathfrak{m},\mathfrak{m}'}^{(l)} \mathfrak{m}'.
\end{array}
$$

Ce morphisme sera utile à travers la caractérisation suivante de l'idéal éliminant.

LEMME 2.3. *Soit I un idéal de B. Pour $f \in B[d]$, il y a équivalence entre*
1. $f \in \mathfrak{U}_d(I)$;
2. *il existe $k \in \mathbf{N}^q$ tel que $\partial(f)\mathcal{M}_k \subset I.\mathfrak{S}B[d]$.*

DÉMONSTRATION. Montrons $1 \Longrightarrow 2$. D'après 1 on peut trouver $k \in \mathbf{N}^q$ tel que $f\mathcal{M}_k \subset I[d]$ donc $\partial(f)\mathcal{M}_k \subset \partial(I[d])$. En outre $\partial(I[d])$ est engendré par I et les éléments $\partial(U_l)$ qui valent ($1 \leq l \leq r$) :

$$
\partial(U_l) = \sum_{\mathfrak{m} \in \mathcal{M}_{d_l}} \left(\sum_{\mathfrak{m}' \in \mathcal{M}_{d_l}} s_{\mathfrak{m},\mathfrak{m}'}^{(l)} \mathfrak{m}' \right) \mathfrak{m} = \sum_{\mathfrak{m} \prec \mathfrak{m}'} \left(s_{\mathfrak{m},\mathfrak{m}'}^{(l)} + s_{\mathfrak{m}',\mathfrak{m}}^{(l)} \right) \mathfrak{m}\mathfrak{m}' = 0
$$

donc $\partial(I[d]) \subset I.\mathfrak{S}B[d]$ et l'assertion 2 en découle. Démontrons maintenant la réciproque. Pour $j \in \mathbf{N}^q$ avec $j_i \leq n_i$, on définit un morphisme de B-algèbres

$$
\begin{array}{ccc}
\mathfrak{S}B[d] & \xrightarrow{\;\sigma_j\;} & B[d][X_{j_1}^{(1)-1}, \ldots, X_{j_q}^{(q)-1}] \\
s_{\mathfrak{m},\mathfrak{m}'}^{(l)} & \longmapsto & \dfrac{\delta_{\mathfrak{m}',\mathfrak{m}_l} u_{\mathfrak{m}}^{(l)} - \delta_{\mathfrak{m},\mathfrak{m}_l} u_{\mathfrak{m}'}^{(l)}}{\mathfrak{m}_l}
\end{array}
$$

où δ désigne le symbole de Kronecker et $\mathfrak{m}_l = X_{j_1}^{(1)^{d_{l,1}}} \ldots X_{j_q}^{(q)^{d_{l,q}}}$. On calcule alors pour $1 \leq l \leq r$ et $\mathfrak{m} \in \mathcal{M}_{d_l}$:

$$
\begin{aligned}
\sigma_j \circ \partial(u_{\mathfrak{m}}^{(l)}) - u_{\mathfrak{m}}^{(l)} &= \sum_{\mathfrak{m}' \in \mathcal{M}_{d_l}} \left(\frac{\delta_{\mathfrak{m}',\mathfrak{m}_l} u_{\mathfrak{m}}^{(l)} - \delta_{\mathfrak{m},\mathfrak{m}_l} u_{\mathfrak{m}'}^{(l)}}{\mathfrak{m}_l} \right) \mathfrak{m}' - u_{\mathfrak{m}}^{(l)} \\
&= u_{\mathfrak{m}}^{(l)} - \frac{\delta_{\mathfrak{m},\mathfrak{m}_l}}{\mathfrak{m}_l} \left(\sum_{\mathfrak{m}' \in \mathcal{M}_{d_l}} u_{\mathfrak{m}'}^{(l)} \mathfrak{m}' \right) - u_{\mathfrak{m}}^{(l)} \\
&= -\delta_{\mathfrak{m},\mathfrak{m}_l} \frac{U_l}{\mathfrak{m}_l}
\end{aligned}
$$

ainsi $\sigma_j \circ \partial(u_{\mathfrak{m}}^{(l)}) - u_{\mathfrak{m}}^{(l)} \in I[d].B[d][X_{j_1}^{(1)-1}, \ldots, X_{j_q}^{(q)-1}]$ puis comme on a affaire à des morphismes de B-algèbres

$$
\sigma_j \circ \partial(f) - f \in I[d].B[d][X_{j_1}^{(1)-1}, \ldots, X_{j_q}^{(q)-1}]
$$

pour tout $f \in B[d]$. L'hypothèse 2 est que $\mathfrak{d}(f)\mathcal{M}_k \subset I.\mathfrak{S}B[d]$ pour un certain $k \in \mathbf{N}^q$; on a donc $\sigma_j \circ \mathfrak{d}(f)\mathcal{M}_k \subset I.B[d][X_{j_1}^{(1)^{-1}}, \ldots, X_{j_q}^{(q)^{-1}}]$ d'où $f\mathcal{M}_k \subset I.B[d][X_{j_1}^{(1)^{-1}}, \ldots, X_{j_q}^{(q)^{-1}}]$ puis en chassant les dénominateurs

$$X_{j_1}^{(1)^{\alpha_1}} \ldots X_{j_q}^{(q)^{\alpha_q}} f\mathcal{M}_k \subset I[d] \ .$$

Ceci étant vrai pour tout j on peut trouver k' tel que $f\mathcal{M}_{k'} \subset I[d]$ et on a donc bien $f \in \mathfrak{U}_d(I)$. $\qquad\square$

Cette caractérisation nous permet de donner le

LEMME 2.4. *Soit I un idéal de B.*

1. $\mathcal{M}_\varepsilon \subset \sqrt{I} \iff \mathfrak{U}_d(I) = B[d] \iff \mathfrak{E}_d(I) = A[d]$.
2. *Si I est premier et si $\mathcal{M}_\varepsilon \not\subset \sqrt{I}$ alors $\mathfrak{U}_d(I)$ et $\mathfrak{E}_d(I)$ sont premiers et de plus $I = \mathfrak{U}_d(I) \cap B$.*
3. *Si $I = \bigcap_{h=1}^p I_h$ alors $\mathfrak{U}_d(I) = \bigcap_{h=1}^p \mathfrak{U}_d(I_h)$ et $\mathfrak{E}_d(I) = \bigcap_{h=1}^p \mathfrak{E}_d(I_h)$.*

DÉMONSTRATION. La première assertion résulte clairement des définitions, par exemple en disant $1 \in \mathfrak{E}_d(I) \iff \exists k \ \mathcal{M}_k \subset I[d] \iff \exists k \ \mathcal{M}_k \subset I \iff \mathcal{M}_\varepsilon \subset \sqrt{I}$. Pour la seconde, sachant que $\mathfrak{S}B[d]$ est un anneau de polynômes sur B, on vérifie que $I.\mathfrak{S}B[d]$ est premier et que $\mathcal{M}_\varepsilon \not\subset \sqrt{I.\mathfrak{S}B[d]}$ (car $I.\mathfrak{S}B[d] \cap B = I$) donc $\mathfrak{d}(f)\mathcal{M}_k \subset I.\mathfrak{S}B[d] \iff \mathfrak{d}(f) \in I.\mathfrak{S}B[d]$. D'après le lemme précédent $\mathfrak{U}_d(I) = \mathfrak{d}^{-1}(I.\mathfrak{S}B[d])$ est bien premier ainsi que $\mathfrak{E}_d(I) = \mathfrak{U}_d(I) \cap A[d]$. Enfin on a toujours $I \subset \mathfrak{U}_d(I)$ et si $f \in \mathfrak{U}_d(I) \cap B$ alors $f = \mathfrak{d}(f) \in I.\mathfrak{S}B[d] \cap B = I$. La dernière assertion découle du même lemme en notant que

$$I.\mathfrak{S}B[d] = \bigcap_{h=1}^p (I_h.\mathfrak{S}B[d]).$$

$\qquad\square$

L'étape suivante consiste à montrer dans certains cas la principalité de l'idéal éliminant pour définir ce que l'on appelle les formes éliminantes. Nous aurons besoin, pour cela, d'une théorie fine du polynôme de Hilbert-Samuel pour les modules multigradués. Remarquons que cette notion (pour les modules de la forme B/I) apparaît pour la première fois dans [VdW][1] où l'auteur traite des degrés d'un idéal bihomogène.

2.2. Polynôme de Hilbert-Samuel multihomogène. Nous citons les résultats d'algèbre commutative multihomogène que nous utiliserons. En premier lieu on a

LEMME 2.5. *Soit M un module multigradué sur un anneau multigradué B. Tous les idéaux premiers associés à M sont multihomogènes. De plus chacun d'entr'eux est de la forme $\mathrm{Ann}(m)$ où m est un élément multihomogène de M.*

DÉMONSTRATION. Soit donc $\mathfrak{p} \in \mathrm{Ass}(M)$. Écrivons $\mathfrak{p} = \mathrm{Ann}(m)$ et $m = \sum_j m_j$ une décomposition en éléments multihomogènes, où j désigne le multidegré de m_j. Clairement $\bigcap_j \mathrm{Ann}(m_j) \subset \mathfrak{p}$; il existe par conséquent un indice j tel que

[1][VdW] B.V.L. Van der Waerden. On Hilbert's functions, series of composition of ideals and a generalization of a theorem of Bézout, *Proc. Royal Acad. Amsterdam* 31, (1928), 749–770.

Ann(m_j) $\subset$ $\mathfrak{p}$ (puisque l'intersection est finie); choisissons j_0 le plus grand *dans l'ordre lexicographique* des tels indices. De la formule

$$\bigcap_{j > j_0} \text{Ann}(m_j) \not\subset \mathfrak{p}$$

on déduit l'existence d'un élément multihomogène λ n'appartenant pas à $\mathfrak{p}$ tel que $\lambda m_j = 0$ si $j > j_0$. Ainsi $\lambda m_{j_0} \neq 0$ est le terme dominant de λm. Soit maintenant $x \in \mathfrak{p}$ que l'on écrit $x = \sum_j x_j$. On note j_1 le plus grand indice tel que $x_{j_1} \neq 0$. Alors $\lambda x m$ n'a pas de terme d'indice supérieur à $j_0 + j_1 + \deg(\lambda)$ et son terme de ce degré est $\lambda x_{j_1} m_{j_0}$. Ainsi $\lambda x m = 0$ donne $\lambda x_{j_1} m_{j_0} = 0$ puis $\lambda x_{j_1} \in \text{Ann}(m_{j_0}) \subset \mathfrak{p} = \text{Ann}(m)$. En utilisant $\lambda \notin \mathfrak{p}$ on a $x_{j_1} \in \mathfrak{p}$. Finalement, par soustractions successives, $x_j \in \mathfrak{p}$ pour tout j. On en conclut que $\mathfrak{p}$ est multihomogène et que $\mathfrak{p} = \text{Ann}(m_{j_0})$. $\qquad\square$

Lorsque l'on impose des conditions de finitude on a un utile résultat de décomposition. Pour un B-module multigradué M et $f \in \mathbf{Z}^q$, on note $M(f)$ le module multigradué de même module sous-jacent et gradué par $M(f)_n = M_{n+f}$.

LEMME 2.6. *Soient B un anneau multigradué nœthérien et M un B-module multigradué de type fini. Il existe une suite de sous-modules multihomogènes $0 = M_0 \subset M_1 \subset \cdots \subset M_t = M$, des idéaux premiers multihomogènes $\mathfrak{p}_1, \ldots, \mathfrak{p}_t$ de B et des éléments $f_1, \ldots, f_t$ de $\mathbf{Z}^q$ de telle sorte que pour tout $1 \leq i \leq t$ on ait un isomorphisme de B-modules multigradués $M_i/M_{i-1} \simeq (B/\mathfrak{p}_i)(f_i)$.*

DÉMONSTRATION. Considérons l'ensemble Φ des sous-modules N multihomogènes qui admettent une suite de la forme voulue. Comme le sous-module 0 est élément de Φ et que M est nœthérien il existe un élément N maximal dans Φ. En particulier il existe une suite $0 = M_0 \subset M_1 \subset \cdots \subset M_t = N$ dont les quotients sont du type $(B/\mathfrak{p})(f)$. Si le module multihomogène M/N n'est pas nul il possède un premier associé $\mathfrak{q}$ (car B est nœthérien) qui, par le lemme précédent, est multihomogène et s'écrit $\text{Ann}(m)$ avec $m \in M/N$ multihomogène; si l'on définit alors N' comme l'image réciproque de Bm dans M on constate que $N'/N \simeq (B/\mathfrak{q})(f')$ où f' est le multidegré de m. Ceci contredit la maximalité de N et l'on avait donc en fait $N = M$ ce qui démontre le lemme. $\qquad\square$

On reprend à présent les notations du premier paragraphe. On impose que A soit un corps K. Notre anneau multigradué B est donc $K[X] = K[X^{(1)}, \ldots, X^{(q)}]$. Pour une partie $J \subset \{1, \ldots, q\}$ on désigne par $X^{(J)}$ la famille des indéterminées $X^{(i)}$ avec $i \in J$ et $n_J = \sum_{i \in J} n_i$; on convient de noter I_J l'intersection $I \cap K[X^{(J)}]$ dès que I est un idéal de B. On considérera aussi une famille d'indéterminées $T_1, \ldots, T_q$ et $T^{(J)}$ en sera la sous-famille $\{T_i \mid i \in J\}$. Lorsque M est un B-module multigradué, on pose les définitions suivantes dans lesquelles $\mathcal{M}_\varepsilon$ a été défini plus haut et $\text{ht}(I)$ est la hauteur d'un idéal I (c'est-à-dire la hauteur algébrique de l'idéal I définie à l'aide des chaînes d'idéaux premiers distincte des hauteurs arithmétiques définies au chapitre 7 et notées h) :

$$e(M) = \max_{\text{Ann}(M) \subset \mathfrak{p}, \mathcal{M}_\varepsilon \not\subset \mathfrak{p}} (n - \text{ht}(\mathfrak{p})) \quad \text{et}$$

$$e_J(M) = \max_{\text{Ann}(M) \subset \mathfrak{p}, \mathcal{M}_\varepsilon \not\subset \mathfrak{p}} (n_J - \text{ht}(\mathfrak{p}_J)).$$

On adopte la convention naturelle de noter $\max(\emptyset) = -\infty$; de manière cohérente le degré d'un polynôme nul est $-\infty$. Remarquons que l'on peut limiter les maxima écrits ci-dessus aux premiers minimaux contenant $\text{Ann}(M)$ (et ne contenant pas $\mathcal{M}_\varepsilon$) et que ces derniers sont multihomogènes d'après le lemme 2.5. On peut reformuler ces définitions en posant d'abord pour un premier multihomogène $e_J(B/\mathfrak{p}) = -\infty$ si $\mathcal{M}_\varepsilon \subset \mathfrak{p}$ et $e_J(B/\mathfrak{p}) = n_J - \mathrm{ht}(\mathfrak{p}_J)$ sinon puis

$$e_J(M) = \max\{e_J(B/\mathfrak{p}) \mid \mathrm{Ann}(M) \subset \mathfrak{p},\ \mathfrak{p}\ \text{multihomogène}\}\ .$$

On déduit alors

LEMME 2.7. *Si $\mathfrak{p}$ est un idéal premier multihomogène de B ne contenant pas $\mathcal{M}_\varepsilon$ et si J_1 et J_2 sont des parties de $\{1, \ldots, q\}$ on a*

$$e_{J_1}(B/\mathfrak{p}) \le e_{J_1 \cup J_2}(B/\mathfrak{p}) \le e_{J_1}(B/\mathfrak{p}) + e_{J_2}(B/\mathfrak{p}) - e_{J_1 \cap J_2}(B/\mathfrak{p}).$$

Lorsque M est un B-module multigradué

$$\max_{1 \le l \le t} e_{J_l}(M) \le e_J(M) \le \sum_{l=1}^{t} e_{J_l}(M)$$

chaque fois que J est la réunion de $J_1, \ldots, J_t$. De plus les nombres $e_J(M)$ sont soit tous égaux à $-\infty$ soit tous positifs.

DÉMONSTRATION. Pour la première inégalité il suffit de montrer $e_{J'}(B/\mathfrak{p}) \le e_J(B/\mathfrak{p})$ si $J = J' \cup \{i\}$, $J' \ne J$. Supposons par l'absurde $e_{J'}(B/\mathfrak{p}) > e_J(B/\mathfrak{p})$ soit $n_{J'} - \mathrm{ht}(\mathfrak{p}_{J'}) > n_J - \mathrm{ht}(\mathfrak{p}_J)$ ce que l'on peut encore écrire $\mathrm{ht}(\mathfrak{p}_J) \ge \mathrm{ht}(\mathfrak{p}_{J'}) + n_i + 1$. Or par multihomogénéité l'idéal $\mathfrak{p}_J$ est inclus dans l'idéal de B engendré par $\mathfrak{p}_{J'}$ et $\mathcal{M}_{\varepsilon_i}$; il est aisé de constater que ce dernier idéal est premier de hauteur $\mathrm{ht}(\mathfrak{p}_{J'}) + n_i + 1$. Alors la condition sur les hauteurs est absurde car elle implique que l'inclusion citée se transforme en égalité et donc que $\mathcal{M}_{\varepsilon_i} \subset \mathfrak{p}$. La seconde inégalité peut s'écrire

$$\mathrm{ht}(\mathfrak{p}_{J_1}) - \mathrm{ht}(\mathfrak{p}_{J_1 \cap J_2}) + \mathrm{ht}(\mathfrak{p}_{J_2}) - \mathrm{ht}(\mathfrak{p}_{J_1 \cap J_2}) \le \mathrm{ht}(\mathfrak{p}_{J_1 \cup J_2}) - \mathrm{ht}(\mathfrak{p}_{J_1 \cap J_2}).$$

Cette formulation montre que l'on peut supposer $J_1 \cap J_2 = \emptyset$ en remplaçant K par le corps des fractions de $K[X^{(J_1 \cap J_2)}]/\mathfrak{p}_{J_1 \cap J_2}$. Dans ce cas $\mathfrak{p}_{J_1 \cup J_2}$ contient l'idéal engendré dans $K[X^{(J_1 \cup J_2)}]$ par $\mathfrak{p}_{J_1}$ et $\mathfrak{p}_{J_2}$ qui est de hauteur $\mathrm{ht}(\mathfrak{p}_{J_1}) + \mathrm{ht}(\mathfrak{p}_{J_2})$.

Ensuite l'encadrement de $e_J(M)$ est clair si $M = B/\mathfrak{p}$; pour M quelconque on prend le maximum sur les $\mathfrak{p}$ multihomogènes tel que $\mathrm{Ann}(M) \subset \mathfrak{p}$.

Si $e(M) = -\infty$, on a certainement $e_J(M) = -\infty$ pour toute partie J car tous les maxima sont pris sur le même ensemble. Si $e(M) \ne -\infty$ alors il est clair que $e_\emptyset(M) = 0$; la première inégalité montre bien que $e_J(M) \ge e_\emptyset(M) = 0$ pour toute partie J.   □

Pour le lemme technique suivant, on étend la notation $\varepsilon_J = \sum_{i \in J} \varepsilon_i$.

LEMME 2.8. *Soient $\mathfrak{p}$ un idéal premier multihomogène de B, P un élément multihomogène de B et J une partie de $\{1, \ldots, q\}$ telle que $P \in K[X^{(J)}]$. Si l'on suppose $\mathcal{M}_\varepsilon \not\subset \sqrt{(\mathfrak{p}, P)}$, il existe un idéal premier $\mathfrak{q}$ de B tel que $\mathcal{M}_\varepsilon \not\subset \mathfrak{q}$ et, pour toute partie J_1 de $\{1, \ldots, q\}$ contenant J, l'idéal $\mathfrak{q}_{J_1}$ est minimal associé à $(\mathfrak{p}_{J_1}, P)$. Pour une telle partie J_1 on a*

$$e_{J_1}(B/(\mathfrak{p}, P)) = e_{J_1}(B/\mathfrak{q}).$$

De plus si $P \in \mathfrak{p}$ alors $\mathfrak{q} = \mathfrak{p}$ et sinon $e_{J_1}(B/\mathfrak{q}) = e_{J_1}(B/\mathfrak{p}) - 1$.

DÉMONSTRATION. Si $P \in \mathfrak{p}$ le lemme est trivial; on suppose donc $P \notin \mathfrak{p}$. Considérons l'idéal $(\mathfrak{p}_J, P)$ de $K[X^{(J)}]$. L'hypothèse $\mathcal{M}_\varepsilon \not\subset \sqrt{(\mathfrak{p}, P)}$ entraîne $\mathcal{M}_{\varepsilon_J} \not\subset \sqrt{(\mathfrak{p}_J, P)}$ donc il existe un idéal premier $\mathfrak{r}$ (de $K[X^{(J)}]$) minimal associé à $(\mathfrak{p}_J, P)$ ne contenant pas $\mathcal{M}_{\varepsilon_J}$. Soit ensuite $j \notin J$ et $J' = J \cup \{j\}$. Nous allons établir qu'il existe un idéal premier $\mathfrak{r}'$ de $K[X^{(J')}]$ minimal associé à $(\mathfrak{p}_{J'}, P)$, ne contenant pas $\mathcal{M}_{\varepsilon_{J'}}$ et tel que $\mathfrak{r}' \cap K[X^{(J)}] = \mathfrak{r}$. Désignons par $\mathfrak{p}_1, \ldots, \mathfrak{p}_s$ les premiers minimaux associés à $(\mathfrak{p}_{J'}, P)$. On a alors

$$(\mathfrak{p}_{J'}, P) \subset \bigcap_{i=1}^{s} \mathfrak{p}_i = \sqrt{(\mathfrak{p}_{J'}, P)};$$

comme $\sqrt{(\mathfrak{p}_{J'}, P)} \cap K[X^{(J)}] = \sqrt{(\mathfrak{p}_J, P)}$ on en déduit

$$(\mathfrak{p}_J, P) \subset \bigcap_{i=1}^{s} (\mathfrak{p}_i \cap K[X^{(J)}]) \subset \mathfrak{r}.$$

Il existe donc un indice i_0 tel que $(\mathfrak{p}_J, P) \subset \mathfrak{p}_{i_0} \cap K[X^{(J)}] \subset \mathfrak{r}$ et par minimalité de $\mathfrak{r}$ on a $\mathfrak{p}_{i_0} \cap K[X^{(J)}] = \mathfrak{r}$. Notons $\mathfrak{r}' = \mathfrak{p}_{i_0}$ de sorte qu'il reste à voir seulement $\mathcal{M}_{\varepsilon_{J'}} \not\subset \mathfrak{r}'$. Or si $\mathcal{M}_{\varepsilon_{J'}} \subset \mathfrak{r}'$ on a $\mathcal{M}_{\varepsilon_j} \subset \mathfrak{r}'$ (puisque, si $i \in J$, $\mathcal{M}_{\varepsilon_i} \not\subset \mathfrak{r}'$ à cause de $\mathcal{M}_{\varepsilon_i} \not\subset \mathfrak{r} = \mathfrak{r}' \cap K[X^{(J)}]$) puis $\mathfrak{p}_{J'} \subset (\mathfrak{p}_J, \mathcal{M}_{\varepsilon_j}) \subset \mathfrak{r}'$. Maintenant le théorème de l'idéal principal de Krull conjointement avec le fait que l'anneau B est caténaire entraîne $\mathrm{ht}(\mathfrak{r}') = \mathrm{ht}(\mathfrak{p}_{J'}) + 1$. Ainsi l'une des deux inclusions serait une égalité : $\mathfrak{p}_{J'} = (\mathfrak{p}_J, \mathcal{M}_{\varepsilon_j})$ est en contradiction avec $\mathcal{M}_\varepsilon \not\subset \mathfrak{p}$ et $\mathfrak{r}' = (\mathfrak{p}_J, \mathcal{M}_{\varepsilon_j})$ donne par intersection $\mathfrak{p}_J = \mathfrak{r}$ également absurde.

En appliquant à $\mathfrak{r}'$ ce qui précède, on peut itérer le procédé et on aboutit à $\mathfrak{q} \in \mathrm{Spec} B$ ayant les propriétés requises. Montrons que $e_{J_1}(B/(\mathfrak{p}, P)) = e_{J_1}(B/\mathfrak{q}) = e_{J_1}(B/\mathfrak{p}) - 1$. La seconde égalité qui s'écrit $\mathrm{ht}(\mathfrak{q}_{J_1}) = \mathrm{ht}(\mathfrak{p}_{J_1}) + 1$ résulte comme précédemment du théorème de l'idéal principal dans un anneau caténaire. La minoration $e_{J_1}(B/(\mathfrak{p}, P)) \geq e_{J_1}(B/\mathfrak{q})$ est claire par $(\mathfrak{p}, P) \subset \mathfrak{q}$ et $\mathcal{M}_\varepsilon \not\subset \mathfrak{q}$. L'inégalité restante suit de ce que $\mathfrak{q}_{J_1}$ est de hauteur minimale parmi les idéaux premiers contenant $(\mathfrak{p}_{J_1}, P) = (\mathfrak{p}, P)_{J_1}$. $\qquad\square$

Lorsque M est un B-module multigradué de type fini sur B on définit la *fonction de Hilbert-Samuel* de M comme étant

$$\mathbf{Z}^q \xrightarrow{\Psi_M} \mathbf{N}$$

$$d \longmapsto \dim_K M_d.$$

Nous allons donner des informations sur Ψ_M avec les $e_J(M)$. Le lemme suivant décrit le cas trivial. On utilise l'ordre produit sur $\mathbf{Z}^q$.

LEMME 2.9. *Soit M un B-module multigradué de type fini. On a équivalence entre*

1. $\mathcal{M}_\varepsilon \subset \sqrt{\mathrm{Ann}(M)}$;
2. $e(M) = -\infty$;
3. *il existe $b \in \mathbf{Z}^q$ tels que si $d \geq b$ alors $\Psi_M(d) = 0$.*

DÉMONSTRATION. L'équivalence de 1 et 2 est immédiate. Montrons $1 \Longrightarrow 3$. On choisit $k \in \mathbf{N}^q$ tel que $\mathcal{M}_k \subset \mathrm{Ann}(M)$ et $b \in \mathbf{N}^q$ tel que M soit engendré par des éléments multihomogènes de multidegrés majorés par $b - k$. En se servant de ces générateurs on vérifie immédiatement que si $d \geq b$ alors $M_d \subset B_k \cdot M = 0$ car B_k est engendré par $\mathcal{M}_k$. Montrons $3 \Longrightarrow 1$. On choisit $k \in \mathbf{N}^q$ tel que M soit engendré par ses éléments multihomogènes de multidegrés minorés par $b - k$. On constate $B_k \cdot M \subset \oplus_{d \geq b} M_d = 0$ et donc $\mathcal{M}_k \subset \mathrm{Ann}(M)$ ce qui entraîne bien $\mathcal{M}_\varepsilon \subset \sqrt{\mathrm{Ann}(M)}$. $\qquad\square$

Le résultat central de ce paragraphe est

THÉORÈME 2.10. *Soit M un B-module multigradué de type fini. Il existe un polynôme (aussitôt unique) dit de Hilbert-Samuel de M noté H_M dans $\mathbf{Q}[T_1, \ldots, T_q]$ pour lequel on peut trouver $b \in \mathbf{Z}^q$ de telle sorte que si $d \geq b$ alors $\Psi_M(d) = H_M(d)$. De plus le polynôme H_M jouit des propriétés ci-dessous.*

1. *H_M est de degré total égal à $e(M)$ et, plus généralement, de degré partiel en $T^{(J)}$ égal à $e_J(M)$ pour toute partie J de $\{1, \ldots, q\}$.*
2. *Les coefficients des monômes de H_M de degré total égal à $e(M)$ sont positifs.*
3. *Si $\mathrm{Ann}(M)$ est de la forme $(\mathfrak{p}, P)$ où $\mathfrak{p}$ est un premier multihomogène de B et P un élément multihomogène de B et si $\mathcal{M}_\varepsilon \not\subset \sqrt{\mathrm{Ann}(M)}$, alors, pour chaque couple de parties $J \subset J_1 \subset \{1, \ldots, q\}$ vérifiant $P \in K[X^{(J)}]$, il existe un monôme de coefficient non nul dans H_M de degré total $e(M)$ et de degrés partiels en $T^{(J)}$ et $T^{(J_1)}$ respectivement $e_J(M)$ et $e_{J_1}(M)$.*

DÉMONSTRATION. On vérifie aisément qu'un polynôme de $\mathbf{Q}[T_1, \ldots, T_q]$ qui s'annule en tous les éléments de $\mathbf{Z}^q$ tels que $d \geq b$ (pour $b \in \mathbf{Z}^q$ donné) est nul. Ce fait donne en particulier l'unicité de H_M. Notons aussi immédiatement que si $e(M) = -\infty$ le théorème découle du lemme précédent.

Lorsque l'on a une suite exacte

$$0 \longrightarrow M' \longrightarrow M \longrightarrow M'' \longrightarrow 0$$

de modules multigradués de type fini sur B, on constate sans peine que d'une part $\Psi_M = \Psi_{M'} + \Psi_{M''}$ et d'autre part que

$$\sqrt{\mathrm{Ann}(M)} = \sqrt{\mathrm{Ann}(M')} \cap \sqrt{\mathrm{Ann}(M'')} \ .$$

Alors par le lemme 2.6 on obtient pour chaque M multigradué de type fini des premiers multihomogènes $\mathfrak{p}_1, \ldots, \mathfrak{p}_t$ et des éléments $f_1, \ldots, f_t$ de $\mathbf{Z}^q$ de telle sorte que

$$\Psi_M = \sum_{i=1}^{t} \Psi_{(B/\mathfrak{p}_i)(f_i)} \qquad \text{et} \qquad \sqrt{\mathrm{Ann}(M)} = \bigcap_{i=1}^{t} \mathfrak{p}_i.$$

De la deuxième égalité on déduit que, pour toute partie J de $\{1, \ldots, q\}$, on a $e_J(M) = \max_{1 \leq i \leq t} e_J(B/\mathfrak{p}_i)$: en effet les minimaux contenant $\mathrm{Ann}(M)$ apparaissent parmi les $\mathfrak{p}_i$ (qui contiennent tous $\mathrm{Ann}(M)$). Vérifions que si l'on connaît le théorème pour chacun des $(B/\mathfrak{p}_i)(f_i)$ alors on le connaît pour M. En effet l'existence de H_M est claire par somme. Par la remarque ci-dessus on peut démontrer les autres propriétés seulement lorsque $e(M) \neq -\infty$. La formule pour les $e_J(M)$ assure que les degrés sont au plus ceux annoncés; comme il existe un $\mathfrak{p}_i$ tel que $e(B/\mathfrak{p}_i) = e(M)$ et que pour chaque tel $\mathfrak{p}_i$ les coefficients dominants de $H_{(B/\mathfrak{p}_i)(f_i)}$ sont positifs, on constate que H_M est effectivement de degré total $e(M)$; on obtient en outre aussitôt que les coefficients dominants de H_M sont positifs. Choisissons ensuite, pour J fixée, un premier $\mathfrak{p}$ minimal contenant $\mathrm{Ann}(M)$ tel que $e_J(B/\mathfrak{p}) = e_J(M)$ et pour lequel $e(B/\mathfrak{p})$ est maximal parmi les tels premiers. En appliquant la troisième assertion à $B/\mathfrak{p}$ (c'est-à-dire avec $P = 0$) on constate que $H_{B/\mathfrak{p}}$ comporte un monôme de degré en $T^{(J)}$ égal à $e_J(M)$ et de degré total $e(B/\mathfrak{p})$; par ailleurs la maximalité impose que les seuls premiers $\mathfrak{p}'$ pour lesquels le coefficient du monôme en question est non nul dans $H_{B/\mathfrak{p}'}$ satisfont aux mêmes contraintes que $\mathfrak{p}$ (à savoir $e_J(B/\mathfrak{p}') = e_J(M)$ et $e(B/\mathfrak{p}') = e(B/\mathfrak{p})$) et on conclut donc, toutes les contributions étant positives, que le coefficient correspondant de H_M est non nul. Ceci donne bien l'assertion sur le degré partiel de H_M en $T^{(J)}$. Si finalement M vérifie les hypothèses de la troisième assertion, le lemme 2.8 montre qu'il existe $\mathfrak{q}$ parmi les $\mathfrak{p}_i$ tel que $e_J(M) = e_J(B/\mathfrak{q})$, $e_{J_1}(M) = e_{J_1}(B/\mathfrak{q})$ et $e(M) = e(B/\mathfrak{q})$. En utilisant cette assertion pour $B/\mathfrak{q}$ on constate qu'il existe dans $H_{B/\mathfrak{q}}$, donc dans un certain $H_{(B/\mathfrak{p}_i)(f_i)}$, un monôme ayant les degrés requis; à nouveau par positivité le coefficient correspondant est non nul dans H_M. Ainsi le théorème est acquis pour M.

Établissons maintenant le théorème pour les modules tels que $e(M) = e$ par récurrence sur $e \geq 0$. Soit donc un tel e et l'on suppose que l'on connaît le résultat pour tous les modules tels que $e(M) < e$ (ce qui est loisible en vertu de la remarque sur le cas où $e(M) = -\infty$). Par le raisonnement précédent (et parce que les $\mathfrak{p}_i$ y intervenant vérifient $e(B/\mathfrak{p}_i) \leq e(M) = e$) il suffit de prouver le théorème lorsque M est de la forme $(B/\mathfrak{p})(f)$ et même, le décalage n'affectant pas les termes dominants, lorsque $M = B/\mathfrak{p}$. L'hypothèse $e(M) \geq 0$ assure que $\mathcal{M}_\varepsilon \not\subset \mathfrak{p}$. Il existe donc $j_1, \ldots, j_q$ tels que $X_{j_1}^{(1)} \ldots X_{j_q}^{(q)} \notin \mathfrak{p}$. Pour une partie J on note alors X_J le monôme $\prod_{i \in J} X_{j_i}^{(i)}$ de degré ε_J. On considère alors la suite exacte

$$0 \longrightarrow B/\mathfrak{p}(-\varepsilon_J) \xrightarrow{\times X_J} B/\mathfrak{p} \longrightarrow B/(\mathfrak{p}, X_J) \longrightarrow 0$$

qui donne pour tout $d \in \mathbf{N}^q$ l'égalité

$$\Psi_M(d) - \Psi_M(d - \varepsilon_J) = \Psi_{B/(\mathfrak{p}, X_J)}(d).$$

Par le lemme 2.8 on a soit $e(B/(\mathfrak{p}, X_J)) = -\infty$ soit $e(B/(\mathfrak{p}, X_J)) = e - 1$. Dans tous les cas l'hypothèse de récurrence assure que le membre de droite de l'égalité ci-dessus est un polynôme pour d assez grand. En considérant la base de $\mathbf{Q}[T_1, \ldots, T_q]$ formée des polynômes $\binom{T_1}{b_1} \cdots \binom{T_q}{b_q}$ pour $b \in \mathbf{N}^q$, on peut, par un raisonnement analogue à celui utilisé dans le cas homogène (voir par exemple [**Har**][2], I.7.3), montrer avec l'égalité pour $J = \{1\}$ que, pour d assez grand, Ψ_M diffère d'un polynôme par une fonction de $d_2, \ldots, d_q$ puis avec l'égalité pour $\{2\}$ que la dite fonction diffère pour $d_2, \ldots, d_q$ assez grand d'un polynôme en $d_3, \ldots, d_q$ et ainsi de suite. Ceci montre l'existence de H_M et il vérifie pour tout J

$$H_M(T_1, \ldots, T_q) - H_M(T - \varepsilon_J) = H_J(T_1, \ldots, T_q)$$

où l'on note pour simplifier $H_J = H_{B/(\mathfrak{p}, X_J)}$. Comme $e(M) \neq -\infty$ le polynôme H_M est non nul.

Examinons d'abord le cas où il est constant. Alors chaque $H_{\{i\}}$ est nul donc $\mathcal{M}_\varepsilon \subset (\mathfrak{p}, X_{j_i})$. Par multihomogénéité (et $\mathcal{M}_\varepsilon \not\subset \mathfrak{p}$) ceci entraîne $\mathcal{M}_{\varepsilon_i} \subset (\mathfrak{p}_{\{i\}}, X_{j_i})$ pour tout $1 \leq i \leq q$. On a alors $\mathrm{ht}(\mathfrak{p}_{\{i\}}) = n_i$ puis $e_{\{i\}}(M) = 0$ et enfin, par le lemme 2.7, $e_J(M) = 0$ pour toute partie J. L'assertion 1 sur les degrés est donc vérifiée; l'assertion 3 est immédiate (avec $H_M \neq 0$) et l'assertion 2 découle de la positivité de Ψ_M.

Supposons maintenant H_M non constant. Soit i un indice tel qu'il existe un monôme de degré total maximal de H_M de degré partiel non nul en T_i. D'une part $H_{\{i\}}$ est non nul par la formule ce qui implique qu'il est de degré total $e-1$. D'autre part la même formule montre qu'il y a bijection entre les monômes de H_M de degré total maximal et de degré non nul en T_i et les monômes de degré total maximal (c'est-à-dire $e - 1$) de $H_{\{i\}}$. Ceci entraîne, par le choix de i, que H_M est de degré

[2][**Har**] R. Hartshorne. *Algebraic Geometry*, Springer, Berlin, (1977).

total $e = e(M)$. Nous allons établir la troisième assertion pour M, ce qui donnera les degrés partiels. Soient donc $J \subset J_1 \subset \{1, \ldots, q\}$. Supposons en premier lieu $e(B/(\mathfrak{p}, X_J)) \neq -\infty$. Alors il découle du lemme 2.8 que $e_J(B/(\mathfrak{p}, X_J)) = e_J(M) - 1$ et $e_{J_1}(B/(\mathfrak{p}, X_J)) = e_{J_1}(M) - 1$. Ainsi H_J a par la troisième assertion un terme de degrés total, en $T^{(J)}$ et en $T^{(J_1)}$ respectivement égaux à $e - 1$, $e_J(M) - 1$ et $e_{J_1}(M) - 1$. Par la formule donnant H_J ceci prouve que H_M contient effectivement un terme de la forme voulue. Examinons maintenant le cas $e(B/(\mathfrak{p}, X_J)) = -\infty$. On a alors $\mathcal{M}_\varepsilon \subset (\mathfrak{p}, X_J) \subset (\mathfrak{p}, X_{j_i})$ dès que $i \in J$. Ainsi pour $i \in J$ on déduit $H_{\{i\}} = 0$; comme dans le cas où H_M est constant, ceci montre $e_{\{i\}}(M) = 0$ si $i \in J$ puis $e_J(M) = 0$ par le lemme 2.7 (même si $J = \emptyset$ car $e(M) \neq -\infty$). Par ailleurs H_M est effectivement de degré partiel nul en $X^{(J)}$ car sinon on aurait un terme non nul dans H_J. Il suffit donc de trouver un monôme de bon degré total et en $T^{(J_1)}$. Si $H_{J_1} \neq 0$ cela résulte du premier cas; sinon $e_{J_1}(M) = 0$ et n'importe quel monôme degré total e convient.

Il reste à montrer l'assertion 2. La positivité du coefficient d'un monôme de degré total e dans lequel le degré en T_i est non nul (pour chaque tel monôme on peut trouver un i vérifiant cela) se déduit de l'assertion correspondante pour $H_{\{i\}}$. □

On peut remarquer que la troisième assertion du théorème est énoncée sous une condition assez restrictive sur $\mathrm{Ann}(M)$ (bien qu'elle contienne, c'est essentiel, le cas des annulateurs premiers). L'exemple suivant montre qu'une condition de ce type est nécessaire. On choisit $q = 3$, $n_1 = n_2 = 1$ et $n_3 = 2$ ce qui consiste à travailler avec

$$B = K[X_{0,0}, X_{0,1}, X_{1,0}, X_{1,1}, X_{2,0}, X_{2,1}, X_{2,2}] \ ;$$

on définit ensuite l'idéal multihomogène de B suivant (intersection de deux premiers) $I = (X_{0,1}, X_{1,1}) \cap (X_{2,1}, X_{2,2})$; le calcul montre alors que le polynôme de Hilbert-Samuel de $M = B/I$ est

$$H_M = \frac{1}{2}(T_3 + 1)(T_3 + 2) + (T_1 + 1)(T_2 + 1) - 1;$$

on trouve aussi $e_{\{2\}}(M) = 1$, $e_{\{2,3\}}(M) = 2$ et $e_{\{1,2,3\}}(M) = 2$ pourtant aucun monôme de H_M n'a ces degrés partiels (autrement dit le coefficient de $T_2 T_3$ est nul).

Nous introduisons maintenant quelques notations supplémentaires. Tout d'abord nous allons quelque peu formaliser la notion de termes dominants d'un polynôme de $\mathbf{Q}[T_1, \ldots, T_q]$; on associe à $P \in \mathbf{Q}[T_1, \ldots, T_q]$ l'élément $\delta(P) \in \mathbf{Q}^{\mathbf{N}^q}$ défini comme suit : si $\alpha_1 + \cdots + \alpha_q = \deg P$ alors $\frac{1}{\alpha_1! \ldots \alpha_q!} \delta(P)_\alpha$ est le coefficient de $T_1^{\alpha_1} \ldots T_q^{\alpha_q}$ dans P et $\delta(P)_\alpha = 0$ sinon. Par exemple l'assertion 2 du théorème se traduit par : pour tout module M et tout $\alpha \in \mathbf{N}^q$ on a $\delta(H_M)_\alpha \geq 0$. En fait on peut même montrer, en reprenant pas à pas la preuve, que $\delta(H_M) \in \mathbf{N}^{\mathbf{N}^q}$ (voir aussi plus bas proposition 3.4). Lorsque I est un idéal multihomogène de B on définit — voir [**VdW**] (*loc. cit.*, p. 57) — les degrés (ou multidegrés) de I comme $\deg(I) = \delta(H_{B/I})$. Notons que l'on a la formule de décomposition

$$\deg(I) = \sum_{\substack{\mathfrak{p} \supset I \\ \mathrm{ht}(\mathfrak{p}) = \mathrm{ht}(I)}} \ell(B_\mathfrak{p}/I_\mathfrak{p}) \deg(\mathfrak{p}).$$

En effet ceci découle du lemme 2.6 car, de même que dans la preuve du théorème, on écrit $H_{B/I}$ comme somme des $H_{B/\mathfrak{p}_i(f_i)}$; on constate ensuite que, d'une part, seuls les $\mathfrak{p}_i$ de hauteur minimale contribuent aux termes dominants donc à $\deg(I)$ et d'autre part (en localisant) que le nombre d'occurrence d'un tel $\mathfrak{p}$ donné parmi les $\mathfrak{p}_i$ est égal à $\ell(B_\mathfrak{p}/I_\mathfrak{p})$; ceci donne la formule.

On notera $\mathrm{Supp}(k)$ le support d'un élément k de $\mathbf{N}^q$ c'est-à-dire $\{i \mid k_i \neq 0\}$. Pour un polynôme $P \in \mathbf{Q}[T_1, \ldots, T_q]$ et $k \in \mathbf{Q}^q$ on définit un nouveau polynôme $\Delta_k(P)$ par $\Delta_k(P)(T_1, \ldots, T_q) = P(T_1, \ldots, T_q) - P(T_1 - k_1, \ldots, T_q - k_q)$. De manière analogue à [**Bro3**] on introduit un opérateur $*\colon \mathbf{Q}^{\mathbf{N}^q} \times \mathbf{Q}^q \to \mathbf{Q}^{\mathbf{N}^q}$ par la formule

$$(\delta * k)_\alpha = \sum_{i=1}^{q} k_i.\delta_{\alpha+\varepsilon_i}.$$

L'usage de cet opérateur est illustré par l'exemple fondamental suivant :

LEMME 2.11. *Si $\mathfrak{p}$ est un idéal multihomogène premier de B et P un élément de B_k (multihomogène de multidegré k) n'appartenant pas à $\mathfrak{p}$ alors*

$$\deg((\mathfrak{p}, P)) = \deg(\mathfrak{p}) * k.$$

Si $k \neq 0$, les deux termes sont nuls si et seulement si $e_{\mathrm{Supp}(k)}(B/\mathfrak{p}) = 0$.

DÉMONSTRATION. On suppose $k \neq 0$ car sinon le résultat est immédiat; on a donc $e(B/\mathfrak{p}) \neq -\infty$. L'hypothèse $P \notin \mathfrak{p}$ montre que l'on a une suite exacte de B-modules

$$0 \longrightarrow B/\mathfrak{p}(-k) \xrightarrow{\times P} B/\mathfrak{p} \longrightarrow B/(\mathfrak{p}, P) \longrightarrow 0$$

et donc que $H_{B/(\mathfrak{p}, P)} = \Delta_k H_{B/\mathfrak{p}}$. En revenant aux définitions on constate qu'un terme de degré $e(B/\mathfrak{p})$ de la forme $\frac{1}{\alpha_1! \ldots \alpha_q!} \deg(\mathfrak{p})_\alpha T_1^{\alpha_1} \ldots T_q^{\alpha_q}$ de $H_{B/\mathfrak{p}}$ donne naissance à q termes de $H_{B/(\mathfrak{p}, P)}$ de degré $e(B/\mathfrak{p}) - 1$ à savoir

$$\frac{1}{\alpha_1! \ldots (\alpha_i - 1)! \ldots \alpha_q!} \deg(\mathfrak{p})_\alpha k_i T_1^{\alpha_1} \ldots T_i^{\alpha_i - 1} \ldots T_q^{\alpha_q}$$

pour $1 \leq i \leq q$. Ainsi les coefficients des termes de degré $e(B/\mathfrak{p}) - 1$ de $H_{B/(\mathfrak{p}, P)}$ sont $\deg(\mathfrak{p}) * k$. Ceci prouve la formule sur les degrés lorsque $H_{B/(\mathfrak{p}, P)}$ est de degré total $e(B/\mathfrak{p}) - 1$. Si ce n'est pas le cas, d'une part $\deg(\mathfrak{p}) * k = 0$ et d'autre part (voir démonstration précédente) $e(B/(\mathfrak{p}, P)) = -\infty$ soit $H_{B/(\mathfrak{p}, P)} = 0$ d'où $\deg((\mathfrak{p}, P)) = 0$. Mais, comme il n'y a pas de simplifications par positivité, cela signifie que tous les termes cités ci-dessus sont nuls; ceci n'est possible que lorsque $e_J(B/\mathfrak{p}) = 0$ si $J = \{i \mid k_i \neq 0\}$. □

Nous donnons enfin un lemme technique qui sera au cœur de la preuve du paragraphe suivant car il fournit un résultat de dévissage qui permet de faire des raisonnements par récurrence en gardant trace des informations véhiculées par le polynôme de Hilbert-Samuel.

LEMME 2.12. *Soient $\mathfrak{p}$ un idéal multihomogène premier de B et $d \in (\mathbf{N}^q)^r$ avec $r \geq 1$. On note $C = K[(d_1)]$ et L le corps des fractions de C. Alors si $\mathfrak{E}_{(d_1)}(\mathfrak{p}) = 0$ l'idéal $\mathfrak{U}_{(d_1)}(\mathfrak{p})$ de $C[X]$ donne naissance à un idéal premier $\mathfrak{q}$ de $L[X]$ et on a*

- $H_{L[X]/\mathfrak{q}} = \Delta_{d_1} H_{B/\mathfrak{p}}$,
- $\deg(\mathfrak{q}) = \deg(\mathfrak{p}) * d_1$,
- $\mathfrak{U}_{d'}(\mathfrak{q}) = \mathfrak{U}_d(\mathfrak{p}).L[d'][X]$ *et*
- $\mathfrak{E}_{d'}(\mathfrak{q}) = \mathfrak{E}_d(\mathfrak{p}).L[d']$

si d' désigne $(d_2, \ldots, d_r)$. De plus si $J \subset J_1 \subset \{1, \ldots, q\}$ et $e_J(B/\mathfrak{p}) = e_{J_1}(B/\mathfrak{p})$ alors $e_J(L[X]/\mathfrak{q}) = e_{J_1}(L[X]/\mathfrak{q})$.

DÉMONSTRATION. L'idéal $\mathfrak{p}'$ engendré par $\mathfrak{p}$ dans $C[X]$ est encore un idéal premier (passage à un anneau de polynômes). Par le lemme 2.4, $\mathfrak{E}_{(d_1)}(\mathfrak{p}) = 0 \neq K[(d_1)]$ implique que $\mathcal{M}_\varepsilon \not\subset \mathfrak{p}$ et l'idéal $\mathfrak{U}_{(d_1)}(\mathfrak{p})$ de $C[X]$ est donc premier. De sa définition même il découle que c'est le seul idéal premier minimal contenant $(\mathfrak{p}', U_1)$ qui ne contient pas $\mathcal{M}_\varepsilon$ car $\mathcal{M}_\varepsilon \not\subset \mathfrak{U}_{(d_1)}(\mathfrak{p})$ (sinon $1 \in \mathfrak{U}_{(d_1)}(\mathfrak{p})$). L'hypothèse $\mathfrak{E}_{(d_1)}(\mathfrak{p}) = 0$ signifie exactement $\mathfrak{U}_{(d_1)}(\mathfrak{p}) \cap C = 0$. Alors les idéaux $\mathfrak{p}''$ et $\mathfrak{q}$ engendrés respectivement par $\mathfrak{p}'$ et $\mathfrak{U}_{(d_1)}(\mathfrak{p})$ dans $L[X]$ sont premiers : on voit $L[X]$ comme localisé de $C[X]$ en le système multiplicatif $C \setminus 0$. Dans $L[X]$ l'idéal $\mathfrak{q}$ est encore le seul premier minimal contenant $(\mathfrak{p}'', U_1)$ et non $\mathcal{M}_\varepsilon$. En écrivant la suite exacte de $L[X]$-modules suivante

$$0 \longrightarrow N \longrightarrow L[X]/(\mathfrak{p}'', U_1) \longrightarrow L[X]/\mathfrak{q} \longrightarrow 0$$

on constate que le $L[X]$-module N vérifie $\mathcal{M}_\varepsilon \subset \sqrt{\mathrm{Ann}(N)}$. On a par conséquent $H_{L[X]/(\mathfrak{p}'', U_1)} = H_{L[X]/\mathfrak{q}}$. Par primalité de $\mathfrak{p}''$ et $U_1 \not\in \mathfrak{p}''$ (le contraire impliquerait $\mathcal{M}_\varepsilon \subset \mathfrak{p}$) on a d'après la démonstration du lemme précédent $H_{L[X]/\mathfrak{q}} = H_{L[X]/(\mathfrak{p}'', U_1)} = \Delta_{d_1} H_{L[X]/\mathfrak{p}''}$ car $U_1 \in C[X]_{d_1}$. Par ailleurs on a un isomorphisme $L[X]/\mathfrak{p}'' \simeq (K[X]/\mathfrak{p}) \underset{K}{\otimes} L$ qui montre que $\dim_L(L[X]/\mathfrak{p}'')_b = \dim_L(K[X]/\mathfrak{p})_b \underset{K}{\otimes} L = \dim_K(K[X]/\mathfrak{p})_b$ pour tout $b \in \mathbf{N}^q$ puis $H_{L[X]/\mathfrak{p}''} = H_{B/\mathfrak{p}}$. Ceci prouve $H_{L[X]/\mathfrak{q}} = \Delta_{d_1} H_{B/\mathfrak{p}}$ ainsi que la formule sur les degrés d'après le lemme précédent.

Si $f \in \mathfrak{U}_d(\mathfrak{p})$ on a $f\mathcal{M}_k \subset (\mathfrak{p}, U_1, \ldots, U_r) = (\mathfrak{p}', U_1)[d']$ dans $B[d]$ donc $f\mathcal{M}_k \subset \mathfrak{q}[d']$ soit $f \in \mathfrak{U}_{d'}(\mathfrak{q})$. Réciproquement si $f \in \mathfrak{U}_{d'}(\mathfrak{q})$ on écrit $f\mathcal{M}_k \subset (\mathfrak{q}, U_2, \ldots, U_r)$ dans $L[d'][X]$ puis $fg\mathcal{M}_k \subset (\mathfrak{U}_{(d_1)}(\mathfrak{p}), U_2, \ldots, U_r)$ dans $C[d'][X] = B[d]$ pour un certain $g \in C \setminus 0$. Ainsi chaque élément de $fg\mathcal{M}_k$ est de la forme $p + \sum_{l=2}^r Q_l U_l$ avec $p\mathcal{M}_{k'} \subset (\mathfrak{p}, U_1)$ (on peut choisir k' indépendant de l'élément) donc $fg\mathcal{M}_{k+k'} \subset (\mathfrak{p}, U_1, U_2, \ldots, U_r)$. Finalement on trouve $fg \in \mathfrak{U}_d(\mathfrak{p})$ ce qui montre bien que $f \in \mathfrak{U}_d(\mathfrak{p}).L[d'][X]$ (puisque $\frac{1}{g} \in L[d'][X]$). La formule pour $\mathfrak{E}_{d'}(\mathfrak{q})$ s'obtient immédiatement par intersection.

Pour notre dernière assertion il·importe de remarquer que $\mathfrak{q} \cap K[X] = \mathfrak{p}$: en effet par la deuxième assertion du lemme 2.4 on a $\mathfrak{U}_{(d_1)}(\mathfrak{p}) \cap K[X] = \mathfrak{p}$ et il suffit de localiser en $C \setminus 0$. On a donc a fortiori $\mathfrak{q}_J \cap K[X^{(J)}] = \mathfrak{p}_J$ et ceci permet de considérer l'application

$$(K[X^{(J)}]/\mathfrak{p}_J)[X^{(J_0)}] \hookrightarrow (L[X^{(J)}]/\mathfrak{q}_J)[X^{(J_0)}]$$

(où l'on a noté $J_0 = J_1 \setminus J$) par laquelle l'image réciproque de $\mathfrak{q}_{J_1}/\mathfrak{q}_J$ est $\mathfrak{p}_{J_1}/\mathfrak{p}_J$. Cette flèche en induit une autre $K_1[X^{(J_0)}] \hookrightarrow L_1[X^{(J_0)}]$ où K_1 et L_1 sont les corps des fractions respectifs de $K[X^{(J)}]/\mathfrak{p}_J$ et $L[X^{(J)}]/\mathfrak{q}_J$. Comme on a évidemment $(\mathfrak{q}_{J_1}/\mathfrak{q}_J) \cap (L[X^{(J)}]/\mathfrak{q}_J) = 0$ l'idéal $\mathfrak{q}_{J_1}/\mathfrak{q}_J$ engendre un idéal $\mathfrak{q}_1$ de $L_1[X^{(J_0)}]$ de hauteur $\mathrm{ht}(\mathfrak{q}_1) = \mathrm{ht}(\mathfrak{q}_{J_1}) - \mathrm{ht}(\mathfrak{q}_J) = n_{J_0} + e_J(L[X]/\mathfrak{q}) - e_{J_1}(L[X]/\mathfrak{q})$; de même $\mathfrak{p}_{J_1}/\mathfrak{p}_J$ donne naissance à $\mathfrak{p}_1 \in \mathrm{Spec}(K_1[X^{(J_0)}])$ avec $\mathrm{ht}(\mathfrak{p}_1) = n_{J_0} + e_J(M) - e_{J_1}(M) = n_{J_0}$ par hypothèse. Par ailleurs $\mathfrak{p}_1$ est l'image réciproque de $\mathfrak{q}_1$ donc $\mathrm{ht}(\mathfrak{q}_1) \geq n_{J_0}$ ce qui se lit $e_{J_1}(L[X]/\mathfrak{q}) \leq e_J(L[X]/\mathfrak{q})$; l'autre inégalité découlant de $J \subset J_1$ d'après le lemme 2.7 on a bien $e_{J_1}(L[X]/\mathfrak{q}) = e_J(L[X]/\mathfrak{q})$ comme escompté. $\square$

2.3. Formes éliminantes. Voici le résultat de principalité annoncé.

THÉORÈME 2.13. *Soient K un corps infini, r un entier, $\mathfrak{p}$ un idéal premier multihomogène de $K[X]$ et d un élément de $(\mathbf{N}^q \setminus 0)^r$. Pour une partie J de $\{1, \ldots, q\}$ on note r_J le nombre d'indices j ($1 \leq j \leq r$) tels que $\mathrm{Supp}(d_j) \subset J$ (c'est-à-dire $d_{j,i} = 0$ si $i \not\in J$).*

1. $\mathfrak{E}_d(\mathfrak{p}) = 0 \iff \mathrm{ht}(\mathfrak{p}_J) < n_J - r_J + 1$ *pour toute partie* J.
2. *Si* $\mathrm{ht}(\mathfrak{p}_J) \leq n_J - r_J + 1$ *pour toute partie* J *alors* $\mathfrak{E}_d(\mathfrak{p})$ *est principal*.

DÉMONSTRATION. Pour tout entier s tel que $0 \leq s \leq r$, on notera s_J le nombre d'indices j $(1 \leq j \leq s)$ tels que $d_{j,i} = 0$ pour $i \notin J$. Le corps K étant infini, un polynôme de $K[d]$ est nul si et seulement si sont nulles toutes ses spécialisations dans K c'est-à-dire ses images par tous les morphismes $\rho: K[d] \to K$ de K-algèbres. La condition $\mathfrak{E}_d(\mathfrak{p}) = 0$ est donc équivalente à $\rho(\mathfrak{E}_d(\mathfrak{p})) = 0$ pour tous ces ρ. Le théorème de l'élimination montre qu'elle est encore équivalente à : pour tout r-uplet $(P_1, \ldots, P_r)$ de polynômes multihomogènes de multidegrés respectifs $d_1, \ldots, d_r$ il existe un zéro non trivial de l'idéal $(\mathfrak{p}, P_1, \ldots, P_r)$ dans $\bar{K}^{n_1+1} \times \cdots \times \bar{K}^{n_q+1}$. En réutilisant le théorème de l'élimination dans le cas le plus simple $(r = 0)$ on traduit finalement $\mathfrak{E}_d(\mathfrak{p}) = 0$ par : pour tout r-uplet $(P_1, \ldots, P_r)$ de polynômes multihomogènes de B de multidegrés respectifs $d_1, \ldots, d_r$ on a $\mathcal{M}_\varepsilon \not\subset \sqrt{(\mathfrak{p}, P_1, \ldots, P_r)}$.

Lorsque $P_1, \ldots, P_s$ sont donnés on note $M_s = B/(\mathfrak{p}, P_1, \ldots, P_s)$; en particulier M_0 que l'on notera M n'est autre que $B/\mathfrak{p}$. Démontrons tout d'abord le sens direct de la première assertion. On suppose donc $\mathfrak{E}_d(\mathfrak{p}) = 0$. Montrons par récurrence sur s $(0 \leq s \leq r)$ qu'il existe des polynômes $P_1, \ldots, P_s$ multihomogènes de multidegrés respectifs $d_1, \ldots, d_s$ de telle sorte que $e_J(M_s) \leq e_J(M) - s_J$ pour tout J. Le cas $s = 0$ étant limpide, on suppose $s \geq 1$ et $P_1, \ldots, P_{s-1}$ construits. Notons alors $\mathfrak{q}_1, \ldots, \mathfrak{q}_t$ les premiers minimaux contenant $\mathrm{Ann}(M_{s-1})$ et ne contenant pas $\mathcal{M}_\varepsilon$. Pour bâtir P_s, on considère l'espace vectoriel B_{d_s} des polynômes multihomogènes de multidegré d_s. Chaque $B_{d_s} \cap \mathfrak{q}_i$ est un sous-espace vectoriel strict de B_{d_s} (puisque $\mathcal{M}_{d_s} \not\subset \mathfrak{q}_i$); par conséquent, K étant infini, $B_{d_s} \neq \bigcup_{i=1}^t B_{d_s} \cap \mathfrak{q}_i$ c'est-à-dire $B_{d_s} \not\subset \bigcup_{i=1}^t \mathfrak{q}_i$. On choisit maintenant $P_s \in B_{d_s}$ tel que $P_s \notin \mathfrak{q}_i$ pour $1 \leq i \leq t$. Ce polynôme convient car si $\mathfrak{q}$ est un premier contenant $\mathrm{Ann}(M_s)$ et non $\mathcal{M}_\varepsilon$ il existe un indice i tel que $\mathfrak{q}_i \subset \mathfrak{q}$; ainsi $\mathrm{ht}(\mathfrak{q}_J) \geq \mathrm{ht}(\mathfrak{q}_{iJ})$ pour toute partie J. Maintenant si d_s contribue à s_J (c'est-à-dire si $(s-1)_J < s_J$) on a $\mathrm{ht}(\mathfrak{q}_J) \geq \mathrm{ht}(\mathfrak{q}_{iJ}) + 1$: en effet $\mathfrak{q}_J = \mathfrak{q}_{iJ}$ est absurde car $P_s \in \mathfrak{q}_J$ $(B_{d_s} \subset K[X_J])$ et $P_s \notin \mathfrak{q}_i$; en considérant le maximum sur tous les tels $\mathfrak{q}$ on démontre notre récurrence. L'hypothèse que $\mathfrak{E}_d(\mathfrak{p}) = 0$ assure que $e_J(M_r) \geq 0$. Le cas $s = r$ donne donc $e_J(M) \geq r_J$. Ceci est exactement notre conclusion car $e(M) \geq 0$ entraîne $e_J(M) = n_J - \mathrm{ht}(\mathfrak{p}_J)$.

Démontrons maintenant la réciproque. On procède par récurrence sur r. Le cas $r = 0$ est clair car si $\mathrm{ht}(\mathfrak{p}_{\{i\}}) < n_i + 1$ pour tout i alors $\mathcal{M}_\varepsilon \not\subset \mathfrak{p}$. Étudions ensuite le cas $r = 1$. On suppose donc que $e_J(M) \geq r_J$ et imaginons (par l'absurde) que $\mathfrak{E}_d(\mathfrak{p}) \neq 0$ c'est-à-dire qu'il existe P_1 de multidegré d_1 tel que $\mathcal{M}_\varepsilon \subset \sqrt{(\mathfrak{p}, P_1)}$. Alors le lemme 2.11 assure que $\deg(\mathfrak{p}) * d_1 = 0$ donc $e_{\mathrm{Supp}(d_1)}(M) = 0$. Ceci est absurde car on a clairement $r_{\mathrm{Supp}(d_1)} = 1$. Ceci démontre le cas $r = 1$.

Supposons donc maintenant $r \geq 2$ et l'assertion établie pour les indices d' de longueur $r - 1$, tous les premiers et tous les corps infinis. Nous noterons ici C l'anneau $K[(d_1)]$ et L son corps des fractions qui est infini. Posons également $d' = (d_2, \ldots, d_r)$. Le cas $r = 1$ montre que les conditions de hauteurs sur $\mathfrak{p}$ impliquent en particulier que $\mathfrak{E}_{(d_1)}(\mathfrak{p}) = 0$. Par le lemme 2.12 l'idéal $\mathfrak{U}_{(d_1)}(\mathfrak{p})$ de $C[X]$ donne donc naissance à $\mathfrak{q}$, idéal premier de $L[X]$, et l'on a $H_{L[X]/\mathfrak{q}} = \Delta_{d_1} H_M$. Montrons que l'idéal $\mathfrak{q}$ de $L[X]$ satisfait les conditions de hauteurs pour d'. Notons $r' = r - 1$ et définissons les r'_J grâce à d'. On veut vérifier que, pour toute partie J,

$$e_J(L[X]/\mathfrak{q}) \geq r'_J;$$

on a $r'_J = r_J - 1$ si $\mathrm{Supp}(d_1) \subset J$ et $r'_J = r_J$ sinon. Considérons d'abord le cas où $\mathrm{Supp}(d_1) \subset J$. Alors (troisième assertion du théorème 2.10 avec $P = 0$) H_M a un terme de degré total $e(M)$, de degré en $T^{(J)}$ égal à $e_J(M)$ et de degré $e_{\mathrm{Supp}(d_1)}(M) \geq r_{\mathrm{Supp}(d_1)} \geq 1$ en $T^{(\mathrm{Supp}(d_1))}$. Un tel terme donne naissance dans $\Delta_{d_1} H_M$ à plusieurs termes (qui ne peuvent se simplifier par positivité) de degré en $T^{(J)}$ égal à $e_J(M) - 1 \geq r_J - 1 = r'_J$. Ainsi on a bien $e_J(L[X]/\mathfrak{q}) \geq r'_J$.

Supposons à présent $\mathrm{Supp}(d_1) \not\subset J$. Notons $J_1 = J \cup \mathrm{Supp}(d_1)$ et $J_0 = J_1 \setminus J$ de sorte que J_1 est l'union disjointe de J et de J_0 et que J_0 est non vide. Par le théorème 2.10, H_M a un terme de degré total $e(M)$, de degré en $T^{(J_1)}$ égal à $e_{J_1}(M)$ et de degré en $T^{(J)}$ égal à $e_J(M)$. Dans ce monôme le degré en $T^{(J_0)}$ est exactement $e_{J_1}(M) - e_J(M)$. Si cette quantité est non nulle, il y a donc au moins une variable d'indice appartenant à $\mathrm{Supp}(d_1)$ et non à J dans ce monôme et, par suite, il donne naissance dans $H_{L[X]/\mathfrak{q}} = \Delta_{d_1} H_M$ à au moins un monôme non nul de degré en $T^{(J)}$ égal à $e_J(M) \geq r_J = r'_J$. Dans ce cas ceci prouve bien $e_J(L[X]/\mathfrak{q}) \geq r'_J$. Reste à montrer cette égalité lorsque $e_{J_1}(M) = e_J(M)$; or, d'après le lemme 2.12, elle impose $e_{J_1}(L[X]/\mathfrak{q}) = e_J(L[X]/\mathfrak{q})$. Comme $\mathrm{Supp}(d_1) \subset J_1$ le premier raisonnement a montré $e_{J_1}(L[X]/\mathfrak{q}) \geq e_{J_1}(M) - 1$; on en déduit donc $e_J(L[X]/\mathfrak{q}) \geq r_{J_1} - 1$. Enfin on a $r_{J_1} \geq r_J + 1$ car $J \subset J_1$ montre que si $\mathrm{Supp}(d_i) \subset J$ alors $\mathrm{Supp}(d_i) \subset J_1$ tandis que $\mathrm{Supp}(d_1) \not\subset J$ mais $\mathrm{Supp}(d_1) \subset J_1$. Ceci montre à nouveau $e_J(L[X]/\mathfrak{q}) \geq r'_J$ et on a prouvé que dans tous les cas $\mathfrak{q}$ vérifie les conditions de hauteurs relatives à d'.

Par hypothèse de récurrence on a donc $\mathfrak{E}_{d'}(\mathfrak{q}) = 0$ c'est-à-dire par le lemme 2.12 $\mathfrak{E}_d(\mathfrak{p}).L[d'] = 0$. On en déduit bien $\mathfrak{E}_d(\mathfrak{p}) = 0$ et ceci conclut la démonstration de cette réciproque par récurrence.

Démontrons à présent la seconde assertion. Le cas $r = 0$ est trivial car $K[d] = K$ est alors un anneau principal. D'un autre côté si $\mathcal{M}_\varepsilon \subset \mathfrak{p}$ on sait que $\mathfrak{E}_d(\mathfrak{p})$ est égal à $K[d]$ qui est un idéal principal. On suppose désormais que $r \geq 1$ et $\mathcal{M}_\varepsilon \not\subset \mathfrak{p}$. Montrons dans un premier temps qu'il existe une façon de réordonner d pour que $\mathfrak{E}_{d'}(\mathfrak{p}) = 0$ si l'on note $d' = (d_1, \ldots, d_{r-1})$. Notre hypothèse étant que pour toute partie J on a $e_J(M) \geq r_J - 1$ (M est encore $B/\mathfrak{p}$), on introduit $\mathcal{J} = \{J \subset \{1, \ldots, q\} \mid e_J(M) = r_J - 1\}$. Si $\mathcal{J}$ est vide, on a tout de suite par la première assertion $\mathfrak{E}_d(\mathfrak{p}) = 0$ principal. Supposons $\mathcal{J} \neq \emptyset$. Montrons que $\mathcal{J}$ est stable par intersection : si $J_1, J_2 \in \mathcal{J}$ alors on a $e_{J_1 \cup J_2}(M) \leq e_{J_1}(M) + e_{J_2}(M) - e_{J_1 \cap J_2}(M)$ par le lemme 2.7 et $r_{J_1 \cup J_2} \geq r_{J_1} + r_{J_2} - r_{J_1 \cap J_2}$ immédiatement d'après la définition des r_J. Ainsi

$$e_{J_1}(M) + e_{J_2}(M) \geq e_{J_1 \cup J_2}(M) + e_{J_1 \cap J_2}(M)$$
$$\geq r_{J_1 \cup J_2} + r_{J_1 \cap J_2} - 2 \geq r_{J_1} + r_{J_2} - 2$$

(vrai sans hypothèses) et $J_1, J_2 \in \mathcal{J}$ montre $e_{J_1}(M) + e_{J_2}(M) = r_{J_1} + r_{J_2} - 2$ ce qui entraîne que toutes les inégalités ci-dessus sont des égalités et en particulier $e_{J_1 \cap J_2}(M) = r_{J_1 \cap J_2} - 1$ soit $J_1 \cap J_2 \in \mathcal{J}$. Ainsi il existe un plus petit élément dans $\mathcal{J}$; notons-le J_0. Comme $e_{J_0}(M) = r_{J_0} - 1$ on a $r_{J_0} \geq 1$ donc il existe i tel que $\mathrm{Supp}(d_i) \subset J_0$. On choisit alors de réordonner d de sorte que $\mathrm{Supp}(d_r) \subset J_0$. Montrons que pour les r'_J associés à d' on a $e_J(M) \geq r'_J$ ce qui donnera bien, par la première assertion, $\mathfrak{E}_{d'}(\mathfrak{p}) = 0$. Or si $J \notin \mathcal{J}$ on a $e_J(M) \geq r_J \geq r'_J$ tandis que si $J \in \mathcal{J}$ on trouve $e_J(M) = r_J - 1 = r'_J$ car l'indice r intervient dans le décompte de r_J d'après $\mathrm{Supp}(d_r) \subset J_0 \subset J$.

En utilisant $\mathcal{M}_\varepsilon \not\subset \mathfrak{p}$ on choisit $0 \leq j_i \leq n_i$ pour $1 \leq i \leq q$ tels que $X_{j_1}^{(1)} \ldots X_{j_q}^{(q)} \notin \mathfrak{p}$. Pour la suite, on note Z l'indéterminée $u_{X_{j_1}^{(1)d_{r,1}} \ldots X_{j_q}^{(q)d_{r,q}}}^{(r)}$ et S le sous-anneau de $B[d]$ engendré sur K par toutes les autres indéterminées (ou encore l'anneau des polynômes de degré nul en Z); en particulier on a $B[d] = S[Z]$. Nous allons prouver que $\mathfrak{E}_d(\mathfrak{p}) \cap S = 0$. On se fixe donc $f \in \mathfrak{E}_d(\mathfrak{p}) \cap S$. On considère le morphisme de S-algèbres $S[Z] \to S[X_{j_1}^{(1)^{-1}}, \ldots, X_{j_q}^{(q)^{-1}}]$ qui donne pour image à Z l'élément

$$-X_{j_1}^{(1)^{-d_{r,1}}} \ldots X_{j_q}^{(q)^{-d_{r,q}}} \sum_{\mathrm{m} \in \mathcal{M}_{d_r}, u_{\mathrm{m}}^{(r)} \neq Z} u_{\mathrm{m}}^{(r)} \mathrm{m}.$$

L'intérêt de ce morphisme est que l'image de U_r est nulle. On déduit alors de $f\mathcal{M}_k \subset \mathfrak{p}[d]$ que $f\mathcal{M}_k$ est inclus dans l'idéal engendré dans $S[X_{j_1}^{(1)^{-1}}, \ldots, X_{j_q}^{(q)^{-1}}]$ par $\mathfrak{p}$ et $U_1, \ldots, U_{r-1}$. En chassant les dénominateurs on trouve $b \in \mathbf{N}^q$ tel que $X_{j_1}^{(1)^{b_1}} \ldots X_{j_q}^{(q)^{b_q}} f\mathcal{M}_k$ est inclus dans l'idéal engendré dans S par $\mathfrak{p}$ et $U_1, \ldots, U_{r-1}$. On considère maintenant f comme un polynôme en les indéterminées $u_{\mathrm{m}}^{(r)}$ et soit $g \in B[d']$ l'un de ses coefficients. L'inclusion précédente montre, en passant aux coefficients dans $B[d']$, que $X_{j_1}^{(1)^{b_1}} \ldots X_{j_q}^{(q)^{b_q}} g \in \mathfrak{U}_{d'}(\mathfrak{p})$. Comme $\mathcal{M}_\varepsilon \not\subset \mathfrak{p}$ l'idéal $\mathfrak{U}_{d'}(\mathfrak{p})$ est premier et $X_{j_1}^{(1)^{b_1}} \ldots X_{j_q}^{(q)^{b_q}} \notin \mathfrak{U}_{d'}(\mathfrak{p})$ car $X_{j_1}^{(1)} \ldots X_{j_q}^{(q)} \notin \mathfrak{p} = \mathfrak{U}_{d'}(\mathfrak{p}) \cap B$; ainsi $g \in \mathfrak{U}_{d'}(\mathfrak{p})$ c'est-à-dire $g \in \mathfrak{E}_{d'}(\mathfrak{p})$. La première étape ayant assuré que ce dernier idéal est nul on a $g = 0$. Tous les coefficients de f sont donc nuls et on a ainsi établi $\mathfrak{E}_d(\mathfrak{p}) \cap S = 0$. Ceci donne la conclusion en combinant le fait que $\mathfrak{E}_d(\mathfrak{p})$ est premier avec la factorialité de S (on vérifie immédiatement que si un irréductible de $S[Z]$ appartient à $\mathfrak{E}_d(\mathfrak{p})$ il l'engendre). $\square$

L'intérêt d'un résultat de principalité est de remplacer la manipulation d'un idéal par celui d'un élément de l'anneau, son générateur. Cependant ce dernier n'est pas uniquement déterminé mais seulement à une constante près. Si cela n'est pas gênant pour les applications, il est toutefois plus commode de privilégier arbitrairement un générateur afin de pouvoir écrire facilement certaines formules. Pour cela on fixe une fois pour toutes un ensemble $\mathrm{Irr}(K[d])$ de représentants des éléments irréductibles de $K[d]$ modulo les éléments inversibles (on peut par exemple ne retenir que les irréductibles dont le coefficient dominant est 1 dans un certain ordre lexicographique sur les variables). Comme exemple d'application, sachant que l'on appelle générateur de la partie principale de J (idéal de $K[d]$) tout p.g.c.d. des éléments de J, on privilégiera l'unique générateur qui s'écrit

$$\prod_{\pi \in \mathrm{Irr}(K[d])} \pi^{n_\pi}$$

et on le notera $\mathrm{ppr}(J)$. Dans le même ordre d'idées on note par la suite

$$\sqrt{f} = \prod_{\pi \in \mathrm{Irr}(K[d]), \pi | f} \pi$$

pour $f \in K[d] \setminus 0$.

DÉFINITION 2.14. Soient I un idéal multihomogène de $K[X]$, $r \in \mathbf{N}$ et $d \in (\mathbf{N}^q)^r$. On appelle *forme éliminante d'indice d de I* tout générateur de la partie principale de $\mathfrak{E}_d(I)$. On note en particulier $\mathrm{élim}_d(I) = \mathrm{ppr}(\mathfrak{E}_d(I))$.

L'intérêt de cette définition est essentiellement résumé dans la conséquence suivante du théorème.

COROLLAIRE 2.15. *Sous les hypothèses et notations du théorème, nous noterons* $f = \text{élim}_d(\mathfrak{p})$. *Alors*

1. $f = 0 \iff \text{ht}(\mathfrak{p}_J) < n_J - r_J + 1$ *pour toute partie* J.
2. *Si* $\text{ht}(\mathfrak{p}_J) \le n_J - r_J + 1$ *pour toute partie* J *alors* f *engendre* $\mathfrak{E}_d(\mathfrak{p})$.
3. *Sinon* $f = 1$.

DÉMONSTRATION. Les assertions 1 et 2 découlent clairement du théorème. Montrons 3. Si $\mathcal{M}_\varepsilon \subset \mathfrak{p}$ on a $f = 1$ car $\mathfrak{E}_d(\mathfrak{p}) = K[d]$. Supposons donc $\mathcal{M}_\varepsilon \not\subset \mathfrak{p}$. Notons alors $d' = (d_1, \dots, d_{r-1})$. On a $\mathfrak{E}_{d'}(\mathfrak{p}) \ne 0$ car la diminution de (au plus) un de chacun des r_J montre que la condition de 1 ne peut être satisfaite pour $r - 1$. Aussi, comme $\mathfrak{E}_{d'}(\mathfrak{p}) = \mathfrak{E}_d(\mathfrak{p}) \cap K[d']$, si f engendrait $\mathfrak{E}_d(\mathfrak{p})$ on aurait $f \in K[d']$. Par symétrie on peut supprimer chacun des d_l avec $1 \le l \le r$ et on conclut que $f \in K$. Supposons maintenant que f n'engendre pas $\mathfrak{E}_d(\mathfrak{p})$. Comme $\mathfrak{E}_d(\mathfrak{p})$ est premier et non principal, il contient deux éléments distincts de $\text{Irr}(K[d])$ donc f, qui les divise tous deux, vaut 1. $\qquad\square$

2.4. Spécialisation. Une propriété utile des formes éliminantes est que l'on a un résultat qui donne la forme de leurs spécialisations.

PROPOSITION 2.16. *Soient* K *un corps,* $\mathfrak{p}$ *un premier multihomogène de* $K[X]$, $r \in \mathbf{N}$ *un entier et* $d \in (\mathbf{N}^q)^r$. *On pose* $f = \text{élim}_d(\mathfrak{p})$. *Si* $\rho\colon K[(d_2, \dots, d_r)] \to L$ *est un morphisme d'anneaux dans un corps algébriquement clos* L, *il existe des points* $z_1, \dots, z_t$ *de* $(L^{n_1+1} \setminus 0) \times \cdots \times (L^{n_q+1} \setminus 0)$ *non nécessairement distincts et* $\lambda \in L$ *tels que*

$$\rho(f) = \lambda \prod_{i=1}^{t} U_1(z_i).$$

DÉMONSTRATION. Si $\rho(f) = 0$ le résultat est immédiat avec $\lambda = 0$. Nous supposerons donc que $\rho(f) \ne 0$. De même on suppose que $d_l \ne 0$ pour tout $1 \le l \le r$. En effet dans le cas contraire f divise $u_1^{(l)}$ (1 est ici le monôme de multidegré $(0, \dots, 0)$) et le résultat est clair. Il en est de même si $f = 1$. On se restreint donc au cas où f engendre $\mathfrak{E}_d(\mathfrak{p})$.

Notons I l'idéal multihomogène $(\rho(\mathfrak{p}), \rho(U_2), \dots, \rho(U_r))$ de $L[X]$ et $\mathfrak{q}_1, \dots, \mathfrak{q}_{s'}$ les premiers minimaux contenant I et $0 \le s \le s'$ tel que $\mathcal{M}_\varepsilon \subset \mathfrak{q}_i \iff i > s$. Considérons un morphisme de L-algèbres $\rho'\colon L[(d_1)] \to L$ et notons $\tilde{\rho}$ le morphisme $K[d] \to L$ obtenu à partir de ρ et ρ'. On a ainsi $\tilde{\rho}(f) = \rho'(\rho(f))$. Par le théorème de l'élimination on a : $\tilde{\rho}(f) = 0 \iff \tilde{\rho}(\mathfrak{E}_d(\mathfrak{p})) = 0 \iff \tilde{\rho}(\mathfrak{p}[d])$ a un zéro non trivial dans $L^{n_1+1} \times \cdots \times L^{n_q+1}$. Or $\tilde{\rho}(\mathfrak{p}[d]) = (I, \rho'(U_1)) = \rho'(I[(d_1)])$ par définition de $\tilde{\rho}$; en outre $(I, \rho'(U_1)) \subset (\sqrt{I}, \rho'(U_1)) \subset \sqrt{(I, \rho'(U_1))}$ donc $\rho'(I[(d_1)])$ et $\rho'(\sqrt{I}[(d_1)])$ ont mêmes zéros. En appliquant le théorème de l'élimination à ρ' on a $\rho'(\rho(f)) = 0 \iff \rho'(\mathfrak{E}_{(d_1)}(\sqrt{I})) = 0$. Ceci est valable pour tout ρ' et montre donc, L étant algébriquement clos, que $V(\rho(f)) = V(\mathfrak{E}_{(d_1)}(\sqrt{I}))$ dans $\text{Spec} L[(d_1)]$ d'où $\sqrt{\rho(f)} = \sqrt{\mathfrak{E}_{(d_1)}(\sqrt{I})}$. Par le lemme 2.4 $\mathfrak{E}_{(d_1)}(\sqrt{I}) = \mathfrak{E}_{(d_1)}(\bigcap_{i=1}^{s'} \mathfrak{q}_i) = \bigcap_{i=1}^{s'} \mathfrak{E}_{(d_1)}(\mathfrak{q}_i) = \bigcap_{i=1}^{s} \mathfrak{E}_{(d_1)}(\mathfrak{q}_i)$, la dernière égalité étant justifiée par $i > s \implies \mathcal{M}_\varepsilon \subset \mathfrak{q}_i \implies \mathfrak{E}_{(d_1)}(\mathfrak{q}_i) = L[(d_1)]$. Par le même lemme chaque $\mathfrak{E}_{(d_1)}(\mathfrak{q}_i)$ est premier pour $1 \le i \le s$ et donc $\sqrt{\rho(f)} = \bigcap_{i=1}^{s} \mathfrak{E}_{(d_1)}(\mathfrak{q}_i)$. Maintenant si $1 \le i \le s$ on a ht$(\mathfrak{q}_{iJ}) \le n_J$ pour toute partie J (puisque $\mathcal{M}_\varepsilon \not\subset \mathfrak{q}_i$ donne $e_J(B/\mathfrak{q}) \geq 0$); par conséquent $\mathfrak{E}_{(d_1)}(\mathfrak{q}_i)$ est engendré par $g_i = \text{élim}_{(d_1)}(\mathfrak{q}_i)$. Par factorialité on a alors $(\sqrt{\rho(f)}) = \sqrt{g_1 \cdots g_s}$ puis $\rho(f) = \lambda \prod_{i=1}^{s} g_i^{l_i}$ avec $\lambda \in L$ et $l_i \geq 0$. Pour obtenir la proposition il suffit de voir que chaque g_i peut se mettre sous la forme $U_1(z)$ avec $z \in (L^{n_1+1} \setminus 0) \times \cdots \times (L^{n_q+1} \setminus 0)$; on ne se préoccupe pas des constantes car, L étant algébriquement clos, on peut faire rentrer par multihomogénéité toute constante dans un $U_1(z)$: pour tous $\lambda \neq 0$ et z non trivial il existe z' non trivial tel que $U_1(z') = \lambda U_1(z)$. L'hypothèse $\rho(f) \neq 0$ montre que $\mathfrak{E}_{(d_1)}(\mathfrak{q}_i)$ est non nul donc on a $\text{ht}(\mathfrak{q}_{iJ}) \geq n_J - 1_J + 1$ pour une certaine partie J. Comme $\text{ht}(\mathfrak{q}_{iJ}) \leq n_J$ ceci impose $\text{Supp}(d_1) \subset J$ et $\text{ht}(\mathfrak{q}_{iJ}) = n_J$. Par ailleurs $\mathfrak{q}_{iJ}$ a un zéro non trivial x dans $\prod_{j \in J} L^{n_j+1}$ donc est inclus dans l'idéal engendré par les formes $x_\alpha^{(j)} X_\beta^{(j)} - x_\beta^{(j)} X_\alpha^{(j)}$ pour $j \in J$ et $0 \leq \alpha < \beta \leq n_j$. Ce dernier idéal est clairement premier de hauteur n_J donc il est égal à $\mathfrak{q}_i$. Choisissons $z \in L^{n_1+1} \times \cdots \times L^{n_q+1}$ non trivial coïncidant avec x sur les facteurs correspondant à J. Le théorème de l'élimination montre alors que $\rho'(g_i) = 0$ si et seulement si $\rho'(U_1)$ s'annule en un zéro non trivial de $\mathfrak{q}_i$ (pour $\rho' : L[(d_1)] \to L$ comme précédemment). Comme $\text{Supp}(d_1) \subset J$ ceci se teste sur $\prod_{j \in J} L^{n_j+1}$ c'est-à-dire sur les zéros non triviaux de $\mathfrak{q}_{iJ}$. Vu sa forme explicite, ce dernier a un unique zéro non trivial à savoir x. On conclut $\rho'(g_i) = 0 \iff \rho'(U_1)(z) = 0$ et, par primalité, on a bien $g_i = \lambda U_1(z)$. $\qquad\square$

2.5. Géométrie. Nous n'avons jusqu'ici considéré que des idéaux de $K[X]$. Faisons à présent le lien avec les sous-schémas fermés de $\mathbf{P}_K^{n_1} \times \cdots \times \mathbf{P}_K^{n_q}$. Nous utiliserons le cas dégénéré de l'élimination multihomogène; ainsi $\emptyset$ ci-dessous désigne le 0-uplet, l'unique élément de $(\mathbf{N}^q)^0$.

Un idéal de $K[X]$ sera dit multisaturé si $I = \mathfrak{U}_\emptyset(I)$. Plus généralement on appelle multisaturation de I l'idéal $\bar{I} = \mathfrak{U}_\emptyset(I)$.

Considérons la collection Z des indéterminées Z_j pour $j \in \mathbf{N}^q$, $j_i \leq n_i$ et notons

$$\theta : K[Z] \longrightarrow K[X]$$
$$Z_j \longmapsto X_{j_1}^{(1)} \cdots X_{j_q}^{(q)}.$$

Le noyau de θ est engendré par les $Z_j Z_k - Z_l Z_{j+k-l}$ et l'on sait (voir [**Har**][3], II, Ex.5.12, pour le cas $q = 2$) que l'on a un isomorphisme canonique

$$\text{Proj}(K[Z]/\text{Ker}(\theta)) \simeq \mathbf{P}_K^{n_1} \times \cdots \times \mathbf{P}_K^{n_q}.$$

Si à présent I est un idéal multihomogène de $K[X]$, il est clair que $\theta^{-1}(I)$ est un idéal homogène de $K[Z]$ (gradué par $\deg(Z_i) = 1$). Ce dernier donne donc naissance à un idéal homogène de $K[Z]/\text{Ker}(\theta)$ puis à un sous-schéma fermé de $\text{Proj}(K[Z]/\text{Ker}(\theta))$. Finalement l'isomorphisme ci-dessus permet d'attacher à I un sous-schéma fermé $Y(I)$ de $\mathbf{P}_K^{n_1} \times \cdots \times \mathbf{P}_K^{n_q}$. L'intérêt du procédé est donné par la

PROPOSITION 2.17. *L'application $I \mapsto Y(I)$ décrite ci-dessus induit une bijection décroissante entre les idéaux multihomogènes multisaturés de $K[X]$ et les sous-schémas fermés de $\mathbf{P}_K^{n_1} \times \cdots \times \mathbf{P}_K^{n_q}$. L'adjectif décroissante signifie que si $I \subset J$ l'immersion fermée $Y(J) \hookrightarrow \mathbf{P}_K^{n_1} \times \cdots \times \mathbf{P}_K^{n_q}$ se factorise en $Y(J) \hookrightarrow Y(I) \hookrightarrow \mathbf{P}_K^{n_1} \times \cdots \times \mathbf{P}_K^{n_q}$.*

---

[3][**Har**] R. Hartshorne. *Algebraic Geometry*, Springer, Berlin, (1977).

DÉMONSTRATION. Le résultat dans le cas homogène (voir [**Har**] (*loc. cit.*), II, Ex.5.10) appliqué à $K[Z]$ montre qu'il suffit de prouver que θ^{-1} induit une bijection entre les idéaux multihomogènes multisaturés de $K[X]$ et les idéaux homogènes saturés de $K[Z]$ qui contiennent $\mathrm{Ker}(\theta)$. Le fait que $\theta^{-1}(I)$ est saturé est assez simple : si $fZ_j^N \in \theta^{-1}(I)$ pour tout j alors

$$X_{j_1}^{(1)\,N} \ldots X_{j_q}^{(q)\,N} \theta(f) \in I$$

pour tout j donc $\theta(f) \in \mathfrak{U}_\emptyset(I) = I$ par hypothèse soit $f \in \theta^{-1}(I)$. Exhibons alors l'application inverse : à J idéal homogène de $K[Z]$ on associe $\mathfrak{U}_\emptyset(\theta(J).K[X])$ qui est clairement multihomogène et multisaturé. Montrons que si $\mathrm{Ker}(\theta) \subset J$ et J saturé alors $J = \theta^{-1}(\mathfrak{U}_\emptyset(\theta(J).K[X]))$: comme $\theta(J) \subset \theta(J).K[X] \subset \mathfrak{U}_\emptyset(\theta(J).K[X])$ on a immédiatement l'inclusion $J \subset \theta^{-1}(\mathfrak{U}_\emptyset(\theta(J).K[X]))$. Si à présent $\theta(f') \in \mathfrak{U}_\emptyset(\theta(J).K[X])$ avec $f' \in K[Z]$ homogène, on a $\mathcal{M}_k\theta(f') \subset \theta(J).K[X]$. Quitte à augmenter k on peut le remplacer par $N\varepsilon = (N, \ldots, N)$ avec $N \in \mathbf{N}$; les éléments de $\mathcal{M}_k\theta(f')$ sont alors des sommes de polynômes multihomogènes de multidegrés de la forme $N'\varepsilon$. De tels polynômes appartiennent à l'image de θ donc $\mathcal{M}_{N\varepsilon}\theta(f') \subset \theta(J)$. En particulier $\theta(Z_j^N f') \in \theta(J)$ pour tout j. Comme $\mathrm{Ker}(\theta) \subset J$ cela donne $Z_j^N f' \in J$ puis, J étant saturé, $f' \in J$. Reste à voir que si I est multihomogène et multisaturé on a $I = \mathfrak{U}_\emptyset(\theta(\theta^{-1}(I)).K[X])$. L'inclusion $\supset$ découle de $\theta(\theta^{-1}(I)) \subset I$ et de la multisaturation de I. Dans l'autre sens si $f \in I$ est multihomogène il existe $k \in \mathbf{N}^q$ tel que $f\mathcal{M}_k$ est dans l'image de θ (on choisit k de la forme $N\varepsilon - \deg f$ pour N assez grand). Donc $f\mathcal{M}_k \subset \theta(\theta^{-1}(I)) = I \cap \mathrm{Im}(\theta)$ par conséquent on a bien $f \in \mathfrak{U}_\emptyset(\theta(\theta^{-1}(I)).K[X])$. Le caractère décroissant est clair. $\square$

Le seul idéal multisaturé contenant $\mathcal{M}_\varepsilon$ est B lui-même. Un idéal premier multihomogène de B qui ne contient pas $\mathcal{M}_\varepsilon$ est multisaturé. On constate aussi que l'on a pour un idéal multisaturé $\dim Y(I) = e(B/I)$; en effet le premier nombre est obtenu, d'après la proposition ci-dessus, en considérant la longueur des chaînes d'idéaux premiers multihomogènes; le second se calcule sur les chaînes de premiers quelconques; l'égalité s'obtient alors en combinant le lemme 2.5 et le fait que B est un anneau caténaire. De même $e_J(B/I)$ s'interprète comme la dimension de la projection (sur les facteurs $\mathbf{P}_K^{n_i}$, $i \in J$) de $Y(I)$: il suffit de constater que si I est multisaturé alors aucun des premiers minimaux associés à I ne contient $\mathcal{M}_\varepsilon$.

3. Formes résultantes des idéaux multihomogènes

Dans le cadre homogène la considération des formes éliminantes était suffisante pour définir les hauteurs projectives (voir [**PPh7**]). Pour le cas multihomogène nous devrons passer par les formes résultantes. Ces dernières (voir ci-dessous) ont les mêmes facteurs irréductibles que les formes éliminantes correspondantes mais avec en général des exposants différents. Ainsi elles permettent encore d'utiliser le théorème de l'élimination — qui ne concerne que l'idéal $\sqrt{\mathfrak{E}_d(I)}$ — mais sont de plus beaucoup mieux adaptées à notre étude, en particulier au cas des idéaux non premiers pour lesquels on aura une formule naturelle de décomposition en termes des idéaux premiers associés ainsi qu'aux résultats de spécialisation. La différence de traitement avec le cas homogène tient à ce que l'on pouvait alors se restreindre dans la définition des hauteurs aux idéaux premiers (et compléter par linéarité pour les cycles) pour lesquels formes éliminantes et résultantes coïncidaient. Cette propriété tombe en défaut dans notre situation multihomogène et explique donc l'usage des formes résultantes, y compris pour les idéaux premiers.

3.1. Forme associée à un module. On conserve les notations K, B, $K[d]$, ... de la première partie. Pour ce paragraphe on pourrait essentiellement remplacer $K[d]$ par un anneau factoriel plus quelconque.

Lorsque M est un $K[d]$-module de type fini et de torsion $(\mathrm{Ann}(M) \neq 0)$ on lui associe la forme

$$\chi(M) = \prod_{\pi \in \mathrm{Irr}(K[d])} \pi^{\ell(M_{(\pi)})} \in K[d] \setminus 0$$

où $\ell(M_{(\pi)})$ est la longueur du $K[d]_{(\pi)} = K[d]_{\pi K[d]}$-module $M_{\pi K[d]}$. Cette définition est légitime car d'une part $M_{(\pi)}$ étant un module de type fini et de torsion sur un anneau principal (en fait $K[d]_{(\pi)}$ est même de valuation discrète) est de longueur finie et d'autre part on a

$$
\begin{aligned}
M_{(\pi)} \neq 0 \quad &\Longleftrightarrow \quad \mathrm{Ann}(M_{(\pi)}) \neq K[d]_{(\pi)} \\
&\Longleftrightarrow \quad (\mathrm{Ann}(M))_{(\pi)} \neq K[d]_{(\pi)} \\
&\Longleftrightarrow \quad \mathrm{Ann}(M) \subset (\pi)
\end{aligned}
$$

et il y a ainsi un nombre fini de facteurs différents de 1 à savoir exactement ceux qui correspondent aux π divisant $\mathrm{ppr}(\mathrm{Ann}(M))$. On complète cette définition de façon naturelle en posant $\chi(M) = 0$ si M n'est pas de torsion c'est-à-dire si $\mathrm{Ann}(M) = 0$; on écrira de manière cohérente $\pi^\infty = 0$. Par la théorie élémentaire des longueurs on a immédiatement $\chi(M) = \chi(M')\chi(M'')$ pour toute suite exacte $0 \to M' \to M \to M'' \to 0$ de $K[d]$-modules de type fini. Lorsque d' est un indice plus long débutant par d (ie $d = (d_1, \ldots, d_r)$ et $d' = (d_1, \ldots, d_r, d_{r+1}, \ldots, d_s)$) on a une inclusion naturelle $K[d] \hookrightarrow K[d']$ et pour tout module M de type fini sur $K[d]$

$$\chi\left(M \underset{K[d]}{\otimes} K[d']\right) = \chi(M).$$

En effet $\mathrm{Ann}(M \underset{K[d]}{\otimes} K[d']) = \mathrm{Ann}(M).K[d']$ donc les seuls irréductibles de $K[d']$ intervenant dans $\chi(M \underset{K[d]}{\otimes} K[d'])$ appartiennent à $K[d]$. Pour ceux-là

$$\left(M \underset{K[d]}{\otimes} K[d']\right)_{\pi K[d']} \simeq M_{\pi K[d]} \underset{K[d]_{(\pi)}}{\otimes} K[d']_{(\pi)};$$

si $M_{\pi K[d]}$ s'écrit $\bigoplus_{i=1}^t K[d]_{(\pi)}/\pi^{m_i} K[d]_{(\pi)}$ alors $\ell(M_{(\pi)}) = \sum_{i=1}^t m_i$ et

$$\left(M \underset{K[d]}{\otimes} K[d']\right)_{\pi K[d']} \simeq \bigoplus_{i=1}^t K[d']_{(\pi)}/\pi^{m_i} K[d']_{(\pi)}$$

donc $\ell((M \underset{K[d]}{\otimes} K[d'])_{\pi K[d']}) = \sum_{i=1}^t m_i = \ell(M_{(\pi)})$. Le lemme suivant fournit une utile caractérisation de la longueur sur les anneaux de valuation discrète (tels les $K[d]_{(\pi)}$ ci-dessus).

LEMME 3.1. *Soient A un anneau de valuation discrète, M un A-module de type fini et*

$$A^t \overset{\varphi}{\longrightarrow} A^s \longrightarrow M \longrightarrow 0$$

une présentation finie avec $t \geq s$. Alors $\ell(M)$ est le minimum des valuations des déterminants $s \times s$ extraits de la matrice de φ.

DÉMONSTRATION. Notons que le minimum dont il est question est invariant par les opérations consistant à permuter les lignes ou les colonnes et à soustraire un multiple d'une ligne à une autre (ceci correspond à un changement de base de A^s). Ainsi on peut supposer en premier lieu que la dite matrice, notée

$$\left(a_{i,j}\right)_{1\le i\le s,\, 1\le j\le t},$$

vérifie $v(a_{i,j}) \ge v(a_{1,1})$ pour tous i,j. Comme cela implique $a_{i,j} \in a_{1,1}A$ on peut soustraire un multiple de la première ligne aux suivantes de telle sorte que $a_{i,1} \ne 0 \Rightarrow i = 1$. En itérant le procédé on met la matrice sous la forme

$$\begin{pmatrix} a_{1,1} & \cdots & \cdots & a_{1,s} & \cdots & a_{1,t} \\ 0 & \ddots & & \vdots & & \vdots \\ \vdots & \ddots & \ddots & \vdots & & \vdots \\ 0 & \cdots & 0 & a_{s,s} & \cdots & a_{s,t} \end{pmatrix}$$

avec $v(a_{i,j}) \ge v(a_{i,i})$ pour tous i,j. Comme tout produit du type $a_{1,\sigma(1)} \ldots a_{s,\sigma(s)}$ (avec $\sigma \colon \{1,\ldots,s\} \hookrightarrow \{1,\ldots,t\}$) appartient à $a_{1,1} \ldots a_{s,s}A$, on vérifie que le minimum de l'énoncé vaut $\sum_{i=1}^s v(a_{i,i})$. En faisant ensuite des manipulations de colonnes (changement de base de A^t) on peut remplacer φ par φ' dont la matrice est

$$\begin{pmatrix} a_{1,1} & 0 & \cdots & 0 & 0 & \cdots & 0 \\ 0 & \ddots & \ddots & \vdots & \vdots & & \vdots \\ \vdots & \ddots & \ddots & 0 & \vdots & & \vdots \\ 0 & \cdots & 0 & a_{s,s} & 0 & \cdots & 0 \end{pmatrix}$$

avec les mêmes $a_{i,i}$. Ceci montre que $M \simeq \bigoplus_{i=1}^t A/a_{i,i}$ et on obtient la conclusion car $\ell(A/a_{i,i}) = v(a_{i,i})$. $\qquad\square$

3.2. Définition des formes résultantes. Si I est un idéal multihomogène de $B = K[X]$ alors, pour chaque $k \in \mathbf{Z}^q$, $(B[d]/I[d])_k$ est un $K[d]$-module de type fini. On s'intéresse au comportement de $\chi((B[d]/I[d])_k)$ pour k assez grand. Notons d'abord que

$$\begin{aligned} \mathrm{Ann}_{K[d]}((B[d]/I[d])_k) &= \{f \in K[d] \mid f.B[d]_k \subset I[d]\} \\ &= \{f \in K[d] \mid f\mathcal{M}_k \subset I[d]\}. \end{aligned}$$

Par ailleurs $\mathfrak{E}_d(I) = \bigcup_{k\in\mathbf{N}^q}\{f \in K[d] \mid f\mathcal{M}_k \subset I[d]\}$. Par nœthérianité on a $\mathfrak{E}_d(I) = \{f \in K[d] \mid f\mathcal{M}_k \subset I[d]\}$ pour tout $k \ge k_0$. Ainsi pour tout $k \ge k_0$ on a $\mathrm{Ann}_{K[d]}((B[d]/I[d])_k) = \mathfrak{E}_d(I)$ puis $\sqrt{\chi((B[d]/I[d])_k)} = \sqrt{\text{élim}_d(I)}$. Dorénavant on suppose que $d \in (\mathbf{N}^q \setminus 0)^r$.

LEMME 3.2. *Si $\mathfrak{p}$ est un premier multihomogène de B de hauteur au moins $n - r + 1$ alors $\chi((B[d]/\mathfrak{p}[d])_k)$ est indépendant de k pour k assez grand.*

DÉMONSTRATION. Soient $d_{r+1} \in \mathbf{N}^q \setminus 0$ et $d' = (d, d_{r+1})$. On considère la suite exacte de $B[d']$-modules

$$0 \longrightarrow N \longrightarrow (B[d']/\mathfrak{p}[d])(-d_{r+1}) \overset{\times U_{r+1}}{\longrightarrow} B[d']/\mathfrak{p}[d] \longrightarrow B[d']/\mathfrak{p}[d'] \longrightarrow 0.$$

En en prenant les termes de multidegré k on obtient une suite exacte de $K[d']$-modules et en utilisant $B[d']/\mathfrak{p}[d] \simeq B[d]/\mathfrak{p}[d] \underset{K[d]}{\otimes} K[d']$ on a pour tout $k \in \mathbf{Z}^q$

$$\chi(N_k)\chi((B[d]/\mathfrak{p}[d])_k) = \chi((B[d]/\mathfrak{p}[d])_{k-d_{r+1}})\chi((B[d']/\mathfrak{p}[d'])_k).$$

Comme $\mathrm{ht}(\mathfrak{p}) > n - (r+1) + 1$ le corollaire 2.15 montre que $\mathrm{élim}_{d'}(\mathfrak{p}) = 1$ et, *a fortiori*, $\chi((B[d']/\mathfrak{p}[d'])_k) = 1$ pour $k \geq k_0$. En outre si $f \in \mathfrak{E}_{d'}(\mathfrak{p})$ et $b, k_1 \in \mathbf{N}^q$ sont tels que $f\mathcal{M}_b \subset \mathfrak{p}[d']$ et N est engendré par ses éléments de multidegrés au plus $k_1 - b$ alors un élément P de N_k pour $k \geq k_1$ appartient à $\mathcal{M}_b N_{k-b}$ donc $fP \in f\mathcal{M}_b.N_{k-b} \subset \mathfrak{p}[d'].N_{k-b} = \mathfrak{p}[d].N_{k-b} + U_{r+1}N_{k-b} = 0$. Finalement si $k \geq k_1$ on a $\mathfrak{E}_{d'}(\mathfrak{p}) \subset \mathrm{Ann}(N_k)$ et donc par l'argument précédent $\chi(N_k) = 1$. En prenant le maximum des k_0 et k_1 obtenus pour $d_{r+1} = \varepsilon_i$ on a le résultat. $\qquad\square$

Sous les hypothèses du lemme la valeur commune de $\chi((B[d]/\mathfrak{p}[d])_k)$ pour k assez grand se note $\mathrm{rés}_d(\mathfrak{p})$ et tout multiple (par un élément de $K^\times$) s'appelle une *forme résultante d'indice d de* $\mathfrak{p}$. Comme $\mathfrak{E}_d(\mathfrak{p})$ est premier s'il est différent de $K[d]$ on a toujours $\mathrm{rés}_d(\mathfrak{p}) = \mathrm{élim}_d(\mathfrak{p})^\nu$ pour $\nu > 0$ (qui est uniquement déterminé si $\mathrm{rés}_d(\mathfrak{p}) \neq 1$). C'est encore une conséquence facile du corollaire 2.15 que $\mathrm{rés}_d(\mathfrak{p}) = 1$ si $\mathrm{ht}(\mathfrak{p}) > n - r + 1$ ou si $\mathcal{M}_\varepsilon \subset \mathfrak{p}$. Nous allons voir que le lemme est encore valable pour un idéal quelconque. On définit en fait *a priori* pour un idéal I multihomogène de B de hauteur au moins $n - r + 1$

$$\mathrm{rés}_d(I) = \prod_{\mathfrak{p} \supset I} \mathrm{rés}_d(\mathfrak{p})^{\ell(B_\mathfrak{p}/I_\mathfrak{p})}.$$

Ceci est légitime car d'une part la remarque sur la hauteur montre que l'on peut se limiter aux premiers contenant I de hauteur $n - r + 1$ qui sont en nombre fini (ils font partie des minimaux de $\mathrm{Ass}(B/I)$) et d'autre part pour de tels $\mathfrak{p}$ on a $\sqrt{I_\mathfrak{p}} = \mathfrak{p}B_\mathfrak{p}$ donc $B_\mathfrak{p}/I_\mathfrak{p}$ est de longueur finie (il est artinien comme anneau donc comme $B_\mathfrak{p}$-module). Le théorème suivant (qui donne le lemme pour I quelconque) montre que cette définition est la bonne.

THÉORÈME 3.3. *Soit I un idéal multihomogène de B de hauteur au moins $n - r + 1$. Alors il existe $k_1 \in \mathbf{N}^q$ tel que si $k \geq k_1$ alors*

$$\chi((B[d]/I[d])_k) = \mathrm{rés}_d(I).$$

DÉMONSTRATION. On applique tout d'abord le lemme 2.6 au $B[d]$-module multigradué $B[d]/I[d]$ (la graduation est celle donnée par les $X^{(i)}$). On a ainsi une suite de sous-modules multihomogènes $0 = M_0 \subset M_1 \subset \cdots \subset M_t = B[d]/I[d]$ telle que $M_i/M_{i-1} \simeq (B[d]/\mathfrak{q}_i)(b_i)$ pour $1 \leq i \leq t$ où $\mathfrak{q}_i$ est premier et $b_i \in \mathbf{Z}^q$. En regardant ce que donnent les composantes de multidegré k on obtient

$$\chi((B[d]/I[d])_k) = \prod_{i=1}^{t} \chi((B[d]/\mathfrak{q}_i)_{k+b_i}).$$

Cherchons pour quels $\mathfrak{q}_i$ la contribution du membre de droite n'est pas triviale. Si $\mathfrak{q}_i$ contient $\mathcal{M}_\varepsilon$ on a $\chi((B[d]/\mathfrak{q}_i)_k) = 1$ dès que $k \geq \varepsilon$. Soit donc $\mathfrak{q}$ l'un des $\mathfrak{q}_i$ ne contenant pas $\mathcal{M}_\varepsilon$. On a $I[d] \subset \mathfrak{q}$ donc $\mathfrak{p} = \mathfrak{q} \cap B$ est un premier de B contenant I. De plus $\mathfrak{U}_d(\mathfrak{p}) \subset \mathfrak{q}$ puisque $\mathcal{M}_\varepsilon \not\subset \mathfrak{q}$. Imaginons $\mathfrak{U}_d(\mathfrak{p}) \neq \mathfrak{q}$; alors la suite d'idéaux premiers

$$\mathfrak{p}.B[d] \subset \mathfrak{U}_{(d_1)}(\mathfrak{p}).B[d] \subset \mathfrak{U}_{(d_1,d_2)}(\mathfrak{p}).B[d] \subset \cdots \subset \mathfrak{U}_d(\mathfrak{p}) \subset \mathfrak{q}$$

montre que $\mathrm{ht}(\mathfrak{q}) > \mathrm{ht}(\mathfrak{p}) + r$: en effet les inclusions de la chaîne sont strictes car $\mathfrak{U}_{(d_1,\ldots,d_l)}(\mathfrak{p}).B[d] = \mathfrak{U}_{(d_1,\ldots,d_{l+1})}(\mathfrak{p}).B[d]$ implique que $U_{l+1} \in \mathfrak{U}_{(d_1,\ldots,d_l)}(\mathfrak{p}).B[d]$ c'est-à-dire que les coefficients de U_{l+1} en les $u_\mathrm{m}^{(l+1)}$ appartiennent à $\mathfrak{U}_{(d_1,\ldots,d_l)}(\mathfrak{p})$; ceci s'écrit $\mathcal{M}_{d_l} \subset \mathfrak{U}_{(d_1,\ldots,d_l)}(\mathfrak{p})$ puis $\mathcal{M}_\varepsilon \subset \mathfrak{p}$ qui est absurde. Comme $I \subset \mathfrak{p}$ on a $\mathrm{ht}(\mathfrak{q}) > n + 1$. Par ailleurs $\mathfrak{q}$ donne naissance à un idéal de $(K[d]/\mathfrak{q} \cap K[d])[X]$ de hauteur $\text{ht}(\mathfrak{q}) - \text{ht}(\mathfrak{q} \cap K[d])$. En prenant le corps des fractions L de $K[d]/\mathfrak{q} \cap K[d]$ on voit que $\mathfrak{q}$ engendre dans $L[X]$ un idéal de même hauteur. Comme $\mathcal{M}_\varepsilon \not\subset \mathfrak{q}$ cette hauteur est au plus n. Ainsi

$$n + 1 \leq \text{ht}(I) + r \leq \text{ht}(\mathfrak{p}) + r < \text{ht}(\mathfrak{q}) \leq n + \text{ht}(\mathfrak{q} \cap K[d]).$$

Ceci donne $\text{ht}(\mathfrak{q} \cap K[d]) > 1$. En outre comme $\mathcal{M}_\varepsilon \not\subset \mathfrak{q}$ on a $\text{Ann}((B[d]/\mathfrak{q})_k)$ $= \mathfrak{q} \cap K[d]$ pour k assez grand. Enfin $\text{ppr}(\mathfrak{q} \cap K[d]) = 1$ car c'est un idéal premier non principal et par suite $\chi((B[d]/\mathfrak{q})_k) = 1$ pour k assez grand.

Ainsi les seuls $\mathfrak{q}_i$ qui nous intéressent sont les $\mathfrak{U}_d(\mathfrak{p})$ avec $I \subset \mathfrak{p}$ et $\mathcal{M}_\varepsilon \not\subset \mathfrak{p}$. Dans ce cas $\mathfrak{U}_d(\mathfrak{p})_k = \mathfrak{p}[d]_k$ pour k assez grand (par nœthérianité on peut trouver k_0 tel que $\mathcal{M}_{k_0}\mathfrak{U}_d(\mathfrak{p}) \subset \mathfrak{p}[d]$) donc $\chi((B[d]/\mathfrak{U}_d(\mathfrak{p}))_k) = \text{rés}_d(\mathfrak{p})$ pour $k \geq k_1$ que l'on choisit indépendant de $\mathfrak{p}$ et tel que $\chi((B[d]/\mathfrak{q})_k) = 1$ pour les autres $\mathfrak{q}_i$ précédemment éliminés. Si $\text{ht}(\mathfrak{p}) > n - r + 1$ on a $\text{rés}_d(\mathfrak{p}) = 1$ sinon $\mathfrak{U}_d(\mathfrak{p})$ est un idéal premier minimal contenant $I[d]$ ($I[d] \subset \mathfrak{q} \subset \mathfrak{U}_d(\mathfrak{p})$ donne $\mathfrak{q} \cap B = \mathfrak{p}$ par minimalité de $\mathfrak{p}$ puis $\mathfrak{U}_d(\mathfrak{p}) \subset \mathfrak{q}$ comme ci-dessus car $\mathcal{M}_\varepsilon \subset \mathfrak{q}$ est absurde) et ainsi le nombre d'occurrence de $\mathfrak{U}_d(\mathfrak{p})$ parmi les $\mathfrak{q}_i$ est $\ell((B[d]/I[d])_{\mathfrak{U}_d(\mathfrak{p})})$ (il suffit de localiser en $\mathfrak{U}_d(\mathfrak{p})$). On a donc prouvé que si $k \geq k_1$

$$\chi((B[d]/I[d])_k) = \prod_{\mathfrak{p} \supset I} \text{rés}_d(\mathfrak{p})^{\ell((B[d]/I[d])_{\mathfrak{U}_d(\mathfrak{p})})}.$$

Il reste à voir que $\ell(B_\mathfrak{p}/I_\mathfrak{p}) = \ell((B[d]/I[d])_{\mathfrak{U}_d(\mathfrak{p})})$ pour $\mathfrak{p}$ minimal contenant I et non $\mathcal{M}_\varepsilon$. Choisissons donc une suite de composition de $B_\mathfrak{p}/I_\mathfrak{p}$ notée $0 = N_0 \subset N_1 \subset \cdots \subset N_s = B_\mathfrak{p}/I_\mathfrak{p}$. Ainsi $s = \ell(B_\mathfrak{p}/I_\mathfrak{p})$ et pour chaque $1 \leq i \leq s$ le $B_\mathfrak{p}$-module N_i/N_{i+1} est simple donc (anneau local) isomorphe à $B_\mathfrak{p}/\mathfrak{p}B_\mathfrak{p}$. Montrons que $B_\mathfrak{p} \to (B[d]/(U_1, \ldots, U_r))_{\mathfrak{U}_d(\mathfrak{p})}$ est plat. Comme une flèche de localisation est plate, il suffit par transitivité de vérifier que $B_\mathfrak{p}[d]/(U_1, \ldots, U_r)$ est plat sur $B_\mathfrak{p}$. Il est en fait libre car si $S = B_\mathfrak{p}[(u_\mathfrak{m}^{(l)})_{\mathfrak{m} \neq \mathfrak{m}_l}]$ où pour chaque l le monôme $\mathfrak{m}_l$ est un élément de $\mathcal{M}_{d_l}$ inversible dans $B_\mathfrak{p}$ (il en existe car $\mathcal{M}_\varepsilon \not\subset \mathfrak{p}$) alors le morphisme de S-algèbres $B_\mathfrak{p}[d] \to S$ qui associe à chaque $u_{\mathfrak{m}_l}^{(l)}$ l'élément $-\frac{1}{\mathfrak{m}_l} \sum_{\mathfrak{m} \in \mathcal{M}_{d_l} \setminus \mathfrak{m}_l} u_\mathfrak{m}^{(l)}\mathfrak{m}$ a clairement pour noyau $(U_1, \ldots, U_r)$. On tensorise alors notre suite de composition de $B_\mathfrak{p}/I_\mathfrak{p}$ par $(B[d]/(U_1, \ldots, U_r))_{\mathfrak{U}_d(\mathfrak{p})}$. Par platitude on obtient une suite de longueur s aboutissant à $(B[d]/I[d])_{\mathfrak{U}_d(\mathfrak{p})}$ de quotients isomorphes à $(B[d]/\mathfrak{p}[d])_{\mathfrak{U}_d(\mathfrak{p})}$. Or $(B[d]/\mathfrak{p}[d])_{\mathfrak{U}_d(\mathfrak{p})} \simeq B[d]_{\mathfrak{U}_d(\mathfrak{p})}/\mathfrak{U}_d(\mathfrak{p})B[d]_{\mathfrak{U}_d(\mathfrak{p})}$ puisque $\mathfrak{p}[d]$ engendre $\mathfrak{U}_d(\mathfrak{p})$ dans le localisé car $\mathcal{M}_\varepsilon$ contient un élément inversible. La suite est donc une suite de composition et l'on a bien $\ell(B_\mathfrak{p}/I_\mathfrak{p}) = \ell((B[d]/I[d])_{\mathfrak{U}_d(\mathfrak{p})})$. $\square$

3.3. Degrés des formes résultantes. Notons que le théorème précédent montre que la formule $\sqrt{\text{élim}_d(I)} = \sqrt{\text{rés}_d(I)}$ est valable pour tout I (de hauteur au moins $n - r + 1$). On a maintenant un résultat sur le degré des formes résultantes.

PROPOSITION 3.4. *Si* $\text{ht}(I) \geq n - r + 1$ *alors* $\text{rés}_d(I)$ *est une forme multihomogène de* $K[d] = K[u^{(1)}, \ldots, u^{(r)}]$. *Son degré en le groupe de variables* $u^{(1)}$ *est* $\deg(I) * d_2 * \cdots * d_r$.

DÉMONSTRATION. On procède par récurrence sur $r \geq 1$. Notons que grâce à la formule de décomposition

$$\deg(I) = \sum_{\substack{\mathfrak{p} \supset I \\ \text{ht}(\mathfrak{p}) = \text{ht}(I)}} \ell(B_\mathfrak{p}/I_\mathfrak{p}) \deg(\mathfrak{p})$$

on peut, pour chaque r et chaque corps K, ne démontrer l'énoncé que pour $I = \mathfrak{p}$ premier. Traitons le cas $r = 1$. Il faut montrer que $\text{rés}_d(\mathfrak{p})$ est homogène de degré $H_{B/\mathfrak{p}}(0)$ lorsque $\text{ht}(\mathfrak{p}) \geq n$. Si $\text{ht}(\mathfrak{p}) > n$ on a aussi bien $H_{B/\mathfrak{p}} = 0$ que $\text{rés}_d(\mathfrak{p}) = 1$ ce qui donne le résultat. Supposons donc que $\text{ht}(\mathfrak{p}) = n$. En premier lieu, si K est algébriquement clos, on a (voir démonstration de la proposition 2.16) $B/\mathfrak{p} \simeq K[X_{j_1}^{(1)}, \ldots, X_{j_q}^{(q)}]$ pour certains indices $j_1, \ldots, j_q$. Ainsi $H_{B/\mathfrak{p}} = 1$. Par ailleurs la même démonstration montre que $\pi = \text{élim}_d(\mathfrak{p})$ est de degré 1. Reste à voir que $\text{rés}_d(\mathfrak{p})$ est également de degré 1 c'est-à-dire $\text{rés}_d(\mathfrak{p}) = \text{élim}_d(\mathfrak{p})$ ou encore que

$$(B[d]/\mathfrak{p}[d])_k \underset{K[d]}{\otimes} K[d]_{(\pi)}$$

est de longueur 1 pour k assez grand. Or la forme explicite de $\mathfrak{p}$ montre que dans $B[d]/\mathfrak{p} \simeq K[d][X_{j_1}^{(1)}, \ldots, X_{j_q}^{(q)}]$ on a $U_1 = \pi X_{j_1}^{(1)}{}^{d_{1,1}} \ldots X_{j_q}^{(q)}{}^{d_{1,q}}$ à unité près. Par conséquent si $k \geq d_1$ on trouve

$$(B[d]/\mathfrak{p}[d])_k \simeq K[d]/\pi$$

ce qui prouve l'assertion voulue sur les longueurs et démontre le cas $r = 1$ si K est algébriquement clos.

Si K est un corps quelconque (avec $r = 1$ et $\text{ht}(\mathfrak{p}) = n$) on considère une clôture algébrique $\bar{K}$ de K et $I = \mathfrak{p}.\bar{K}[X]$. Alors $\text{ht}(I) = n$ ("going down"). Par extension des scalaires $\bar{K}[X]/I \simeq B/\mathfrak{p} \underset{K}{\otimes} \bar{K}$ donc $H_{\bar{K}[X]/I} = H_{B/\mathfrak{p}}$ d'où $\deg(I) = \deg(\mathfrak{p})$. Fixons ensuite k assez grand tel que $\text{rés}_d(\mathfrak{p}) = \chi((B[d]/\mathfrak{p}[d])_k)$ et $\text{rés}_d(I) = \chi((\bar{K}[X][d]/I[d])_k)$. On a donc une égalité

$$\text{rés}_d(\mathfrak{p}) = \pi^{\ell\left((B[d]/\mathfrak{p}[d])_k \underset{K[d]}{\otimes} K[d]_{(\pi)}\right)}$$

où $\pi = \text{élim}_d(\mathfrak{p})$. Écrivons $\pi = \lambda \prod_{\omega \in \text{Irr}(\bar{K}[d])} \omega^{\mu_\omega}$ avec $\lambda \in \bar{K}^\times$. Si l'on a

$$(B[d]/\mathfrak{p}[d])_k \underset{K[d]}{\otimes} K[d]_{(\pi)} \simeq \bigoplus_{i=1}^{t} K[d]_{(\pi)}/\pi^{m_i}$$

d'une part on voit que $\text{rés}_d(\mathfrak{p}) = \pi^{\sum_{i=1}^{t} m_i}$ et d'autre part pour chaque $\omega | \pi$, comme on dispose de $K[d]_{(\pi)} \to \bar{K}[d]_{(\omega)}$, on trouve

$$(B[d]/\mathfrak{p}[d])_k \underset{K[d]}{\otimes} \bar{K}[d]_{(\omega)} \simeq \bigoplus_{i=1}^{t} \bar{K}[d]_{(\omega)}/\omega^{m_i \mu_\omega}.$$

Par ailleurs $(B[d]/\mathfrak{p}[d])_k \underset{K[d]}{\otimes} \bar{K}[d]_{(\omega)} \simeq (\bar{K}[X][d]/I[d])_k \underset{\bar{K}[d]}{\otimes} \bar{K}[d]_{(\omega)}$. Ainsi, en calculant les longueurs, on constate immédiatement que l'exposant de ω dans $\text{rés}_d(I)$ est égal à $\mu_\omega \sum_{i=1}^{t} m_i$. En outre si ω ne divise pas π il est clair que π, inversible dans $\bar{K}[d]_{(\omega)}$, annule $(B[d]/\mathfrak{p}[d])_k$ et donc $(\bar{K}[X][d]/I[d])_k$. Aussi a-t-on finalement :

$$\text{rés}_d(I) = \lambda^{-\sum_{i=1}^{t} m_i} \text{rés}_d(\mathfrak{p})$$

ce qui montre bien que $\text{rés}_d(\mathfrak{p})$ est homogène de degré $\deg(\mathfrak{p})$ et conclut dans le cas $r = 1$.

On suppose donc à présent que $r \geq 2$ et que l'on connaît l'énoncé pour $r - 1$ (et tous les idéaux, tous les corps). On se restreint encore au cas $I = \mathfrak{p}$ premier.

Si $\mathfrak{E}_{(d_r)}(\mathfrak{p}) \neq 0$ alors le degré en $u^{(1)}$ de rés$_d(\mathfrak{p})$ est nul : en effet $r \neq 1$ et soit rés$_d(\mathfrak{p}) = 1$ soit rés$_d(\mathfrak{p}) \in K[(d_r)]$. Dans ce cas on vérifie qu'il en est de même de deg$(\mathfrak{p}) * d_2 * \cdots * d_r$. Si à présent $\mathfrak{E}_{(d_r)}(\mathfrak{p}) = 0$ on peut appliquer le lemme 2.12 qui définit C, L, $d' = (d_1, \ldots, d_{r-1})$ et $\mathfrak{q}$. Si ht$(\mathfrak{q}) < n - (r-1) + 1$ la formule deg$(\mathfrak{q}) = $ deg$(\mathfrak{p}) * d_r$ impose deg$(\mathfrak{q}) = 0$ donc $\mathcal{M}_\varepsilon \subset \mathfrak{q}$. Ceci est absurde à cause de $\mathfrak{E}_{(d_r)}(\mathfrak{p}) = 0$. Donc ht$(\mathfrak{q}) \geq n - (r-1) + 1$ et l'hypothèse de récurrence montre que rés$_{d'}(\mathfrak{q})$ est homogène en $u^{(1)}$ de degré deg$(\mathfrak{q}) * d_2 * \cdots * d_{r-1} = $ deg$(\mathfrak{p}) * d_2 * \cdots * d_r$. De plus par le même lemme 2.12 on a $\mathfrak{U}_{d'}(\mathfrak{q}) = \mathfrak{U}_d(\mathfrak{p}).L[d'][X]$ ce qui, à travers le diagramme suivant dans lequel les flèches verticales sont des isomorphismes pour k assez grand

$$
\begin{array}{ccc}
(B[d]/\mathfrak{p}[d])_k \underset{K[d]}{\otimes} L[d'] & & (C[d']/\mathfrak{q}[d'])_k \\
\downarrow & & \downarrow \\
(B[d]/\mathfrak{U}_d(\mathfrak{p}))_k \underset{K[d]}{\otimes} L[d'] & \overset{\sim}{\longrightarrow} & (C[d']/\mathfrak{U}_{d'}(\mathfrak{q}))_k,
\end{array}
$$

implique pour k assez grand

$$
(B[d]/\mathfrak{p}[d])_k \underset{K[d]}{\otimes} L[d'] \simeq (C[d']/\mathfrak{q}[d'])_k
$$

et donc rés$_{d'}(\mathfrak{q}) = $ rés$_d(\mathfrak{p})$. Ceci prouve bien que rés$_d(\mathfrak{p})$ est homogène du degré voulu en $u^{(1)}$. $\qquad\square$

La proposition précédente, qui généralise les résultats sur les degrés que l'on connaît dans le cas homogène (voir par exemple [**PPh2**, p. 15]), sera utile pour comparer entr'elles différentes formes résultantes.

3.4. Spécialisation. Le premier résultat que l'on va déduire de l'étude précédente des degrés consiste à spécialiser une forme en produit de q formes (séparation des variables). Pour cela on considère $d \in (\mathbf{N}^q)^r$ puis $d^{(i)} = (\varepsilon_i, d_2, \ldots, d_r)$ pour $1 \leq i \leq q$. Introduisons $\tilde{d} = (\varepsilon_1, d_2, \ldots, d_r, \varepsilon_2, \ldots, \varepsilon_q)$ et désignons dans $K[\tilde{d}]$ les indéterminées $u^{(r+i-1)}_{X^{(i)}_j}$ par $u^{(1)}_{X^{(i)}_j}$ ($2 \leq i \leq q$) de façon à voir chaque $K[d^{(i)}]$ comme sous-anneau de $K[\tilde{d}]$. Pour éviter les confusions on réserve la notation U_1 à l'élément correspondant de $B[d]$ et l'on note L_i le U_1 de $B[d^{(i)}]$; ainsi dans $B[\tilde{d}]$ les symboles $L_1, L_2, \ldots, L_q$ remplacent $U_1, U_{r+1}, \ldots, U_{r+q-1}$. Définissons ensuite un morphisme de K-algèbres

$$
\begin{array}{rcl}
K[d] & \overset{\omega}{\longrightarrow} & K[\tilde{d}] \\
u^{(l)}_{\mathfrak{m}} & \longmapsto & u^{(l)}_{\mathfrak{m}} \qquad\qquad\qquad\qquad \text{si } l > 1 \\
u^{(1)}_{\mathfrak{m}} & \longmapsto & \displaystyle\binom{d_1}{\mathfrak{m}} \prod_{i=1}^{q}\prod_{j=0}^{n_i} (u^{(1)}_{X^{(i)}_j})^{\alpha_{i,j}} \qquad \text{si } \mathfrak{m} = \prod_{i=1}^{q}\prod_{j=0}^{n_i}(X^{(i)}_j)^{\alpha_{i,j}}.
\end{array}
$$

La propriété de spécialisation de ω est que $\omega(U_1) = \prod_{i=1}^{q} L_i^{d_{1,i}}$. Son intérêt est expliqué par la

PROPOSITION 3.5. *Si I est un idéal multihomogène de B de hauteur au moins $n - r + 1$ on a*

$$
\omega(\text{rés}_d(I)) = \lambda \prod_{i=1}^{q} \text{rés}_{d^{(i)}}(I)^{d_{1,i}}
$$

avec $\lambda \in K^\times$.

DÉMONSTRATION. On peut immédiatement se ramener au cas où $I = \mathfrak{p}$ est premier. Notons $f = \text{rés}_d(\mathfrak{p})$ et $f_i = \text{rés}_{d^{(i)}}(\mathfrak{p})$. En vertu de la proposition 3.4 les formes $f, f_1, \ldots, f_q$ ont toutes le même degré en $u^{(1)}$; notons e ce degré commun. Si $e = 0$, on a, en notant $\hat{d} = (d_2, \ldots, d_r)$, $f \in K[\hat{d}]$ et $f_i \in K[\hat{d}]$ pour tout i. Ainsi d'une part $\omega(f) = f$ et d'autre part $\mathfrak{E}_d(\mathfrak{p}) = \mathfrak{E}_{\hat{d}}(\mathfrak{p}).K[d] = \mathfrak{E}_{d^{(i)}}(\mathfrak{p})$ pour tout i. Les formes f et $f_1, \ldots, f_q$ sont donc toutes puissances de $\pi = \text{élim}_{\hat{d}}(\mathfrak{p}) = \text{élim}_d(\mathfrak{p}) = \text{élim}_{d^{(i)}}(\mathfrak{p})$. Si $\pi = 0$ ou $\pi = 1$ le résultat est clair. Sinon il existe j tel que le degré de π en $u^{(j)}$ est non nul. Par symétrie supposons $j = r$. Il nous suffit de montrer que f et $f_1^{d_{1,1}} \ldots f_q^{d_{q,1}}$ ont même degré en $u^{(r)}$. Or, par la proposition 3.4, ces degrés sont respectivement $\deg(\mathfrak{p}) * d_2 * \cdots * d_{r-1} * d_1$ et

$$\sum_{i=1}^{q} d_{1,i} \deg(\mathfrak{p}) * d_2 * \cdots * d_{r-1} * \varepsilon_i.$$

Ceci donne le résultat par linéarité de l'opérateur $*$ et $\sum_{i=1}^{q} d_{1,i}\varepsilon_i = d_1$.

Supposons maintenant $e \neq 0$. Comme aucune des formes f ou $f_1, \ldots, f_q$ ne peut valoir 1, on sait que $\sqrt{f}$ engendre $\mathfrak{E}_d(\mathfrak{p})$ et, de même, que $\sqrt{f_i}$ engendre $\mathfrak{E}_{d^{(i)}}(\mathfrak{p})$ pour tout i. On considère un morphisme de K-algèbres $\rho \colon K[\hat{d}] \to \bar{K}$. On a alors

$$\rho(\omega(f)) = 0 \iff \rho \circ \omega(\mathfrak{E}_d(\mathfrak{p})) = 0$$

$$\iff (\mathfrak{p}, \prod_{i=1}^{q} \rho(L_i)^{d_{1,i}}, \rho(U_2), \ldots, \rho(U_r)) \text{ a un zéro non trivial}$$
$$\text{dans } \prod_{i=1}^{q} \bar{K}^{n_i+1}$$

$$\iff \text{il existe } i \text{ tel que } (\mathfrak{p}, \rho(L_i)^{d_{1,i}}, \rho(U_2), \ldots, \rho(U_r)) \text{ a un zé-}$$
$$\text{ro non trivial dans } \prod_{i=1}^{q} \bar{K}^{n_i+1}$$

$$\iff \text{il existe } i \text{ tel que } d_{i,1} \neq 0 \text{ et } (\mathfrak{p}, \rho(L_i), \rho(U_2), \ldots, \rho(U_r))$$
$$= \rho(\mathfrak{p}[d^{(i)}]) \text{ a un zéro non trivial dans } \prod_{i=1}^{q} \bar{K}^{n_i+1}$$

$$\iff \text{il existe } i \in \text{Supp}(d_1) \text{ tel que } \rho(\mathfrak{E}_{d^{(i)}}(\mathfrak{p})) = 0$$
$$\iff \exists i \in \text{Supp}(d_1) \ \rho(f_i) = 0$$
$$\iff \exists i \ \rho(f_i^{d_{i,1}}) = 0$$
$$\iff \rho(\prod_{i=1}^{q} f_i^{d_{1,i}}) = 0 \ .$$

Ceci donne

$$\sqrt{\omega(f)} = \sqrt{\prod_{i=1}^{q} f_i^{d_{1,i}}}.$$

On obtient le résultat en comparant les degrés en $u_{X^{(i)}}^{(1)}$: celui de $\omega(f)$ est $d_{1,i}e$ et celui de f_j est $\delta_{i,j}e$. □

C'est surtout à travers le résultat suivant que les formes résultantes sont plus adaptées que les formes éliminantes.

PROPOSITION 3.6. *Soient I un idéal multihomogène de B, r un entier non nul tel que $\text{ht}(I) \geq n - r + 1$ et $d \in (\mathbf{N}^q \setminus 0)^r$. On suppose que $P \in B_{d_1}$ n'est pas diviseur de zéro dans B/I et on note $J = (I, P)$, $d' = (d_2, \ldots, d_r)$ et $\rho \colon K[d] \to K[d']$ le* *morphisme de $K[d']$-algèbres caractérisé par $\rho(U_1) = P$. Alors* $\mathrm{ht}(J) \geq n-(r-1)+1$ *et il existe* $\lambda \in K^\times$ *tel que*

$$\rho(\mathrm{rés}_d(I)) = \lambda.\mathrm{rés}_{d'}(J).$$

DÉMONSTRATION. Comme P n'est pas diviseur de zéro dans B/I, il n'appartient à aucun des premiers minimaux contenant I et par conséquent un premier contenant I et P contient strictement l'un de ces premiers. Ainsi $\mathrm{ht}(J) > n - r + 1$ ce qui nous autorise à écrire une forme éliminante de J d'indice d'.

Il suit immédiatement des définitions que l'on a un isomorphisme

$$B[d']/J[d'] \simeq B[d]/I[d] \underset{K[d]}{\otimes} K[d']$$

où $K[d']$ est un $K[d]$-module *via* ρ. Fixons maintenant $k \in \mathbf{N}^q$ assez grand tel que $\mathrm{rés}_{d'}(J) = \chi((B[d']/J[d'])_k)$ et $\mathrm{rés}_d(I) = \chi((B[d]/I[d])_k)$, notons $M = (B[d]/I[d])_k$ et choisissons une présentation finie du $K[d]$-module M

$$K[d]^t \xrightarrow{\varphi} K[d]^s \longrightarrow M \longrightarrow 0$$

avec $t \geq s$. On considère encore $\mathfrak{I}$ l'idéal de $K[d]$ engendré par les déterminants $s \times s$ extraits de la matrice de φ. D'après le lemme 3.1 pour chaque $\pi \in \mathrm{Irr}(K[d])$ on a $\mathfrak{I}_{(\pi)} = \pi^{\ell(M_{(\pi)})} K[d]_{(\pi)}$: il suffit de localiser la présentation finie de M en (π). Ainsi il apparaît que $\mathrm{rés}_d(I)$ divise tous les éléments de $\mathfrak{I}$. Par conséquent $\rho(\mathrm{rés}_d(I))$ divise tous les éléments de $\rho(\mathfrak{I})$. Or ce dernier idéal (de $K[d']$) est engendré par les déterminants $s \times s$ extraits de la matrice de $\varphi \otimes K[d']$ dans

$$K[d']^t \xrightarrow{\varphi \otimes K[d']} K[d']^s \longrightarrow M \underset{K[d]}{\otimes} K[d'] \longrightarrow 0.$$

De manière analogue à l'argument précédent on a donc

$$\rho(\mathfrak{I})_{(\omega)} = \omega^{\ell(M \otimes K[d']_{(\omega)})} K[d']_{(\omega)}$$

pour $\omega \in \mathrm{Irr}(K[d'])$. Ceci prouve que l'exposant d'un tel ω dans $\rho(\mathrm{rés}_d(I))$ est au plus $\ell(M \otimes K[d']_{(\omega)})$ et ainsi $\rho(\mathrm{rés}_d(I))$ divise $\mathrm{rés}_{d'}(J) = \chi(M \otimes K[d'])$.

Si $\mathrm{rés}_{d'}(J) = 1$ on a $\rho(\mathrm{rés}_d(I)) \in K^\times$ qui est le résultat. Pour conclure dans le cas où $\mathrm{rés}_{d'}(J) \neq 1$ calculons les degrés de $\rho(\mathrm{rés}_d(I))$ et de $\mathrm{rés}_{d'}(J)$. D'après la proposition 3.4 ces formes sont toutes deux multihomogènes en $u^{(2)}, \ldots, u^{(r)}$ de degrés en $u^{(l)}$ égaux à $\deg(I) * d_1 * d_2 * \cdots * d_{l-1} * d_{l+1} * \cdots * d_r$ et $\deg(J) * d_2 * \cdots * d_{l-1} * d_{l+1} * \cdots * d_r$ respectivement. Enfin l'hypothèse que P ne divise pas 0 dans B/I se traduit comme dans le lemme 2.11 par $H_{B/J} = \Delta_{d_1} H_{B/I}$. En écrivant les termes dominants on trouve que $\deg(J) = \deg(I) * d_1$ à condition que $\deg H_{B/J} = \deg H_{B/I} - 1$ (degrés totaux). Or $\mathrm{rés}_{d'}(J) \neq 1$ entraîne $e(B/J) = r - 1$ et la hauteur de J force $e(B/I) \leq r$. Ceci donne bien la condition sur les degrés totaux des polynômes de Hilbert-Samuel puis $\deg(J) = \deg(I) * d_1$. On en déduit que $\rho(\mathrm{rés}_d(I))$ et $\mathrm{rés}_{d'}(J)$, liées par une relation de divisibilité et de mêmes degrés partiels, sont égales à inversible près ce qui termine la démonstration. $\qquad\square$

3.5. Cas de l'idéal nul.
On donne ici un calcul explicite de forme résultante qui correspond à ce que l'on appelle classiquement le *résultant* (voir aussi [**GKZ**][4],

---

[4][**GKZ**] I.M. Gelfand, M.M. Kapranov, A.V. Zelevinsky. *Discriminants, resultants and multidimensional determinants*, Birkhäuser, Boston, (1994).

chap. 13). Précisément nous évaluons $\mathrm{rés}_d(0)$ lorsque

$$d = (\underbrace{\varepsilon_1, \dots, \varepsilon_1}_{n_1 \text{ fois}}, \dots, \underbrace{\varepsilon_q, \dots, \varepsilon_q}_{n_q \text{ fois}}, \delta) \in (\mathbf{N}^q)^{n+1}$$

avec $\delta \in \mathbf{N}^q \setminus 0$. On travaille donc dans $K[d]$ et l'on considère les matrices

$$M_i = \begin{pmatrix} u^{(n_i'+1)}_{X_0^{(i)}} & \cdots & u^{(n_i'+1)}_{X_{n_i}^{(i)}} \\ \vdots & \ddots & \vdots \\ u^{(n_i'+n_i)}_{X_0^{(i)}} & \cdots & u^{(n_i'+n_i)}_{X_{n_i}^{(i)}} \end{pmatrix}$$

de tailles respectives $n_i \times (n_i + 1)$ où $n_i' = n_1 + \cdots + n_{i-1}$. On introduit ensuite un élément

$$\Delta = (\Delta_{1,0}, \dots, \Delta_{1,n_1}, \dots, \Delta_{q,0}, \dots, \Delta_{q,n_q}) \in K[d]^{n+q}$$

où $(-1)^j \Delta_{i,j}$ pour $1 \leq i \leq q$ et $0 \leq j \leq n_i$ est le déterminant de la matrice $n_i \times n_i$ obtenue en retirant la $(j+1)^{\text{ème}}$ colonne de M_i. L'intérêt de cette définition est que pour $n_i' + 1 \leq l \leq n_i' + n_i$ on a

$$\sum_{j=0}^{n_i} u^{(l)}_{X_j^{(i)}} \Delta_{i,j} = 0$$

ce que l'on peut résumer par $U_l(\Delta) = 0$ pour tout $1 \leq l \leq n$. On peut alors énoncer

LEMME 3.7. *La forme $U_{n+1}(\Delta) \in K[d]$ est une forme à la fois éliminante et résultante d'indice d de l'idéal 0 de $K[X]$.*

DÉMONSTRATION. Considérons un morphisme de spécialisation $\rho \colon K[d] \to K$. Si f est une forme éliminante de l'idéal nul d'indice d, le théorème de l'élimination montre que $\rho(f) = 0$ si et seulement si les formes $\rho(U_l)$ pour $1 \leq l \leq n+1$ ont un zéro commun non trivial (c'est-à-dire dans $(K^{n_1+1} \setminus 0) \times \cdots \times (K^{n_q+1} \setminus 0)$). Notons J l'ensemble des $i \in \{1, \dots, q\}$ tels que $\rho(\Delta_i) = (\rho(\Delta_{i,0}), \dots, \rho(\Delta_{i,n_i})) \neq 0$. Si $i \in J$ la matrice $\rho(M_i)$ est de rang maximal et le seul zéro commun non trivial des formes $\rho(U_{n_i'+1}), \dots, \rho(U_{n_i'+n_i})$ est $\rho(\Delta_i)$. A contrario, si $i \notin J$, le rang de $\rho(M_i)$ est $< n_i$ donc il existe k_i avec $1 \leq k_i \leq n_i$ tel que $\rho(U_{n_i'+k_i})$ est combinaison linéaire des $\rho(U_{n_i'+k})$ pour $k \neq k_i$. Dans ce cas nous noterons $\Delta_i' \in K^{n_i+1} \setminus 0$ un zéro commun non trivial de $\rho(U_{n_i'+1}), \dots, \rho(U_{n_i'+n_i})$.

Si $\rho(f) = 0$ il existe un zéro commun non trivial des $\rho(U_l)$ pour $1 \leq l \leq n+1$. Sa composante suivant un indice $i \in J$ est, par ce qui précède, $\rho(\Delta_i)$. Pour $i \notin J$ notons Δ_i'' sa composante selon cet indice. De

$$\rho(U_{n+1})\left((\rho(\Delta_i))_{i \in J}, (\Delta_i'')_{i \notin J}\right) = 0$$

on tire *a fortiori* que

$$\rho(U_{n+1})(\rho(\Delta)) = \rho(U_{n+1})\left((\rho(\Delta_i))_{i \in J}, (0)_{i \notin J}\right) = 0.$$

Si $\rho(f) \neq 0$ on a

$$\rho(U_{n+1})\left((\rho(\Delta_i))_{i \in J}, (\Delta_i')_{i \notin J}\right) \neq 0$$

car $\rho(U_1), \dots, \rho(U_{n+1})$ n'ont pas de zéro commun non trivial. Par relations de dépendance linéaire, il en est de même si l'on exclut les formes $\rho(U_{n_i'+k_i})$ pour $i \notin J$. Par le théorème de l'élimination avec l'indice $d' \in (\mathbf{N}^q)^{n+1+\mathrm{Card}(J)-q}$ correspondant, on obtient $\mathfrak{E}_{d'}(0) \neq 0$. Ainsi soit $f = 1$ soit $f \in K[d']$. En tout état de cause le degré de f en $u^{(n'_i+k_i)}$ pour $i \notin J$ est nul. Or ce degré, calculé à l'aide de la proposition 3.4, vaut δ_i. Par conséquent $\delta_i = 0$ pour $i \notin J$ donc le degré de $\rho(U_{n+1})$ en la composante $i \notin J$ est nul d'où

$$\rho(U_{n+1})(\rho(\Delta)) = \rho(U_{n+1})\left((\rho(\Delta_i))_{i \in J}, (\Delta'_i)_{i \notin J}\right) \neq 0.$$

Finalement on a

$$\rho(f) = 0 \iff \rho(U_{n+1})(\rho(\Delta)) = 0$$

ce qui prouve que f et $U_{n+1}(\Delta)$ ont les mêmes zéros. Alors, comme f est irréductible, $U_{n+1}(\Delta)$ est (à une constante près) une puissance de f. C'est également le cas de $\text{rés}_d(0)$ et ces deux formes ont le même degré en $u^{(n+1)}$ (à savoir 1) d'après la proposition 3.4. Ceci montre que f, $\text{rés}_d(0)$ et $U_{n+1}(\Delta)$ sont proportionnelles et prouve le lemme. $\qquad\square$

CHAPTER 6

Diophantine geometry

Let us start this chapter with a short dictionary. We denote by K a commutative field and take $m, n, d \in \mathbf{N}^*$, $h \in \{0, \ldots, n+1\}$.

Dictionary	
Geometry	Algebra
$\mathbf{P}_n(K)$	$A = K[X_0, \ldots, X_n]$
closed subscheme S	homogeneous ideal I
closed algebraic subset E	radical ideal $\sqrt{I} = I$
projective variety V	prime ideal $\mathfrak{p}$
dimension of V is $h - 1$	rank of $\mathfrak{p}$ is $n - h + 1$
hypersurface H	principal ideal (F)
subscheme S of $\mathbf{P}_m(K) \times \mathbf{P}_n(K)$	bi-homogeneous ideal I of $K[X_0, \ldots, X_n; Y_0, \ldots, Y_m]$
projection of S on $\mathbf{P}_n(K)$	ideal $I \cap A$ or $\mathrm{Fitt}_0^A(A[Y_0, \ldots, Y_m]/I)$
"dual" projective space $\mathbf{P}_N^{\vee}(K)$ where $N = \binom{n+d}{d}$	K-vector space of forms of degree d

We will call *cycle* a formal linear combination Z of varieties, with coefficients in $\mathbf{N}$: $Z = \sum_{j=1}^s m_j.V_j$. We will say that Z is *equidimensional* iff all the varieties involved have the same dimension. To a scheme S we may associate an equidimensional cycle $Z(S)$ as a linear combination of its components of higher dimension, endowed with multiplicities. We stress that when the base field K is not algebraically closed, a variety over K may no longer be a variety, but only an algebraic subset over $\overline{K}$.

1. Elimination theory

Let us now recall some basic facts from projective elimination theory, these are established in chapter 5, as a special case of the multiprojective version of elimination theory (which allows to consider varieties in products of projective spaces).

Let's take positive integers $d_1, \ldots, d_h$, a drawing shows the principle of elimination in projective spaces and how it is used to construct Chow forms of varieties.

(∗) Chapter's author : Patrice PHILIPPON.

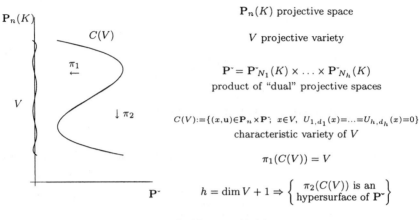

$\mathbf{P}_n(K)$ projective space

V projective variety

$\mathbf{P}^{\check{}} = \mathbf{P}^{\check{}}_{N_1}(K) \times \ldots \times \mathbf{P}^{\check{}}_{N_h}(K)$
product of "dual" projective spaces

$C(V) := \{(x,u) \in \mathbf{P}_n \times \mathbf{P}^{\check{}}; \; x \in V, \; U_{1,d_1}(x) = \ldots = U_{h,d_h}(x) = 0\}$
characteristic variety of V

$$\pi_1(C(V)) = V$$

$$h = \dim V + 1 \Rightarrow \left\{ \begin{array}{l} \pi_2(C(V)) \text{ is an} \\ \text{hypersurface of } \mathbf{P}^{\check{}} \end{array} \right\}$$

NOTATIONS. We set $\mathbf{u} = (\underline{u}^{(1,d_1)}, \ldots, \underline{u}^{(h,d_h)})$ with

$$\underline{u}^{(i,d_i)} = \{u_\alpha^{(i,d_i)}; \alpha \in \mathbf{N}^{n+1}, |\alpha| = d_i\}$$

and

$$U_{i,d_i} = \sum_{|\alpha|=d_i} u_\alpha^{(i,d_i)} \binom{d_i}{\alpha}^{1/2} X_0^{\alpha_0} \ldots X_n^{\alpha_n}$$

where $\binom{d_i}{\alpha} = \binom{d_i!}{\alpha_0!\ldots\alpha_n!}$, $N_i = \binom{d_i+n}{n} - 1$.

If $I(S)$ is the ideal of S, we have $I(C(S)) = I(S).A[\mathbf{u}] + U_{1,d_1}.A[\mathbf{u}] + \cdots + U_{h,d_h}.A[\mathbf{u}]$ and $(f_{S,\mathbf{d}}) = \overline{I(C(S))} \cap K[\mathbf{u}]$ where

$$\overline{I(C(S))} = \{x \in A[\mathbf{u}]; \exists k \in \mathbf{N}, \forall i \quad x.X_i^k \in I(C(S))\}$$

is the *saturation of* $I(C(S))$.

More generally, we can state

BASIC FACTS. *Elimination theory associates to a cycle $Z = \sum_j m_j.V_j$, equidimensional of dimension $h - 1$, in $\mathbf{P}_n(K)$ and integers $d_1, \ldots, d_h$, an hypersurface in $\mathbf{P}^{\check{}}(K)$, that is a form $f_{Z,\mathbf{d}} \in K[\mathbf{u}]$ (unique up to multiplication by a non zero scalar). Furthermore $f_{Z,\mathbf{d}} = \prod_j f_{V_j,\mathbf{d}}^{m_j}$ and $f_{V,\mathbf{d}}$ is irreducible iff V is a variety. We call such a form an* eliminating form of index $\mathbf{d}$ of Z or a Chow form of Z if $\mathbf{d} = \mathbf{1}$.

PROOF. See §§ 2 and 3 from chapter 5.   □

ELIMINATION THEOREM (ET). *In the above notations, let L be an extension of K and $\rho : L[\mathbf{u}] \to L$ an homomorphism of L-algebras, then the two following assertions are equivalent :*

1. *$\rho(f_{S,\mathbf{d}}) = (0)$ (i.e. $\rho(\mathbf{u})$ is a zero of $f_{S,\mathbf{d}}$);*
2. *there exists an extension L' of L and $x = (x_0, \ldots, x_n) \in L'^{n+1} \setminus \{0\}$ such that for all $F \in I(C(S))$ we have $\rho(F)(x) = 0$ (i.e. x is a zero of $\rho(I(C(S)))$).*

PROOF. See Théorème 2.2 from chapter 5.   □

EXAMPLE. For $V = \{x\}$ we have $h = \dim V + 1 = 1$ and any Chow form of V is proportional to $U(x)$.

For $V = \mathcal{Z}(F)$ an hypersurface, any Chow form of V is proportional to the multihomogeneous form $F(\Delta_0, \ldots, \Delta_n)$ where $\Delta_i = \det \left(u_\alpha^{(j,1)} \right)_{j=1,\ldots,n;0 \leq \alpha \leq n, \alpha \neq i}$.

We will now discuss various quantities related to projective varieties through their Chow forms, this discussion extend easily to arbitrary cycles. See the next chapter for the extension to varieties in products of projective spaces.

2. Degree

A fondamental invariant asociated to a projective variety V is its degree. Suppose V is of dimension $h - 1$, for almost all set of hypersurfaces $H_1, \ldots, H_{h-1}$ of given degrees the set of points lying on the intersection $V \cap H_1 \cap \cdots \cap H_{h-1}$ is finite and when so it has a fixed cardinality. We call *degree* and denote $d(V)$ the cardinality obtained in intersecting with hyperplanes.

Taking $L = K(\underline{u}^{(1,d_1)}, \ldots, \underline{u}^{(h-1,d_{h-1})})$ in (ET) we find that $f_{V,\mathbf{d}}$ vanishes on the same set as $\prod_{x \in V \cap H_1 \cap \cdots \cap H_{h-1}} U_{h,d_h}(x)$ with $H_i = \mathcal{Z}(U_{i,d_i})$ (notice that $V \cap H_1 \cap \cdots \cap H_{h-1}$ is finite, whatever $\mathbf{d}$, since it is finite for almost all specialisations of these hypersurfaces). But $f_{V,\mathbf{d}}$ is irreducible and must be proportional to the latter form which is therefore a Chow form of V, the ratio belonging to L^*. In particular, $d(V) = d^{\circ}_{\underline{u}^{(i,d_i)}} f_{V,\mathbf{1}}$.

Now, specializing U_{h,d_h} to $(U_{h,1})^{d_h}$ we see that the specialisation of $f_{V,\mathbf{d}}$ is proportional to $(f_{V,\mathbf{d}'})^{d_h}$ where $\mathbf{d}' = (d_1, \ldots, d_{h-1}, 1)$, the ratio belonging *a priori* to L^*. But, thanks to (ET), an element from

$$K[\underline{u}^{(1,d_1)}, \ldots, \underline{u}^{(h-1,d_{h-1})}]$$

cannot divide $f_{V,\mathbf{d}}$ nor $(f_{V,\mathbf{d}'})^{d_h}$ and we conclude that the ratio actually lays in K^*. Iterating we eventually get that $\rho(f_{V,\mathbf{d}})$ is proportional to $(f_{V,\mathbf{1}})^{d_1 \cdots d_h}$ where ρ is the specialisation of all U_{i,d_i} to $(U_{i,1})^{d_i}$. But, ρ preserves homogeneity and is of degree d_i in $\underline{u}^{(i,d_i)}$, since $f_{V,\mathbf{1}}$ is obviously symmetric (of degree $d(V)$) in each set of variables $\underline{u}^{(i,1)}$ we have proved

$$d^{\circ}_{\underline{u}^{(i,d_i)}} f_{V,\mathbf{d}} = d(V) . \frac{d_1 \ldots d_h}{d_i} \ .$$

REMARK. If $Z = \sum_j m_j . V_j$ is equidimensional, then

$$d(Z) = \sum_j m_j . d(V_j) = d^{\circ}_{\underline{u}^{(i,1)}} f_{Z,\mathbf{1}}$$

for $i = 1, \ldots, h$.

3. Height

Let us now consider our variety V defined over a number field, we want to define a *height* $h(V)$ of V. Of course, we put $h(V) = h(f_{V,\mathbf{1}})$ for a suitable height on $K[\mathbf{u}]$, but there are many possible variants of heights for forms and polynomials. Here, we select a height which is nicer than the others with respect to projective transformations (all other reasonable choice would lead to a notion of height comparable to that one).

As usual our height is made of local factors for the places of K (say that a place is a set of equivalence classes of absolute values on K). For each place we fix a normalized absolute value representing v, in the archimedean case this fixes an embedding $\sigma_v : K \hookrightarrow \mathbf{C}$ such that the normalized absolute value is just $|\sigma_v(\cdot)|$ and in the ultrametric case we require $|p|_v = p^{-1}$ for p the prime number lying under

v. Also, for $x \in K^\ell$ we put $\|x\|_v = \sqrt{|x_1|^2 + \cdots + |x_\ell|^2}$ if v is archimedean and $\|x\|_v = \max(|x_1|_v; \ldots; |x_\ell|_v)$ if v is ultrametric. Finally, for $f \in K[\mathbf{u}]$ we set

$$h(f) = \sum_v \frac{[K_v : \mathbf{Q}_v]}{[K : \mathbf{Q}]} . \log M_v(f)$$

where v runs over all places of K, $M_v(f)$ is the maximum of the v-adic absolute values of the coefficients of f if v is ultrametric and, if $\sigma_v : K \hookrightarrow \mathbf{C}$ is the embedding associated to v in the archimedean case

$$\log M_v(f) = \int_{S_{N_1+1} \times \cdots \times S_{N_h+1}} \log |\sigma_v(f).\eta_{N_1+1} \wedge \cdots \wedge \eta_{N_h+1} + \sum_{i=1}^h d^\circ_{\underline{u}^{(i,d_i)}} f. \sum_{j=1}^{N_i} \frac{1}{2j}$$

where S_{N+1} is the unit sphere in $\mathbf{C}^{N+1}$ and η_{N+1} the Haar measure on S_{N+1} of total mass 1. In the archimedean case, the measure $M_v(f)$ is invariant by unitary transformations of $\mathbf{C}^{n+1}$. It follows from the product formula that $h(\lambda f) = h(f)$ for all $\lambda \in K^*$ and from [Lel2][1] that $h(f)$ if larger than the *Mahler height* of f (as defined and used in [PPh2] for example), in particular $h(f) \geq 0$ for all $f \in K[\mathbf{u}]$.

A computation (see for example [Lau1], prop. 5) shows $\log M_v(U_{i,d_i}(x)) = d_i. \log \|x\|_v$, from which (and the discussion in the previous section) we deduce

$$h(f_{V,\mathbf{d}}) = d_1 \ldots d_h.h(V) .$$

REMARK. If $Z = \sum_j m_j.V_j$ is equidimensional, then

$$h(Z) = \sum_j m_j.h(V_j) = h(f_{Z,\mathbf{1}}) .$$

EXAMPLE. If $V = \mathcal{Z}(L_1, \ldots, L_{n-h+1})$ where the L_i's are independent linear forms (that we may consider as vectors in K^{n+1}) we have

$$h(V) = \sum_v \frac{[K_v : \mathbf{Q}_v]}{[K : \mathbf{Q}]} . \log \|L_1 \wedge \cdots \wedge L_{n-h+1}\|_v + \sum_{i=1}^{h-1} \sum_{j=1}^i \frac{1}{2j} .$$

In particular, $h(\mathbf{P}_n) = \sum_{i=1}^n \sum_{j=1}^i \frac{1}{2j}$ and $h(x) = \sum_v \frac{[K_v:\mathbf{Q}_v]}{[K:\mathbf{Q}]} . \log \|x\|_v$. If $V = \mathcal{Z}(F)$ then $h(V) = h(F) + d^\circ F. \sum_{i=1}^{n-1} \sum_{j=1}^i \frac{1}{2j}$.

4. Geometric and arithmetic Bézout theorems

An important procedure when using varieties is intersection, given two varieties V and W we want to relate the quantities introduced (degree, height, ...) of $V \cap W$ to those of V and W. We will concentrate on the case when $W = \mathcal{Z}(F)$ is an hypersurface, the general case can be reduced to that one by the use of the *join variety* and the *reduction to the diagonal* (see, for example, [PPh7]-III, § 2.B).

Assume now on that $W = \mathcal{Z}(F)$ is an hypersurface of degree d_h, can we compute a Chow form of $V \cap W$ from a Chow form of V? Answer: Yes, considering the homomorphism of K-algebras $\rho_F : K[\mathbf{u}] \to K[\mathbf{u}']$ defined by $\rho_F(U_{h,d_h}) = F$, where $\mathbf{u}' = (\underline{u}^{(1,d_1)}, \ldots, \underline{u}^{(h-1,d_{h-1})})$. In fact, (ET) tells us that $\rho_F(f_{V,\mathbf{d}})$ has the same set of zeros as $f_{V \cap W, \mathbf{d}'}$ where $\mathbf{d}' = (d_1, \ldots, d_{h-1})$. Therefore $\rho_F(f_{V,\mathbf{d}})$ is an eliminating form associated to a cycle, that we will denote $V.W$, supported by $V \cap W$. In any

[1][Lel2] P. Lelong. Mesure de Mahler des polynômes et majorations par convexité. *C. R. Acad. Sci. Paris* 315, (1992), 139–142; Mesure de Mahler et calcul de constantes universelles pour les polynômes de N variables. *Math. Ann.* 299, (1994), 673–695.

case, $f_{V \cap W, \mathbf{d}'}$ divides $f_{V.W, \mathbf{d}'} = \rho_F(f_{V,\mathbf{d}})$, and if $V \cap W = \emptyset$ then we put $V.W$ equal to the empty cycle, a Chow form of which is 1.

Let us start with the classical geometric Bézout theorem. In the sequel we set $f = f_{V,\mathbf{d}}$, $W = \mathcal{Z}(F)$ an hypersurface.

GEOMETRIC BÉZOUT THEOREM (GBT). *In the above notations, if $V \cap W \neq \emptyset$ one has*

$$d(V \cap W) \leq d(V.W) = d(V).d(W) = d(V).d^\circ F \ .$$

PROOF. The first inequality is clear and taking $\mathbf{d} = (1, \ldots, 1, d^\circ F)$ we have

$$d(V.W) = d^\circ_{U_{1,1}} \rho_F(f) = d^\circ_{U_{1,1}} f = d(V).d^\circ F = d(V).d(W) \ . \quad \square$$

Let us now turn to the arithmetic counterpart. To state more neatly the result it is better to use a slightly modified height h_1 for forms $F \in K[X_0, \ldots, X_n]$. Precisely, $h_1(F) = \sum_v \frac{[K_v : \mathbf{Q}_v]}{[K : \mathbf{Q}]} \cdot \log \|F\|_v$ where $\|F\|_v$ is the maximum of the absolute values of the coefficients F_α of F if v is ultrametric and $\left(\sum_{|\alpha| = d^\circ F} \binom{d^\circ F}{\alpha}^{-1} \cdot |\sigma_v(F_\alpha)|^2 \right)^{1/2}$ if v is archimedean. In particular, for $x \in \mathbf{P}_n(\mathbf{C}_v)$ we have $|F(x)| \leq \|F\|_v \cdot \|x\|_v$ (use Cauchy-Schwarz inequality when v is archimedean). Therefore, $M_v(F) = \|F\|_v$ if v is ultrametric and $M_v(F) \leq \|F\|_v \cdot \exp\left(d^\circ F. \sum_{j=1}^n 1/2j \right)$ if v is archimedean. In the other direction we have $\|F\|_v \leq M_v(F).(n+1)^{d^\circ F/2}$ when v is archimedean (see lemme 3.3 in chapter 7). Also, $\|.\|$ is invariant under unitary transformations of $\mathbf{C}^{n+1}$.

ARITHMETIC BÉZOUT THEOREM (ABT). *In the above notation, if $V \cap W \neq \emptyset$ one has*

$$h(V \cap W) \leq h(V.W) \leq h(V).d^\circ F + h_1(F).d(V) \ .$$

We need a lemma for archimedean places, which is a special case of lemme 3.5 in chapter 7.

LEMMA 4.1. *Fix an embedding $K \subset \mathbf{C}$ and $\mathbf{d} = (1, \ldots, 1, d^\circ F)$ then*

$$\log M(\rho_F(f)) = \log M(f) + \int_{V(\mathbf{C})} \log(|F(x)|/\|x\|^{d^\circ F}).\Omega(x)$$

where Ω is a positive measure on $V(\mathbf{C})$ of total mass equal to $d(V)$.

SKETCH OF PROOF OF THE LEMMA. We have :

$$\frac{\rho_F(f)}{f} = \prod_{x \in V \cap H_1 \cap \cdots \cap H_{h-1}} \frac{F(x)}{U_{h,d_h}(x)}$$

from which follows

$$\log M(\rho_F(f)/f) = \int_{(S_{n+1})^{h-1}} \sum_{x \in V \cap H_1 \cap \cdots \cap H_{h-1}} \log(|F(x)|/\|x\|^{d^\circ F}).\eta_{n+1}^{\wedge(h-1)}$$

since $\log M(U_{h,d^\circ F}(x)) = d^\circ F. \log \|x\|$. A tedious change of variable and integration in fibers transforms the above integral in an integral on $V(\mathbf{C})$ against a positive measure Ω the total mass of which is just the cardinality of $V \cap H_1 \cap \cdots \cap H_{h-1}$, that is $d(V)$ (see, for example, [**PPh7**]-I, Proposition 5). $\quad \square$

REMARK. Actually, the measure Ω is the $(h-1)$-th power of the Fubini-Study form on $\mathbf{P}_n(\mathbf{C})$.

PROOF OF THEOREM (ABT). The first inequality is again clear. We can re-place F by a multiple λF, $\lambda \in \overline{\mathbf{Q}}^*$ so that $M_v(\rho_F(f)) = M_v(f)$ for all finite places. Collecting the formulas given by lemma 4.1 for all archimedean places we get

$$h(V.W) = h(f) + \sum_{v \mid \infty} \frac{[K_v : \mathbf{Q}_v]}{[K : \mathbf{Q}]} \cdot \int_{\sigma_v(V)(\mathbf{C})} \log \left(\frac{|\sigma_v(F)(x)|}{\|x\|^{d^\circ F}} \right) . \Omega(x) \ .$$

But, $h(f) = d^\circ F.h(V)$, $\frac{|\sigma_v(F)(x)|}{\|x\|^{d^\circ F}} \le \|F\|_v$ and the result follows. $\square$

5. Distance from a point to a variety

Another ingredient which is needed in diophantine approximation is a notion of distance. We shall restrict our exposition to the case of varieties defined over $\mathbf{C}$, which apply to the archimedean embeddings of varieties defined over a number field for example. Let now V be a projective variety defined over $K = \mathbf{C}$ and $x = (x_0 : \cdots : x_n) \in \mathbf{P}_n(\mathbf{C})$, we first want general forms vanishing at x. A nice way to write down such forms is to consider the morphism of K-algebras

$$\mathfrak{d}_x : \mathbf{C}[\mathbf{u}] \to \mathbf{C}[\mathbf{s}]$$

defined by $\mathfrak{d}_x(u_\alpha^{(i,d_i)}) = \sum_{|\alpha'|=d_i} \binom{d_i}{\alpha'}^{1/2} . s_{\alpha,\alpha'}^{(i,d_i)} . x_0^{\alpha_0'} \ldots x_n^{\alpha_n'}$ where $s_{\alpha,\alpha'}^{(i,d_i)}$ are new vari-ables linked by the only relations $s_{\alpha,\alpha'}^{(i,d_i)} + s_{\alpha',\alpha}^{(i,d_i)} = 0$. Because of these anti-symmetry relations the forms $\mathfrak{d}_x U_{i,d_i}$ vanishes at x and it follows from (ET) that x belongs to V iff $\mathfrak{d}_x f = 0$. How small $\mathfrak{d}_x f$ is will measure the *distance from x to V*, more precisely we set

$$\mathrm{Dist}_\mathbf{d}(x, V) = \|\mathfrak{p}(x)\|_\mathbf{d} := \frac{M(\mathfrak{d}_x f)}{M(f). \prod_{i=1}^h \|x\|^{d_i d^\circ_{U_{i,d_i}} f}} \ .$$

PROPERTIES 5.1. 1. $\mathrm{Dist}_\mathbf{d}(x, V) \le 1$;

2. *If* $\mathbf{d}' = (d_1, \ldots, d_{h-1}, 1)$, $N_i = \binom{d_i+n}{n} - 1$ *and* $\gamma_{N_i} = \exp\left(\sum_{j=1}^{N_i} \frac{1}{2j}\right)$ *for* $i = 1, \ldots, h$, *then :*

- $\mathrm{Dist}_{\mathbf{d}'}(x, V) \le \mathrm{Dist}_\mathbf{d}(x, V). \left(\gamma_{N_h}^{d^\circ f} . \prod_{i=1}^{h-1} \gamma_{N_i}^{d^\circ_{U_{i,d_i}} f}\right)^{1-\frac{1}{d_h}}$;

- $\mathrm{Dist}_\mathbf{d}(x, V) \le \mathrm{Dist}_{\mathbf{d}'}(x, V). (\sqrt{d_h})^{d^\circ_{U_{h,d_h}} f}$.

This is a special case of proposition 4.1 (points 1 and 2) from chapter 7.

REMARK. If $Z = \sum_i m_i . V_i$ is equidimensional we define

$$\mathrm{Dist}_\mathbf{d}(x, Z) = \prod_i \mathrm{Dist}_\mathbf{d}(x, V_i)^{m_i} = \frac{M(\mathfrak{d}_x f_{Z,\mathbf{d}})}{M(f_{Z,\mathbf{d}}). \prod_{i=1}^h \|x\|^{d_i d^\circ_{U_{i,d_i}} f_{Z,\mathbf{d}}}} \ .$$

The quantity $\mathrm{Dist}(x, Z)$ is invariant by unitary transformations on $\mathbf{C}^{n+1}$.

EXAMPLES. If $V = \mathcal{Z}(L_1, \ldots, L_{n-h+1})$ where the L_i's are linear forms (that we may consider as vectors in K^{n+1}) we have

$$\mathrm{Dist}(x, V) = \frac{\|\pi(x)\|}{\|x\|}$$

where $\pi : \mathbf{C}^{n+1} \to W$ is the orthogonal projection of $\mathbf{C}^{n+1}$ on the space spanned by the L_i's, for the canonical hermitian product.

In particular, if $V = \{y\}$ then $\mathrm{Dist}(x, y)$ defines an actual distance on $\mathbf{P}_n(\mathbf{C})$. In fact, $\mathrm{Dist}(x, y)$ behave like a sine between the directions x and y in $\mathbf{C}^{n+1}$ and

$$1 - \mathrm{Dist}_d^2(x, y) = \left(1 - \mathrm{Dist}^2(x, y)\right)^d .$$

In the affine chart of $\mathbf{P}_n(\mathbf{C})$ centered at x (i.e. $x = 0 \in \mathbf{C}^n$) one has

$$\mathrm{Dist}_d(x, y) = \|y\| \cdot \sqrt{\frac{\sum_{i=0}^{d-1} \binom{d}{i+1} \|y\|^{2i}}{(1 + \|y\|^2)^d}}$$

where $y \in \mathbf{C}^n$, $\|y\| = \sum_{i=1}^n |y_i|^2$.

If $V = \mathcal{Z}(F)$ then $\mathrm{Dist}(x, V) = \frac{|F(x)|}{M(F) \cdot \|x\|^{d^\circ F}}$.

CLOSEST POINT PROPERTY. *In the above notations assume $d_1 = \cdots = d_h = 1$, then there exists $y \in V(\mathbf{C})$ such that*

$$\mathrm{Dist}(x, y) \le \mathrm{Dist}(x, V)^{1/d(V)} \cdot \exp\left(\sum_{i=1}^n \frac{h}{i}\right) .$$

To prove properties 5.1 and the above, we need and assume the following lemma, which is proved in [**Jad**], lemmes 3.7 and 4.5, see also chapter 7, lemme 4.5.

LEMMA 5.2. *Suppose $\|x\| = 1$ and let $\tilde{\partial}_x : \mathbf{C}[\underline{s}] \to \mathbf{C}[\underline{s}^{(h,1)}]$ be the morphism defined by $\tilde{\partial}_x(U_{h,1}) = U_{h,1}$ and $\tilde{\partial}_x(U_{i,1}) = \partial_x(U_{i,1})$ for $i = 1, \ldots, h-1$ then*

$$M(\tilde{\partial}_x f) \le M(f) \le M(\tilde{\partial}_x f) \cdot \exp\left(\sum_{i=1}^{h-1} d_{U_{i,d_i}}^\circ f \cdot \sum_{j=1}^{\binom{d_i+n}{n}-1} \frac{1}{2j} + d^\circ f \cdot \sum_{j=1}^{\binom{d_h+n}{n}-1} \frac{1}{2j}\right) .$$

PROOF OF PROPERTIES 5.1. We first reduce to the case of points. Let f and f' be Chow forms of V of index $\mathbf{d}$ and $\mathbf{d}'$ respectively. Let's normalize x so that $\|x\| = 1$, we write

$$\frac{\partial_x f}{\tilde{\partial}_x f} = \prod_{y \in V \cap H_1 \cap \cdots \cap H_{h-1}} \frac{\partial_x U_{h,d_h}(y)}{U_{h,d_h}(y)} , \quad \frac{\partial_x f'}{\tilde{\partial}_x f'} = \prod_{y \in V \cap H_1 \cap \cdots \cap H_{h-1}} \frac{\partial_x U_{h,1}(y)}{U_{h,1}(y)} ,$$

where $H_i = \mathcal{Z}(\partial_x U_{i,d_i})$ and from which follows by integration

$$\log M(\partial_x f / \tilde{\partial}_x f) = \int_{\left(S_{\frac{n(n+1)}{2}}\right)^{h-1}} \sum_{y \in V \cap H_1 \cap \cdots \cap H_{h-1}} \log \mathrm{Dist}_{d_h}(x, y) \cdot \eta_{\frac{n(n+1)}{2}}^{\wedge(h-1)}$$

$$\log M(\partial_x f' / \tilde{\partial}_x f') = \int_{\left(S_{\frac{n(n+1)}{2}}\right)^{h-1}} \sum_{y \in V \cap H_1 \cap \cdots \cap H_{h-1}} \log \mathrm{Dist}(x, y) \cdot \eta_{\frac{n(n+1)}{2}}^{\wedge(h-1)} .$$

Now, we have seen in section 3 that $M(f) = M(f')^{d_h}$, writing

$$\frac{\tilde{\partial}_x f}{(\tilde{\partial}_x f')^{d_h}} = \prod_{y \in V \cap H_1 \cap \cdots \cap H_{h-1}} \frac{U_{h,d_h}(y)}{U_{h,1}(y)^{d_h}}$$

and integrating we check $M(\tilde{\partial}_x f) = M(\tilde{\partial}_x f')^{d_h}$. Therefore

$$\log \mathrm{Dist}_{\mathbf{d}}(x, V) = \log M(\partial_x f / \tilde{\partial}_x f) - \log M(f / \tilde{\partial}_x f) ,$$

$$\log \mathrm{Dist}_{\mathbf{d}'}(x, V) = \log M(\partial_x f' / \tilde{\partial}_x f') - \frac{1}{d_h} \cdot \log M(f / \tilde{\partial}_x f) ,$$

and lemma 5.2 reduces the proof of properties 1 and 2 to the case of points. But it follows directly from the expressions of these distances in the affine chart of $\mathbf{P}_n$ centered at x that $\mathrm{Dist}_{d_h}(x, y) \leq 1$ and

$$1 \leq \frac{\mathrm{Dist}_{d_h}(x, y)}{\mathrm{Dist}(x, y)} \leq \sqrt{d_h} \ . \quad \square$$

PROOF OF CLOSEST POINT PROPERTY. Put $g = \tilde{\partial}_x f$, by the mean value theorem there exists a morphism $\rho : \mathbf{C}[\underline{s}] \to \mathbf{C}[\underline{s}^{(h,1)}]$ such that $\|\rho(\underline{s}^{(i,1)})\| = 1$ for $i = 1, \dots, h-1$ and

$$\frac{M(\partial_x f)}{M(g)} = \frac{M(\partial_x g)}{M(g)} \geq \frac{M(\rho \circ \partial_x(g))}{M(\rho(g))} = \frac{M(\partial_x \circ \rho(g))}{M(\rho(g))} \ .$$

But, $\rho(g) = \prod_{y \in Z} U_{h,1}(y)$ where $Z = V \cap H_1 \cap \cdots \cap H_{h-1}$ with $H_i = \mathcal{Z}(\rho \circ \partial_x(U_{i,1}))$. Thanks to lemma 5.2 we can write

$$\prod_{y \in Z} \mathrm{Dist}(x, y) = \mathrm{Dist}(x, Z) = \frac{M(\partial_x \circ \rho(g))}{M(\rho(g))} \leq \frac{M(\partial_x f)}{M(g)}$$

$$\leq \frac{M(\partial_x f)}{M(f)} \cdot \exp\left(d^\circ f . \sum_{i=1}^n \frac{1}{i} \right) \leq \mathrm{Dist}(x, V) . \exp\left(d^\circ f . \sum_{i=1}^n \frac{1}{i} \right) \ .$$

Since $\mathrm{card} Z = d(V) = \frac{1}{h} d^\circ f$ there exists $y \in Z$ satisfying the inequality in the property. $\qquad\square$

6. Auxiliary results

We fix a point $x \in \mathbf{P}_n(\mathbf{C})$, a projective variety $V \subset \mathbf{P}_n(\mathbf{C})$ of dimension $h - 1$ (defined over $K = \mathbf{C}$) and a form $F \in \mathbf{C}[X_0, \dots, X_n]$ of degree d. We denote $x^{(d)}$ the projective point in $\mathbf{P}_N(\mathbf{C})$, with $N = \binom{d+n}{n} - 1$, whose coordinates are the monomials $\binom{d}{\alpha}^{1/2} . x_0^{\alpha_0} \dots x_n^{\alpha_n}$, $|\alpha| = d$.

LEMMA 6.1. *Let f (resp. f_1) be an eliminating form of index $\mathbf{d} = (1, \dots, 1, d)$ (resp. of index $(1, \dots, 1)$) of V and $T \in \mathbf{N}^*$. Assume $\|x\| = 1$ and $x \notin V(\mathbf{C})$, then*

$$\log M\left(\frac{\partial_x \circ \rho_F(f)}{\partial_x(f)} \right) = (T-1) . \log M\left(\frac{\partial_x(f_1)}{\tilde{\partial}_x(f_1)} \right)$$

$$+ \int_{V(\mathbf{C})} \log\left(\frac{|F(y)| . \|y\|^{T-1}}{\|x^{(d)} \wedge y^{(d)}\| . \|x \wedge y\|^{T-1}} \right) . \Omega_x(y)$$

where Ω_x is a positive measure on $V(\mathbf{C})$ of total mass $d(V)$ (this is the $(h-1)$th power of the Fubini-Study metric on $x^\perp$). Furthermore, if F vanishes at x with order at least T the integral is bounded above by $d(V) . \log\left(\binom{d}{T}^{1/2} . \|F\| \right)$.

SKETCH OF PROOF. We have

$$\frac{\partial_x \circ \rho_F(f) . \tilde{\partial}_x(f_1)^{T-1}}{\partial_x(f) . \partial_x(f_1)^{T-1}} = \prod_{y \in V \cap H_1 \cap \cdots \cap H_{h-1}} \frac{F(y) . U_{h,1}(y)^{T-1}}{\partial_x U_{h,d}(y) . \partial_x U_{h,1}(y)^{T-1}}$$

where $H_i = \mathcal{Z}(\eth_x U_{i,1})$, from which follows that $\log M\left(\frac{\eth_x \circ \rho_F(f) \cdot \tilde{\eth}_x(f_1)^{T-1}}{\eth_x(f) \cdot \eth_x(f_1)^{T-1}}\right)$ is equal to

$$\int_{\left(S_{\frac{n(n+1)}{2}}\right)^{h-1}} \sum_{y \in V \cap H_1 \cap \cdots \cap H_{h-1}} \log\left(\frac{|F(y)| \cdot \|y\|^{T-1}}{\|x^{(d)} \wedge y^{(d)}\| \cdot \|x \wedge y\|^{T-1}}\right) \cdot \eta_{\frac{n(n+1)}{2}}^{\wedge(h-1)} ,$$

since $M(\eth_x U_{h,1}(y)) = \|y\|$, $M(\eth_x U_{h,1}(y)) = \|x \wedge y\|$ and $M(\eth_x U_{h,d}(y)) = \|x^{(d)} \wedge y^{(d)}\|$. After applying a unitary transformation sending x to $(1 : 0 : \cdots : 0)$, another tedious change of variable and integration in fibers transforms the above integral in an integral on $V(\mathbf{C})$ against a positive measure Ω_x the total mass of which is just the cardinality of $V \cap H_1 \cap \cdots \cap H_{h-1}$, that is $d(V)$ (see [**PPh7**]-II, lemme 1'). We note that each term in the equality of the lemma is invariant under unitary transformations, thus we apply again such a transformation sending x to $(1 : 0 : \cdots : 0)$ in order to bound the integral. Then, assuming $x = (1 : 0 : \cdots : 0)$ and since F vanishes at x with order at least T, we get

$$|F(y)|^2 \leq \sum_{\substack{\alpha \in \mathbf{N}^{n+1} \\ |\alpha| = d}} \frac{|F_\alpha|^2}{\binom{d}{\alpha}} \cdot \sum_{\substack{\alpha \in \mathbf{N}^{n+1} \\ |\alpha| = d; \alpha_0 \leq d - T}} \binom{d}{\alpha} |y_0|^{2\alpha_0} \dots |y_n|^{2\alpha_n}$$

$$\leq \binom{d}{T} \cdot \|F\|^2 \cdot \sum_{\substack{\beta \in \mathbf{N}^n \\ |\beta| = T}} \binom{T}{\beta} |y_1|^{2\beta_1} \dots |y_n|^{2\beta_n} \cdot \sum_{\substack{\alpha \in \mathbf{N}^{n+1} \\ |\alpha| = d - T}} \binom{d-T}{\alpha} |y_0|^{2\alpha_0} \dots |y_n|^{2\alpha_n}$$

$$\leq \binom{d}{T} \cdot \|F\|^2 \cdot \|x \wedge y\|^{2T} \cdot \|y\|^{2(d-T)}$$

$$\leq \binom{d}{T} \cdot \|F\|^2 \cdot \|x^{(d)} \wedge y^{(d)}\|^2 \cdot \|x \wedge y\|^{2(T-1)} \cdot \frac{\|y\|^{2(d-T-1)}}{\|y^{(d)}\|^2}$$

for any $y \in V(\mathbf{C})$, since $\frac{\|x \wedge y\|^2}{\|y\|^2} = 1 - \frac{|y_0|^2}{\|y\|^2} \leq 1 - \frac{|y_0|^{2d}}{\|y\|^{2d}} = \frac{\|x^{(d)} \wedge y^{(d)}\|^2}{\|y^{(d)}\|^2}$. Therefore

$$\frac{|F(y)| \cdot \|y\|^{T-1}}{\|x^{(d)} \wedge y^{(d)}\| \cdot \|x \wedge y\|^{T-1}} \leq \binom{d}{T}^{1/2} \cdot \|F\| \cdot \frac{\|y\|^d}{\|y^{(d)}\|} = \binom{d}{T}^{1/2} \cdot \|F\|$$

and the last assertion of the lemma follows.   □

See chapter 7, § 4.4 for a statement corresponding to lemma 6.1 (with $T = 1$) in products of projective spaces. In lemma 6.1, assuming $\|x\| = 1$ is no restriction and the condition $x \notin V(\mathbf{C})$ is harmless since we can always apply it to a point arbitraily close to x which doesnot belong to $V(\mathbf{C})$. In particular, if F vanishes to order T at x the right-hand side of the first formula in the proof of lemma 6.1 is bounded above near x and since $\tilde{\eth}(f_1)$ doesnot vanish at x this shows that $\mathrm{ord}_x \eth \circ \rho_F(f) \geq (T - 1)\mathrm{ord}_x \eth(f_1) + \mathrm{ord}_x \eth(f)$. In fact, one can show [**PPh5**] that $\mathrm{ord}_x \eth(f_1) = \mathrm{ord}_x \eth(f)$ is just the *Samuel's multiplicity* of the ring $\mathbf{C}[X_0, \dots, X_n]/I(V)$ localized at x.

For $\tau \in \mathbf{N}^{n+1}$ let us denote by $F^{(\tau)}(x)$ the τ-th *divided derivative* of F, that is the coefficient of $Y^{\tau_0} \dots Y^{\tau_n}$ in the expansion of $F(x + Y)$. We further set, for $t \in \mathbf{N}$,

$$\|F^{(t)}(x)\| := \left(\sum_{\substack{\tau \in \mathbf{N}^{n+1} \\ |\tau| = t}} \frac{|F^{(\tau)}(x)|^2}{\binom{t}{\tau} \|x\|^{2(d^\circ F - \tau)}}\right)^{1/2} ,$$

and we remark that this quantity is invariant under unitary transformation of $\mathbf{C}^{n+1}$ (up to a constant it is the L^2-norm on $S_{n+1}(1)$ of the homogeneous part of $F(x+Y)$ of degree t in Y).

In the proof of the following proposition we will use an extension of the norm $\|.\|$ to multi-homogeneous forms. More precisely, for $g \in \mathbf{C}[\mathbf{u}]$ of multi-degree $(\Delta_1,\ldots,\Delta_h)$ we denote $\|g\| = \sum_{\substack{a_i \in \mathbf{N}^{N_i} \\ |\alpha_i|=\Delta_i}} \frac{|g_{\alpha_1,\ldots,\alpha_h}|^2}{\binom{\Delta_1}{\alpha_1}\cdots\binom{\Delta_h}{\alpha_h}}$ with $N_i = \binom{d_i+n}{n} - 1$. In particular, if $\rho : \mathbf{C}[\mathbf{u}] \to \mathbf{C}$ is an homomorphism we have

$$|\rho(g)| \leq \|g\| \cdot \prod_{i=1}^{h} \|\rho(U_{i,d_i})\|^{d^{\circ}_{U_{i,d_i}} g}$$

and $\|g\| \leq M(g) \cdot \prod_{i=1}^{h}(N_i+1)^{d^{\circ}_{U_{i,d_i}} g/2}$.

PROPOSITION 6.2. *Let f (resp. f_1) be an eliminating form of V of index* $\mathbf{d} = (1,\ldots,1,d)$ *(resp. of index $(1,\ldots,1)$). Assume $x \notin V(\mathbf{C})$ and $\|x\| = 1$, then $\frac{M(\partial_x \circ \rho_F(f))}{M(\partial_x(f))}$ is bounded above by*

$$\left(\left(\frac{M(\partial_x(f_1))}{M(\tilde{\partial}_x(f_1))} \right)^{T-1} + \sum_{t=0}^{T-1} \frac{\|F^{(t)}(x)\|}{\|F\|} \cdot \left(\frac{M(\partial_x(f_1))}{M(\tilde{\partial}_x(f_1))} \right)^{t-1} \right) \cdot \left(\|F\|(n+1)^{2dh} \right)^{d(V)} \quad .$$

PROOF. We first make a unitary transformation so that $x = (1 : 0 : \cdots : 0)$. For $t = 0,\ldots,d$ introduce the homomorphism $\rho_t : \mathbf{C}[\mathbf{u}^{(h)}] \to \mathbf{C}[\mathbf{u}^{(h)}]$ defined by $\rho_t(u_\alpha^{(h,d_h)}) = 0$ if $\alpha_0 > d - t$ and $= u_\alpha^{(h,d_h)}$ otherwise. Applying lemma 6.1 to $F = \rho_t(U_{h,d_h})$, which vanishes to order t at x, and then integrating on $U_{h,d_h} \in S_{N+1}(1)$ one gets

$$\frac{M(\partial_x \circ \rho_t(f))}{M(\partial_x(f))} \leq \left(\frac{M(\partial_x(f_1))}{M(\tilde{\partial}_x(f_1))} \right)^{t-1} \cdot \binom{d}{t}^{d(V)} \quad ,$$

(for $t = 0$ note that $\partial_x \circ \rho_0(f) = \tilde{\partial}_x(f)$ and $\frac{\|y^{(d)}\|}{\|x^{(d)} \wedge y^{(d)}\|} \leq \frac{\|y\|}{\|x \wedge y\|}$ which imply $\frac{M(\tilde{\partial}_x(f))}{M(\partial_x(f))} \leq \frac{M(\tilde{\partial}_x(f_1))}{M(\partial_x(f_1))}$). Then we write $G_t = F - \sum_{\substack{|\alpha|=d \\ \alpha_0 > d-t}} F_\alpha \cdot X_0^{\alpha_0} \ldots X_n^{\alpha_n}$, we have $G_0 = F$, $\|G_t\| \leq \|F\|$, and the finite increment theorem implies that $\frac{M(\partial_x \circ \rho_{G_t}(f))}{M(\partial_x(f))}$ is bounded above by

$$\frac{M(\partial_x \circ \rho_{G_{t+1}}(f))}{M(\partial_x(f))} + \left(\sum_{\substack{|\alpha|=d \\ \alpha_0=d-t}} \frac{|F_\alpha|^2}{\binom{d}{\alpha}^2} \right)^{1/2} \cdot \frac{\|\partial_x \circ \rho_t(f)\|}{M(\partial_x(f))} \cdot d(V) \|F\|^{d(V)-1}$$

$$\leq \frac{M(\partial_x \circ \rho_{G_{t+1}}(f))}{M(\partial_x(f))} + \frac{\|F^{(t)}(x)\|}{\|F\|} \cdot \left(\frac{M(\partial_x(f_1))}{M(\tilde{\partial}_x(f_1))} \right)^{t-1} \cdot \left(\|F\|(n+1)^{2dh} \right)^{d(V)}$$

because $F_\alpha = F^{(0,\alpha_1,\ldots,\alpha_n)}(x)$ at $x = (1 : 0 : \cdots : 0)$ and

$$\frac{\|\partial_x \circ \rho_t(f)\|}{M(\partial_x(f))} \leq \left(\frac{n(n+1)}{2} \right)^{dhd(V)} \frac{M(\partial_x \circ \rho_t(f))}{M(\partial_x(f))}$$

$$\leq \left(\frac{n(n+1)}{2} \right)^{dhd(V)} \cdot \binom{d}{t}^{d(V)} \cdot \left(\frac{M(\partial_x(f_1))}{M(\tilde{\partial}_x(f_1))} \right)^{t-1} \quad .$$

Piling up the above inequalities for $t = 0, \ldots, T - 1$ we obtain that $\frac{M(\eth_x \circ \rho_F(f))}{M(\eth_x(f))}$ is bounded above by

$$\frac{M(\eth_x \circ \rho_{G_T}(f))}{M(\eth_x(f))} + \sum_{t=0}^{T-1} \frac{\|F^{(t)}(x)\|}{\|F\|} \cdot \left(\frac{M(\eth_x(f_1))}{M(\tilde{\eth}_x(f_1))} \right)^{t-1} \cdot \left(\|F\|(n+1)^{2dh} \right)^{d(V)} \quad,$$

this conclude the proof of the proposition since by lemma 6.1 one has $\frac{M(\eth_x \circ \rho_{G_T}(f))}{M(\eth_x(f))} \leq$ $\left(\frac{M(\eth_x(f_1))}{M(\tilde{\eth}_x(f_1))} \right)^{T-1} \cdot \left(\|F\| \binom{d}{T} \right)^{d(V)}.$ $\qquad\square$

7. First metric Bézout theorem

FIRST METRIC BÉZOUT THEOREM (FMTB). *Let V a projective subvariety of $\mathbf{P}_n(\mathbf{C})$ defined over a number field K, F a form of degree d in $K[X_0, \ldots, X_n]$, $x \in \mathbf{P}_n(\mathbf{C})$ and $T \in \mathbf{N}^*$. Fix an embedding $K \hookrightarrow \mathbf{C}$, $\Delta = [K : \mathbf{Q}]$ if K is real or $= [K : \mathbf{Q}]/2$ if K is imaginary, and put $W = V.\mathcal{Z}(F)$, then*

$$\frac{1}{\Delta} \cdot \log \mathrm{Dist}(x, W) + h(W) \leq \frac{1}{\Delta} \cdot \log \left(\mathrm{Dist}(x, V)^T + \sum_{t=0}^{T-1} \frac{\|F^{(t)}(x)\|}{\|F\|} . \mathrm{Dist}(x, V)^t \right)$$

$$+ d.h(V) + d(V).(h_1(F) + \frac{(T + 3d)h}{\Delta} \log(n+1)) \quad .$$

PROOF. We may assume $\|x\| = 1$ and we deduce from proposition 6.2 ($\mathbf{d} = (1, \ldots, 1, d)$)

$$\log \left(\frac{\mathrm{Dist}(x, W).M(\rho_F(f))}{\mathrm{Dist}_\mathbf{d}(x, V).M(f)} \right) \leq \log \left(\mathrm{Dist}(x, V)^{T-1} + \sum_{t=0}^{T-1} \frac{\|F^{(t)}(x)\|}{\|F\|} . \mathrm{Dist}(x, V)^{t-1} \right)$$

$$+ \log \left(\frac{M(f_1)}{M(\tilde{\eth}_x(f_1))} \right)^{T-1} + d(V).\log \left(\|F\|(n+1)^{2dh} \right)$$

$$\leq \log \left(\mathrm{Dist}(x, V)^{T-1} + \sum_{t=0}^{T-1} \frac{\|F^{(t)}(x)\|}{\|F\|} . \mathrm{Dist}(x, V)^{t-1} \right)$$

$$+ d(V). (\log \|F\| + (T + 2d)h \log(n+1))$$

since $\frac{M(f_1)}{M(\tilde{\eth}_x(f_1))} \leq (n+1)^{hd(V)}$ by lemma 5.2. For all places v of K it follows from lemma 4.1

$$\log M_v(\rho_F(f)) \leq \log M_v(f) + d(V) \log \|F\|_v \quad .$$

We also have $\mathrm{Dist}_\mathbf{d}(x, V) \leq (\sqrt{d})^{d(V)} . \mathrm{Dist}(x, V)$ by property 5.1-(2). Adding up the first inequality above for the fixed place and the second for all the other places with the weights $\frac{[K_v : \mathbf{Q}_v]}{[K : \mathbf{Q}]}$, gives the inequality in the theorem. $\qquad\square$

8. Second metric Bézout theorem

SECOND METRIC BÉZOUT THEOREM (SMTB). *Let $\sigma \geq 1$ a real, V a projective subvariety of $\mathbf{P}_n(\mathbf{C})$ defined over a number field K, $F \in K[X_0, \ldots, X_n]$ a form* of degree d, $x \in \mathbf{P}_n(\mathbf{C})$ and $T \in \mathbf{N}^*$. Fix an embedding $K \hookrightarrow \mathbf{C}$, $\Delta = [K : \mathbf{Q}]$ if K is real or $= [K : \mathbf{Q}]/2$ if K is imaginary, and put $W = V.\mathcal{Z}(F)$. Assuming

$$\frac{\|F^{(t)}(x)\|}{\|F\|} \leq \min_{y \in V(\mathbf{C})} \left(\mathrm{Dist}(x, y)^{(T-t)/\sigma} \right)$$

for all $t = 0, \ldots, T - 1$ one has

$$\frac{1}{\Delta} . \log \mathrm{Dist}(x, W) + h(W) \leq \frac{T}{\sigma \Delta} . \log \mathrm{Dist}(x, V) + d.h(V) +$$

$$+ d(V). \left(h_1(F) + \left(\frac{Th}{\sigma} + 3d \right) \log(n + 1) \right) \quad .$$

PROOF. We assume $\|x\| = 1$ and make a unitary transformation so that $x = (1 : 0 : \cdots : 0)$. Let $\rho : \mathbf{C}[\underline{s}] \to \mathbf{C}[\underline{s}]$ be an homomorphism such that $\|\rho(\underline{s}^{(i,d_i)})\| = 1$, and $\mathfrak{d}_x$, $\tilde{\mathfrak{d}}_x$ the homomorphisms defined in § 5. We have the following identities, where $H_i = \mathcal{Z}(\rho(U_{i,1}))$, $i = 1, \ldots, h - 1$,

$$M \left(\frac{\rho \circ \mathfrak{d}_x \circ \rho_F(f)}{\rho \circ \tilde{\mathfrak{d}}_x(f)} \right) = \prod \frac{|F(y)|}{\|y\|^d}$$

$$M \left(\frac{\rho \circ \mathfrak{d}_x(f_1)}{\rho \circ \tilde{\mathfrak{d}}_x(f_1)} \right) = \prod \frac{\|x \wedge y\|}{\|y\|} = \prod \mathrm{Dist}(x, y) \quad ,$$

where products run over $y \in V \cap H_1 \cap \cdots \cap H_{h-1}$. But, proposition 6.2 applied to the variety $\{y\}$ tells us ($h = 1$, $d(\{y\}) = 1$, $f = U_{1,d}(y)$ and $\rho_F(f) = F(y)$)

$$\frac{|F(y)|}{\|y\|^d} \leq \left(\mathrm{Dist}(x, y)^T + \sum_{t=0}^{T-1} \frac{\|F^{(t)}(x)\|}{\|F\|} . \mathrm{Dist}(x, y)^t \right) . \|F\|.(n + 1)^{3d} \quad .$$

It results from the hypothesis

$$\frac{|F(y)|}{\|y\|^d} \leq \mathrm{Dist}(x, y)^{T/\sigma} . T . \|F\|.(n + 1)^{3d}$$

and we can write

$$M \left(\frac{\rho \circ \mathfrak{d}_x \circ \rho_F(f)}{\rho \circ \tilde{\mathfrak{d}}_x(f)} \right) \leq M \left(\frac{\rho \circ \mathfrak{d}_x(f_1)}{\rho \circ \tilde{\mathfrak{d}}_x(f_1)} \right)^{T/\sigma} . \left(\|F\|.T.(n + 1)^{3d} \right)^{d(V)} \quad .$$

Then, integrating on ρ gives

$$M(\mathfrak{d}_x \circ \rho_F(f)) \leq \frac{M(\tilde{\mathfrak{d}}_x(f))}{M(\tilde{\mathfrak{d}}_x(f_1))^{T/\sigma}} . M(\mathfrak{d}_x(f))^{T/\sigma} . \left(\|F\|.T.(n + 1)^{3d} \right)^{d(V)}$$

$$\leq M(f). \left(\frac{M(\mathfrak{d}_x f_1)}{M(f_1)} \right)^{T/\sigma} . \left(\|F\|.(n + 1)^{\frac{Th}{\sigma} + 3d} \right)^{d(V)}$$

since $M(\tilde{\mathfrak{d}}_x(f)) \leq M(f)$ and $M(f_1) \leq M(\tilde{\mathfrak{d}}_x(f_1)).(n + 1)^{hd(V)}$ by lemma 5.2. Combining the last inequality for the fixed place and the inequality from lemma 4.1 for all the other places, gives the theorem. $\qquad \square$

CHAPTER 7

Géométrie diophantienne multiprojective

1. Introduction

Ce chapitre utilise, dans un cadre géométrique, le chapitre 5 sur la théorie de l'élimination multihomogène. On rappelle que l'on travaille avec un espace multiprojectif de la forme

$$\mathbf{P} = \mathbf{P}_K^{n_1} \times \cdots \times \mathbf{P}_K^{n_q}$$

où $n_1, \ldots, n_q$ sont des entiers naturels et K un corps de nombres.

L'élimination multihomogène s'applique à la géométrie, en premier lieu, par la définition de hauteurs d'un sous-schéma fermé de $\mathbf{P}$. La construction de hauteurs a été étudiée par de nombreux auteurs soit dans le cadre de la théorie de l'élimination : Nesterenko, Philippon, Jadot (voir [**PPh2, PPh7, Jad**]) soit dans le cadre de la géométrie d'Arakelov : Faltings, Bost, Gillet, Soulé (voir [**BGS**][1]). Le fait que les deux approches définissent essentiellement la même notion a été montré par Soulé [**Sou**] et Philippon [**PPh7**, III]. Comme nous l'avons dit notre démarche s'inscrit dans la théorie des formes de Chow; nous examinerons (proposition 2.5) comment nos définitions trouvent leurs équivalents dans la théorie d'Arakelov (d'après [**BGS**] (*loc. cit.*), 3.1.3).

Ainsi (paragraphe 2) on associe à un sous-schéma fermé $V \hookrightarrow \mathbf{P}$ des hauteurs. En effet, alors que la théorie usuelle attache à $W \hookrightarrow \mathbf{P}_K^n$ un seul nombre réel $h(W)$, on constate que l'analogue de $h(W)$ consiste en une famille de nombres réels $h_\alpha(V)$ où α parcourt les éléments de $\mathbf{N}^q$ tels que $\alpha_1 + \cdots + \alpha_q = \dim V + 1$. On envisage ensuite le comportement des hauteurs à travers les opérations simples sur les variétés (multi-)projectives que représentent les plongements de Segre et de Veronese. On décrit aussi le cas important où V est un produit de la forme

$$V = V_1 \times \cdots \times V_q$$

(avec $V_i \hookrightarrow \mathbf{P}_K^{n_i}$ pour $1 \leq i \leq q$).

Le paragraphe 3 établit une formule d'intersection : on y montre comment varient hauteurs et degrés lorsqu'une variété V est coupée par une hypersurface (c'est-à-dire un sous-schéma fermé de $\mathbf{P}_K^{n_1} \times \cdots \times \mathbf{P}_K^{n_q}$ défini par une seule équation multihomogène). Ceci s'inspire de la proposition 4 de [**PPh7**, III], que l'on généralise, et explicite les formules suggérées dans [**BGS**] (*loc. cit.*), 3.1.3.

Enfin nous conclurons en examinant ce que deviennent les notions de distances dans le cas multiprojectif. La théorie projective a été développée par Philippon [**PPh2**] et Jadot [**Jad**, chap. III]. On remplace pour cette étude (locale) K par

---

(∗) Chapter's author : Gaël RÉMOND.

[1][**BGS**] J.-B. Bost, H. Gillet, C. Soulé. Heights of projective varieties and positive Green forms, *J. Amer. Math. Soc.* 7, (1994), 903–1027.

$\mathbf{C}_v$. Lorsque V est un sous-schéma fermé de $\mathbf{P}$, x un point fermé de $\mathbf{P}$ et d un indice de $(\mathbf{N}^q)^{\dim V + 1}$ on définit une distance $\mathrm{Dist}_d(x, V)$ qui est un nombre réel. On montre même que $0 \leq \mathrm{Dist}_d(x, V) \leq 1$. On étudie les cas particuliers où V est soit un produit soit un point. Dans le cas général, on constate que certains indices se prêtent mal à la définition et donnent au mieux une distance partielle. On se restreint alors aux indices dont toutes les composantes sont non nulles. On montre alors que les distances obtenues sont toutes comparables (en un sens précis) et l'on fait le lien avec les plongements de Segre-Veronese.

Ce travail fournit donc des outils techniques pour l'approximation diophantienne qui sont susceptibles d'applications assez vastes. En effet, outre les illustrations qu'en fournissent déjà dans le cas projectif les articles [**PPh7**], il apparaît que certains problèmes requièrent des notions respectant une situation produit inscrite dans les données. On peut par exemple citer le lemme de zéros de [**PPh9**] : la remarque qui suit l'énoncé des théorèmes 1 et 2 fait intervenir des hauteurs et dans le cadre du paragraphe 7 on pourrait travailler directement dans l'espace produit $\mathbf{P} = \mathbf{P}^{n_1} \times \cdots \times \mathbf{P}^{n_q}$ (par un résultat d'intersection du type du théorème 3.4) pour avoir des majorations de chaque $h_\alpha(V)$ et non seulement de certaines combinaisons linéaires correspondant à un plongement de Segre-Veronese. L'étude des hauteurs et distances semble particulièrement utile en indépendance algébrique : voir les critères de [**PPh2, Jab, Jad**] (cas projectif) et le chapitre 8, § 2 et 3, qui utilise la théorie multiprojective.

Une partie des résultats présentés ici se trouvent déjà dans [**Rem**, chap. III].

2. Hauteurs

2.1. Hauteurs des formes. On définit tout d'abord une mesure locale complexe de la manière suivante. Soit F une forme multihomogène non nulle de $\mathbf{C}[z^{(1)}, \ldots, z^{(p)}]$ où $z^{(i)}$ représente le groupe d'indéterminées $(z_0^{(i)}, \ldots, z_{l_i}^{(i)})$. On pose alors

$$\log M(F) = \int_{S_{l_1+1}(1) \times \cdots \times S_{l_p+1}(1)} \log |F| \sigma_{l_1+1} \wedge \ldots \wedge \sigma_{l_p+1}$$

$$+ \sum_{i=1}^{p} \deg_{z^{(i)}}(F) \sum_{j=1}^{l_i} \frac{1}{2j}$$

où pour $m \geq 1$ on note $S_m(1)$ la sphère unité de $\mathbf{C}^m$ et σ_m la mesure invariante de masse totale 1 sur $S_m(1)$. Remarquons que cette définition est stable par ajout de variables muettes : précisément si $y^{(1)}, \ldots, y^{(p')}$ sont d'autres groupes d'indéterminées $y^{(i)} = (y_0^{(i)}, \ldots, y_{l_i'}^{(i)})$ avec $p' \geq p$ et $l_i' \geq l_i$ si $1 \leq i \leq p$ alors en notant

$$\mathbf{C}[z^{(1)}, \ldots, z^{(p)}] \xrightarrow{\beta} \mathbf{C}[y^{(1)}, \ldots, y^{(p')}]$$
$$z_j^{(i)} \longmapsto y_j^{(i)}$$

on a $M(\beta(F)) = M(F)$ (voir [**PPh7**, III, p. 346] ou [**Jad**, p. 19–23]). Notons aussi qu'un calcul direct donne la formule suivante pour le cas linéaire ($a_0, \ldots, a_{l_1} \in \mathbf{C}$) :

$$M\left(\sum_{j=0}^{l_1} a_j z_j^{(i)}\right) = \sqrt{\sum_{j=0}^{l_1} |a_j|^2}.$$

Soit maintenant un corps de nombres K. Si v est une place de K on note K_v le complété de K pour v, $\mathbf{C}_v$ le complété d'une clôture algébrique de K_v et $\sigma_v \colon K \hookrightarrow \mathbf{C}_v$. Pour F une forme non nulle de $K[z^{(1)}, \ldots, z^{(p)}]$ on note $M_v(F)$ le maximum des valeurs absolues des coefficients de $\sigma_v(F)$ si v est une place finie et $M_v(F) = M(\sigma_v(F))$ sinon. On appelle alors *hauteur de la forme* F le nombre réel

$$h(F) = \sum_v \frac{[K_v : \mathbf{Q}_v]}{[K : \mathbf{Q}]} \log M_v(F)$$

où la somme est prise sur toutes les places v de K. On complète cette définition par $h(0) = 0$. On constate grâce à la formule du produit que $h(F) = 0$ si F est constante. La remarque sur l'ajout de variables muettes s'étend à h (car est évidente pour les places finies). Notons enfin que l'on peut remplacer K par une extension finie sans changer la valeur de $h(F)$. On a ainsi défini en fait une fonction

$$h \colon \bar{\mathbf{Q}}[z^{(1)}, \ldots, z^{(p)}] \longrightarrow \mathbf{R}.$$

Pour l'étude locale il est commode d'introduire une définition commune aux places finies et infinies. On considère donc comme précédemment des entiers naturels $l_1, \ldots, l_p$ puis une fonction $F \colon \mathbf{C}_v^{(l_1+1)+\cdots+(l_p+1)} \to \mathbf{R}_+$. On définit, sous certaines hypothèses, une "moyenne" de la fonction F notée $\mathrm{moy}_v(F)$ ou, s'il faut préciser,

$$\underset{z^{(1)},\ldots,z^{(p)}}{\mathrm{moy}_v} F(z^{(1)}, \ldots, z^{(p)}) \qquad \text{voire} \qquad \underset{\rho \colon z^{(1)},\ldots,z^{(p)}}{\mathrm{moy}_v} \rho(F)$$

si ρ désigne une spécialisation $\mathbf{C}_v[z^{(1)}, \ldots, z^{(p)}] \to \mathbf{C}_v$. Bien que principalement on supposera que F est la valeur absolue d'un polynôme multihomogène, il peut être pratique dans les calculs d'autoriser d'autres fonctions. Ainsi, lorsque v est infinie, on définira

$$\mathrm{moy}_v F = \exp \int_{S_{l_1+1}(1) \times \cdots \times S_{l_p+1}(1)} \log |F| \sigma_{l_1+1} \wedge \ldots \wedge \sigma_{l_p+1}$$

lorsque l'intégrale est définie (et $\mathrm{moy}_v 0 = 0$ s'il y a lieu).

En revanche, si v est finie, on se limitera à des fonctions F beaucoup plus particulières. On adopte ici un point de vue beaucoup plus restrictif, mais plus élémentaire, que celui de [**PPh7**, II, 1B]. Notre définition, différente, copie le cas des polynômes et ne constitue en fait qu'une commodité d'écriture. La condition sur F est la suivante : en notant $\mathcal{O}_v$ et k_v l'anneau des entiers et le corps résiduel de $\mathbf{C}_v$ puis $\varphi \colon \mathcal{O}_v^{(l_1+1)+\cdots+(l_p+1)} \to k_v^{(l_1+1)+\cdots+(l_p+1)}$ la surjection naturelle, on suppose que F est constante sur $\varphi^{-1}(U)$ où U est un ouvert Zariski-dense de $k_v^{(l_1+1)+\cdots+(l_p+1)}$. En ce cas on note $\mathrm{moy}_v(F)$ la valeur de F sur $\varphi^{-1}(U)$ qui est clairement indépendante du choix de U.

Le lien avec ce qui précède est que si $P \in \mathbf{C}_v[z^{(1)}, \ldots, z^{(p)}]$ est multihomogène alors $M_v(P) = \mathrm{moy}_v |P|_v \prod_{i=1}^p \exp\left(\deg_{z^{(i)}}(P) \sum_{j=1}^{l_i} \frac{1}{2j}\right)$ si v est infinie et $M_v(P) = \mathrm{moy}_v |P|_v$ sinon. Ceci est immédiat aux places infinies; pour une place finie v on peut écrire $P = z P_0$ avec $z \in \mathbf{C}_v$, $|z|_v = M_v(P)$ et $P_v \in \mathcal{O}_v[z^{(1)}, \ldots, z^{(p)}]$. Comme $M_v(P_0) = 1$ l'image $\overline{P_0}$ de P_0 dans $k_v[z^{(1)}, \ldots, z^{(p)}]$ est non nulle. Par suite, le complémentaire U du lieu des zéros de $\overline{P_0}$ est un ouvert dense. De plus si $(z^{(1)}, \ldots, z^{(p)}) \in \varphi^{-1}(U)$ alors $\overline{P_0(z^{(1)}, \ldots, z^{(p)})} \neq 0$ donc $|P_0(z^{(1)}, \ldots, z^{(p)})|_v = 1$ puis $|P(z^{(1)}, \ldots, z^{(p)})|_v = M_v(P)$ ce qui prouve bien $M_v(P) = \mathrm{moy}_v |P|_v$.

Avec ces définitions on prouve instantanément que si F, G sont telles que $\text{moy}_v(F)$ et $\text{moy}_v(G)$ sont définies alors il en va de même de $\text{moy}_v(FG)$ et (si F ne s'annule pas) de $\text{moy}_v(F^{-1})$ qui valent respectivement

$$\text{moy}_v(F)\text{moy}_v(G) \qquad \text{et} \qquad \text{moy}_v(F)^{-1}.$$

Par suite si P et Q sont deux polynômes multihomogènes on a aussi $M_v(PQ) = M_V(P)M_v(Q)$.

Enfin les propriétés plus fines que nous utiliserons sont données par le

LEMME 2.1. *Soient deux polynômes multihomogènes*

$$P \in \mathbf{C}_v[z^{(1)}, \ldots, z^{(p)}, z'^{(1)}, \ldots, z'^{(q)}]$$

et

$$Q \in \mathbf{C}_v[z^{(1)}, \ldots, z^{(p)}, z''^{(1)}, \ldots, z''^{(r)}] \ .$$

1. *On suppose que* $\deg_{z^{(i)}} P = \deg_{z^{(i)}} Q$ *pour* $1 \le i \le p$. *Alors*

$$\frac{M_v(P)}{M_v(Q)} = \underset{\rho:\ z^{(1)},\ldots,z^{(p)}}{\text{moy}_v} \frac{M_v(\rho(P))}{M_v(\rho(Q))}.$$

2. *On suppose que pour tout indice* $1 \le i \le p$ *et toute spécialisation*

$$\rho' : \mathbf{C}_v[z^{(1)}, \ldots, z^{(i-1)}, z^{(i+1)}, \ldots, z^{(p)}, z'^{(1)}, \ldots, z'^{(q)}] \to \mathbf{C}_v$$

le polynôme $\rho'(P) \in \mathbf{C}_v[z^{(i)}]$ *est un produit de formes linéaires. Alors pour tout* $\rho : \mathbf{C}_v[z^{(1)}, \ldots, z^{(p)}] \to \mathbf{C}_v$ *on a*

$$M_v(\rho(P)) \le M_v(P) \prod_{i=1}^{p} M_v \left(\sum_{j=0}^{l_i} \rho\left(z_j^{(i)}\right) z_j^{(i)} \right)^{\deg_{z^{(i)}} P} .$$

DÉMONSTRATION. 1. Établissons le cas des places finies (montrant de ce fait que moy_v a un sens). Eu égard à ce qui précède il existe un ouvert dense U de $k_v^{(l_1+1)+\cdots+(l_p+1)+(l'_1+1)+\cdots+(l'_q+1)}$ sur l'image réciproque duquel $|P|_v$ est égal à $M_v(P)$. Sa projection sur $k_v^{(l_1+1)+\cdots+(l_p+1)}$ est encore un ouvert dense. En raisonnant de même pour Q et en prenant l'intersection, on trouve un ouvert dense V de $k_v^{(l_1+1)+\cdots+(l_p+1)}$ tel que si $\rho \in \varphi^{-1}(V)$ alors il existe $\rho' : \mathbf{C}_v[z'^{(1)}, \ldots, z'^{(p)}] \to \mathbf{C}_v$ tel que $(\rho, \rho') \in \varphi^{-1}(U)$ et $\rho'' : \mathbf{C}_v[z''^{(1)}, \ldots, z''^{(r)}] \to \mathbf{C}_v$ pareillement. Pour ρ fixé l'ensemble des tels ρ' est l'image réciproque d'un ouvert dense de $k_v^{(l'_1+1)+\cdots+(l'_q+1)}$ ce qui montre $M_v(\rho(P)) = M_v(P)$; de même $M_v(\rho(Q)) = M_v(Q)$. Par suite si $\rho \in \varphi^{-1}(V)$ on a bien $\frac{M_v(\rho(P))}{M_v(\rho(Q))} = \frac{M_v(P)}{M_v(Q)}$ ce qui montre la formule.

Pour une place infinie l'égalité des degrés montre

$$\log \frac{M_v(P)}{M_v(Q)} = \sum_{i=1}^{q} \deg_{z'^{(i)}} P \sum_{j=1}^{l'_i} \frac{1}{2j} - \sum_{i=1}^{r} \deg_{z''^{(i)}} Q \sum_{j=1}^{l''_i} \frac{1}{2j}$$

$$+ \int_{S_{l_1+1}(1) \times \cdots \times S_{l_p+1}(1)} \left[\int_{S_{l'_1+1}(1) \times \cdots \times S_{l'_q+1}(1)} \log |\rho(P)|_v \sigma_{l'_1+1} \wedge \ldots \wedge \sigma_{l'_q+1} \right.$$

$$\left. - \int_{S_{l''_1+1}(1) \times \cdots \times S_{l''_r+1}(1)} \log |\rho(Q)|_v \sigma_{l''_1+1} \wedge \ldots \wedge \sigma_{l''_r+1} \right] \sigma_{l_1+1} \wedge \ldots \wedge \sigma_{l_p+1}$$

où ρ correspond à la spécialisation en $(z^{(1)}, \ldots, z^{(p)}) \in S_{l_1+1}(1) \times \cdots \times S_{l_p+1}(1)$. Comme $\rho(P)$ est un polynôme multihomogène de degrés partiels $\deg_{z'^{(i)}} \rho(P) = \deg_{z'^{(i)}} P$ (si $\rho(P) \neq 0$) on peut faire rentrer les termes de degré dans l'expression intégrée (puisque $\rho(P) = 0$ seulement sur un ensemble de mesure nulle); en procédant de même pour Q on obtient exactement la formule cherchée.

2. On procède par récurrence sur p. Si $p = 0$ l'inégalité est une tautologie (car $\rho(P) = P$). Si $p > 0$ toute spécialisation de P par un morphisme $\mathbf{C}_v[z^{(1)}, \ldots, z^{(p-1)}] \to \mathbf{C}_v$ vérifie encore la condition; en lui appliquant l'hypothèse de récurrence on constate sans peine qu'il suffit d'établir le cas $p = 1$. Par le résultat de (1) utilisé avec $\rho(P)$ et P on peut encore supposer $q = 0$. Dès lors on doit prouver

$$|P(a_0, \ldots, a_l)|_v \leq M_v(P) M_v \left(\sum_{j=0}^{l} a_j z_j \right)^{\deg P}$$

où P est un polynôme homogène de $\mathbf{C}_v[z_0, \ldots, z_p]$ qui est un produit de formes linéaires et $a_0, \ldots, a_l \in \mathbf{C}_v$. Par multiplicativité on peut supposer $P = \sum_{j=0}^{l} b_j z_j$. *In fine* le résultat suit de

$$\left| \sum_{j=0}^{l} a_j b_j \right|_v \leq \max_{0 \leq j \leq l} |b_j|_v \max_{0 \leq j \leq l} |a_j|_v$$

si v est finie (immédiat) et

$$\left| \sum_{j=0}^{l} a_j b_j \right|_v \leq \sqrt{\sum_{j=0}^{l} |b_j|_v^2} \sqrt{\sum_{j=0}^{l} |a_j|_v^2}$$

si v est infinie (Cauchy-Schwarz). $\qquad\square$

2.2. Application aux formes résultantes. On fixe un corps de nombres K. On considère à nouveau l'élimination multihomogène dans $K[X]$. Soient donc $r \geq 0$ et $d \in (\mathbf{N}^q \setminus 0)^r$. Notre travail de définition des *hauteurs multiprojectives* passe par le calcul des hauteurs des formes résultantes qui sont des éléments multihomogènes de $K[d] = K[u^{(1)}, \ldots, u^{(r)}]$. Cependant nous calculerons les hauteurs après application du morphisme de K-algèbres

$$\alpha_d \colon K[d] \longrightarrow \bar{K}[d]$$
$$u_{\mathbf{m}}^{(l)} \longmapsto \binom{d_l}{\mathbf{m}}^{1/2} u_{\mathbf{m}}^{(l)}.$$

Notons toutefois que si $d_{l,i} \leq 1$ pour tous $1 \leq l \leq r$ et $1 \leq i \leq q$ le morphisme α_d est l'inclusion. On associe donc à d et à un idéal I le nombre $h(\alpha_d(f))$ où f est une forme résultante d'indice d de I. Ce nombre est nul dans les cas inintéressants où f est constante. Le résultat de base suivant montre son bon comportement pour le changement d'indices.

THÉORÈME 2.2. *Soient K un corps de nombres, $r \in \mathbf{N}$, I un idéal multihomogène de $K[X]$ de hauteur au moins $n - r + 1$ et $d \in (\mathbf{N}^q \setminus 0)^r$. On note $d^{(i)} = (\varepsilon_i, d_2, \ldots, d_r)$ puis f et f_i des formes résultantes de I d'indices respectifs d*

et $d^{(i)}$. Alors on a

$$h(\alpha_d(f)) = \sum_{i=1}^{q} d_{1,i} h(\alpha_{d^{(i)}}(f_i)).$$

DÉMONSTRATION. D'après la proposition 3.5 du chapitre 5 il existe $\lambda \in K^\times$ tel que $\omega(f) = \lambda \prod_{i=1}^{q} f_i^{d_{1,i}}$. On conserve les notations ω, $L_1, \ldots, L_q$ et $\tilde{d}$ introduites avant cette proposition. En utilisant la possibilité de rajouter des variables muettes dans le calcul de h et la compatibilité évidente entre $\alpha_{\tilde{d}}$, α_d et $\alpha_{d^{(i)}}$ on a $h(\alpha_{\tilde{d}}(\omega(f))) = \sum_{i=1}^{q} d_{1,i} h(\alpha_{d^{(i)}}(f_i))$. Il nous suffira de prouver que pour toute place v (d'une extension de corps contenant toutes les racines d'entiers qui interviennent dans les morphismes α) on a $M_v(\alpha_d(f)) = M_v(\alpha_{\tilde{d}}(\omega(f)))$. Noter que l'on n'a pas l'égalité $M_v(\alpha_d(F)) = M_v(\alpha_{\tilde{d}}(\omega(F)))$ pour toute forme F de $K[d]$. Le fait que f soit une forme résultante est indispensable : cela servira cependant uniquement à travers le résultat de la proposition 2.16 du chapitre 5. Dans le calcul qui suit on note encore K l'extension finie citée ci-dessus et l'on considère α_d et $\alpha_{\tilde{d}}$ comme des automorphismes de $K[d]$ et $K[\tilde{d}]$.

Fixons une place v ainsi qu'un morphisme d'anneaux

$$\rho \colon K[(d_2, \ldots, d_r)] \to \mathbf{C}_v$$

prolongeant σ_v tel que $\rho(f) \neq 0$. On note aussi ρ les extensions de ce morphisme à $K[d]$ et $K[\tilde{d}]$. Définissons ensuite un morphisme semblable $\rho' \colon K[(d_2, \ldots, d_r)] \to \mathbf{C}_v$ par $\rho'(u_{\mathfrak{m}}^{(l)}) = \left(\dfrac{d_l}{\mathfrak{m}}\right)^{-1/2} \rho(u_{\mathfrak{m}}^{(l)})$. On constate alors (avec la même convention pour les extensions) que $\rho' \circ \alpha_d = \alpha_{(d_1)} \circ \rho$ et $\rho' \circ \alpha_{\tilde{d}} = \rho$. Il suit de la proposition 2.16 du chapitre 5 (car f est produit de puissances de formes élim$_d(\mathfrak{p})$) qu'il existe des éléments $z_1, \ldots, z_t$ de $(\mathbf{C}_v^{n_1+1} \setminus 0) \times \cdots \times (\mathbf{C}_v^{n_q+1} \setminus 0)$ et $\lambda' \in \mathbf{C}_v$ de telle sorte que

$$\rho(f) = \lambda' \prod_{i=1}^{t} U_1(z_i).$$

On en déduit donc

$$\rho'(\alpha_d(f)) = \alpha_{(d_1)}(\rho(f)) = \lambda' \prod_{i=1}^{t} \alpha_{(d_1)}(U_1)(z_i)$$

et

$$\rho'(\alpha_{\tilde{d}}(\omega(f))) = \rho(\omega(f)) = \omega(\rho(f))$$
$$= \lambda' \prod_{i=1}^{t} \omega(U_1)(z_i) = \lambda' \prod_{i=1}^{t} \prod_{j=1}^{q} L_j(z_i)^{d_{1,j}} .$$

On utilise à présent le lemme 2.1 pour écrire

$$\frac{M_v(\alpha_d(f))}{M_v(\alpha_{\tilde{d}}(\omega(f)))} = \operatorname*{moy}_{\rho' \colon u^{(2)}, \ldots, u^{(r)}} \frac{M_v(\rho'(\alpha_d(f)))}{M_v(\rho'(\alpha_{\tilde{d}}(\omega(f))))}.$$

Au vu des calculs précédents, en reconstruisant ρ à l'aide de ρ', il nous suffit de montrer que (si $z \in (\mathbf{C}_v^{n_1+1} \setminus 0) \times \cdots \times (\mathbf{C}_v^{n_q+1} \setminus 0)$)

$$M_v(\alpha_{(d_1)}(U_1)(z)) = M_v(\omega(U_1)(z))$$

(en effet ceci montrera que $M_v(\alpha_d(f)) = M_v(\alpha_{\tilde{d}}(\omega(f))) \operatorname{moy}_v(1)$).

Distinguons maintenant selon que v est finie ou infinie. Dans le premier cas l'égalité à établir est explicitement

$$\max_{\mathfrak{m}\in\mathcal{M}_{d_1}} \left| \binom{d_1}{\mathfrak{m}}^{1/2} \mathfrak{m}(z) \right|_v = \max_{\mathfrak{m}\in\mathcal{M}_{d_1}} \left| \binom{d_1}{\mathfrak{m}} \mathfrak{m}(z) \right|_v ;$$

cette dernière formule est évidente car $\binom{d_1}{\mathfrak{m}} \in \mathbf{Z}$ qui implique $\left| \binom{d_1}{\mathfrak{m}}^{1/2} \right|_v \le 1$ donne la majoration $\ge$ et la minoration s'obtient en notant que le maximum est atteint avec un monôme de la forme $\mathfrak{m} = X_{j_1}^{(1)^{d_{1,1}}} \ldots X_{j_q}^{(q)^{d_{1,q}}}$ pour lequel $\binom{d_1}{\mathfrak{m}} = 1$. Traitons à présent le cas où v est infinie : on a par multiplicativité

$$M(\omega(U_1)(z)) = \prod_{i=1}^{q} M(L_i(z))^{d_{1,i}}$$

puis par la formule pour les formes linéaires

$$M(\alpha_{(d_1)}(U_1)(z))^2 = \sum_{\mathfrak{m}\in\mathcal{M}_{d_1}} \binom{d_1}{\mathfrak{m}} |\mathfrak{m}(z)|^2 \quad \text{et}$$

$$M(L_i(z))^2 = \sum_{\mathfrak{m}\in\mathcal{M}_{\varepsilon_i}} |\mathfrak{m}(z)|^2.$$

Par conséquent une simple application de la formule multinomiale donne le résultat et ceci termine la démonstration. $\qquad\qquad\square$

Des applications répétées du théorème montrent que la hauteur de toute forme résultante s'exprime comme combinaison linéaire à coefficients entiers en les hauteurs des formes résultantes d'indices du type

$$(\varepsilon_1,\ldots,\varepsilon_1,\ldots,\varepsilon_q,\ldots,\varepsilon_q).$$

Ceci explique les définitions que nous allons voir maintenant pour les sous-schémas fermés.

2.3. Application géométrique. Lorsque V est un sous-schéma fermé de $\mathbf{P}_K^{n_1} \times \cdots \times \mathbf{P}_K^{n_q}$ on lui associe par la proposition 2.17 du chapitre 5 un idéal multihomogène et multisaturé I de $K[X]$. On appellera aussi forme éliminante (*resp.* résultante) d'indice d de V (ou du plongement $V \hookrightarrow \mathbf{P}_K^{n_1} \times \cdots \times \mathbf{P}_K^{n_q}$) une forme éliminante (*resp.* résultante) d'indice d de l'idéal I. Soit maintenant $\alpha \in \mathbf{N}^q$ tel que $\alpha_1 + \cdots + \alpha_q = \dim V + 1$. On note $d \in (\mathbf{N}^q)^{\alpha_1+\cdots+\alpha_q}$ l'élément $(\varepsilon_1,\ldots,\varepsilon_1,\ldots,\varepsilon_q,\ldots,\varepsilon_q)$ constitué de α_1 termes ε_1 suivis de α_2 termes ε_2 et ainsi de suite jusqu'à α_q termes ε_q. On définit

$$h_\alpha(V)$$

comme la hauteur d'une forme résultante d'indice d de V. Ce nombre ne dépend pas du choix de la forme par la formule du produit. Lorsqu'il y a lieu on note plus précisément $h_{\psi,\alpha}(V) = h_\alpha(V)$ où $\psi\colon V \hookrightarrow \mathbf{P}_K^{n_1} \times \cdots \times \mathbf{P}_K^{n_q}$ est l'immersion fermée de V dans $\mathbf{P}_K^{n_1} \times \cdots \times \mathbf{P}_K^{n_q}$. Lorsque V est intègre, d'après le corollaire 2.15 du chapitre 5, le nombre $h_\alpha(V)$ ne peut être non nul que lorsque $\alpha_i \le \dim \pi_i(V) + 1$ pour tout i.

De manière analogue on définit les degrés de V à partir de I. Pour α tel que $\alpha_1 + \cdots + \alpha_q = \dim V$ on note

$$d_\alpha(V) = \deg(I)_\alpha.$$

Notons que si l'on associe à nouveau à α un indice $d \in (\mathbf{N}^q)^{\alpha_1 + \cdots + \alpha_q}$ de la forme $(\varepsilon_1, \ldots, \varepsilon_1, \ldots, \varepsilon_q, \ldots, \varepsilon_q)$ alors la proposition 3.4 du chapitre 5 montre que $d_\alpha(V)$ est le degré en $u^{(1)}$ de toute forme $\mathrm{rés}_{d'}(I)$ où d' est un indice de la forme $(\delta, d) \in (\mathbf{N}^q)^{\alpha_1 + \cdots + \alpha_q + 1}$ avec $\delta \in \mathbf{N}^q \setminus 0$.

Examinons à présent comment les plongements de Veronese et de Segre permettent de relier entr'elles différentes formes résultantes. Soient donc $\psi \colon V \hookrightarrow \mathbf{P}_K^{n_1} \times \cdots \times \mathbf{P}_K^{n_q}$ une immersion fermée et $\delta \in (\mathbf{N} \setminus 0)^q$. On note $N_i = \binom{n_i + \delta_i}{\delta_i} - 1$ pour tout i puis $\chi \colon \mathbf{P}_K^{n_1} \times \cdots \times \mathbf{P}_K^{n_q} \hookrightarrow \mathbf{P}_K^{N_1} \times \cdots \times \mathbf{P}_K^{N_q}$ le produit des *plongements de Veronese* associés à $\delta_1, \ldots, \delta_q$. Enfin pour $r \in \mathbf{N}$ on associe à $d \in (\mathbf{N}^q)^r$ dont les composantes vérifient $d_{i,j} \leq 1$ l'élément $d' = \delta.d$ (multiplication terme à terme). Alors on peut identifier $K[d']$ et $K[d]$ si la première écriture se réfère aux variables $X^{(i)}$ de $\mathbf{P}_K^{n_1} \times \cdots \times \mathbf{P}_K^{n_q}$ et la seconde à celles de $\mathbf{P}_K^{N_1} \times \cdots \times \mathbf{P}_K^{N_q}$ disons $Z^{(i)}$. En effet ces dernières s'identifient aux monômes en $X^{(i)}$ de degré δ_i pour i donné et donc, lorsque l'on forme les monômes de multidegré d_l en Z, on trouve les monômes de multidegré $\delta.d_l$ en X. L'intérêt de cette identification est qu'une forme résultante d'indice d' de ψ devient une forme résultante de $\chi \circ \psi$ d'indice d. En effet ceci provient du fait qu'en indiçant $Z^{(i)}$ par les $\alpha \in \mathbf{N}^{n_i+1}$ tels que $|\alpha| = \delta_i$ l'idéal I' de $\chi \circ \psi$ se déduit de celui de ψ (noté I) comme son image réciproque par le morphisme de K-algèbres

$$\begin{aligned}\xi \colon K[Z] &\longrightarrow K[X] \\ Z_\alpha^{(i)} &\longmapsto (X^{(i)})^\alpha.\end{aligned}$$

Si l'on étend ξ par l'identification précédente en $K[d][Z] \to K[d'][X]$ on a $\xi(U_l) = U_l$ et $\xi(\mathcal{M}_k) = \mathcal{M}_{k\delta}$ où $k\delta = (k_1\delta_1, \ldots, k_q\delta_q)$. On en déduit $I'[d] = \xi^{-1}I[d']$: l'inclusion $\xi(I'[d]) \subset I[d']$ est évidente et si $\xi(f)$ s'écrit comme combinaison linéaire d'un élément de I et des U_l on peut faire en sorte, pour raisons de degré, que les coefficients des U_l soient dans l'image de ξ et ainsi $f \in I'[d]$. Par conséquent ξ induit une injection

$$K[d][Z]/I'[d] \hookrightarrow K[d'][X]/I[d'].$$

D'après la formule $\xi(\mathcal{M}_k) = \mathcal{M}_{k\delta}$, ceci donne pour tout $k \in \mathbf{N}^q$ un isomorphisme

$$(K[d][Z]/I'[d])_k \xrightarrow{\sim} (K[d'][X]/I[d'])_{k\delta}$$

et on obtient bien $\mathrm{rés}_d(I') = \mathrm{rés}_{d'}(I)$ car si $k \geq k_0$ on a aussi $k\delta \geq k_0$ puisque $\delta \in (\mathbf{N} \setminus 0)^q$.

Remarquons que si l'on veut interpréter non la forme résultante d'indice d' de ψ mais l'image d'icelle par $\alpha_{d'}$ (qui se prête mieux au calcul des hauteurs) on l'obtient par le même raisonnement comme forme résultante de $\chi' \circ \psi$ où χ' est défini à partir de *plongements de Veronese remodelés* (voir [**Jad**, p. 14]). Ceci signifie que l'on modifie les plongements de Veronese par des racines de coefficients multinomiaux (il faut faire une extension de corps) : sur les idéaux χ' correspond à

$$\begin{aligned}\xi \colon K[Z] &\longrightarrow K[X] \\ Z_\alpha^{(i)} &\longmapsto \binom{\delta_i}{\alpha}^{1/2} (X^{(i)})^\alpha.\end{aligned}$$

On procède de même pour le *plongement de Segre* $s\colon \mathbf{P}_K^{n_1} \times \cdots \times \mathbf{P}_K^{n_q} \hookrightarrow \mathbf{P}_K^N$ où $N = (n_1 + 1) \ldots (n_q + 1) - 1$. Ici encore $\psi\colon V \hookrightarrow \mathbf{P}_K^{n_1} \times \cdots \times \mathbf{P}_K^{n_q}$ est une immersion fermée. On note $d = (\varepsilon, \ldots, \varepsilon) \in (\mathbf{N}^q)^r$ et $d' = (1, \ldots, 1) \in \mathbf{N}^r$. Alors $K[d]$ s'identifie avec $K[d']$ lorsque d correspond aux variables $X_j^{(i)}$ de $\mathbf{P}_K^{n_1} \times \cdots \times \mathbf{P}_K^{n_q}$ et d' à celles de $\mathbf{P}_K^N$ notées Z_j avec $j \in \mathbf{N}^q$, $0 \le j_i \le n_i$. On identifie en effet simplement $u_{X_{j_1}^{(1)} \ldots X_{j_q}^{(q)}}^{(l)} \in K[d]$ à $u_{Z_j}^{(l)} \in K[d']$. Noter que $K[d']$ fait référence à la théorie de l'élimination homogène qui est traitée dans [**PPh2**] mais qui s'obtient aussi comme cas particulier de la théorie multihomogène $(q = 1)$. Modulo cette identification une forme résultante d'indice d de ψ est une forme résultante d'indice d' de $s \circ \psi$. Ce fait s'obtient de manière analogue au cas du plongement de Veronese en considérant l'application déjà introduite

$$\theta\colon K[Z] \longrightarrow K[X]$$
$$Z_j \longmapsto X_{j_1}^{(1)} \ldots X_{j_q}^{(q)}.$$

On peut rassembler une partie de ces faits de la manière suivante. Fixons $\delta \in (\mathbf{N} \setminus 0)^q$. On désignera par v_i $(1 \le i \le q)$ le plongement de Veronese remodelé de $\mathbf{P}_K^{n_i}$ associé à δ_i de but $\mathbf{P}_K^{N_i - 1}$ avec $N_i = \binom{n_i + \delta_i}{n_i}$ et enfin par s le Segre

$$\mathbf{P}_K^{N_1 - 1} \times \cdots \times \mathbf{P}_K^{N_q - 1} \hookrightarrow \mathbf{P}_K^{N_1 \ldots N_q - 1}.$$

Les considérations précédentes montrent que, pour un plongement $\varphi\colon V \hookrightarrow \mathbf{P}_K^{n_1} \times \cdots \times \mathbf{P}_K^{n_q}$, la hauteur du plongement $s \circ (v_1 \times \cdots \times v_q) \circ \varphi$ est égale à la hauteur de l'image par α_d d'une forme résultante f d'indice d de φ où $d = (\delta, \ldots, \delta) \in (\mathbf{N}^q)^r$ avec $r = \dim V + 1$. Par le théorème 2.2 on en déduit

$$h_{s \circ (v_1 \times \cdots \times v_q) \circ \varphi}(V) = \sum_{|\alpha| = r} \binom{r}{\alpha} h_{\varphi, \alpha}(V) \delta_1^{\alpha_1} \ldots \delta_q^{\alpha_q}.$$

En considérant non plus la hauteur mais le degré des formes résultantes, on trouve que $d_{s \circ (v_1 \times \cdots \times v_q) \circ \varphi}(V)$ est égal au degré en $u^{(1)}$ de la forme f précédente. Par la proposition 3.4 du chapitre 5 ceci vaut exactement $\deg(I) * \delta * \cdots * \delta$ où δ est répété $r - 1$ fois. Après calcul on trouve

$$d_{s \circ (v_1 \times \cdots \times v_q) \circ \varphi}(V) = \sum_{|\alpha| = r-1} \binom{r-1}{\alpha} d_{\varphi, \alpha}(V) \delta_1^{\alpha_1} \ldots \delta_q^{\alpha_q}.$$

Nous allons maintenant donner un exemple de calcul de hauteurs multiprojectives. Nous commençons par déterminer les formes résultantes attachées à un produit.

LEMME 2.3. *Soit V_i un sous-schéma fermé intègre de $\mathbf{P}_K^{n_i}$ pour chaque $1 \le i \le q$. A $\alpha \in \mathbf{N}^q$ tel que $\alpha_1 + \cdots + \alpha_q = a_1 + \cdots + a_q + 1$ (avec $a_i = \dim V_i$) on attache comme précédemment un indice d de la forme $(\varepsilon_1, \ldots, \varepsilon_1, \ldots, \varepsilon_q, \ldots, \varepsilon_q)$. Alors une forme résultante d'indice d de $V_1 \times \cdots \times V_q \hookrightarrow \mathbf{P}_K^{n_1} \times \cdots \times \mathbf{P}_K^{n_q}$ est*

- *$f_{V_i}^{\prod_{j \ne i} d(V_j)}$ si $\alpha_j = a_j + \delta_{i,j}$ pour tout j (et f_{V_i} est une forme résultante de V_i d'indice $(1, \ldots, 1) \in \mathbf{N}^{a_i + 1}$);*
- *1 dans tous les autres cas.*

DÉMONSTRATION. On note $d_{(i)} = (1, \ldots, 1) \in \mathbf{N}^{\alpha_i}$. D'après la définition de d on peut faire l'identification (entre anneaux) suivante :

$$K[d] \simeq \bigotimes_{i=1}^{q} K[d_{(i)}].$$

C'est au travers de cet isomorphisme que l'on voit f_{V_i} comme élément de $K[d]$ lorsque $\alpha_i = a_i + 1$.

Si I_i est l'idéal homogène de $K[X^{(i)}]$ correspondant à V_i alors $I = (I_1, \ldots, I_q) \subset K[X]$ est l'idéal multihomogène associé à $V = V_1 \times \cdots \times V_q$. On déduit de l'isomorphisme

$$K[X]/I \simeq \bigotimes_{i=1}^{q} K[X^{(i)}]/I_i$$

que le polynôme de Hilbert-Samuel du module multigradué $K[X]/I$ est le produit des polynômes de Hilbert-Samuel des modules gradués $K[X^{(i)}]/I_i$. En particulier $H_{K[X]/I}$ a un seul terme dominant et, par suite, tous les degrés de I (ou de V) sont nuls à l'exception de

$$d_{(a_1, \ldots, a_q)}(V) = \prod_{i=1}^{q} d(V_i).$$

On en déduit encore que la seule façon d'obtenir un nombre non nul en appliquant $\dim V$ fois à $\deg(I)$ les opérateurs $* \, \varepsilon_i$ (pour différents i) est que ε_i apparaisse exactement a_i fois. En vertu de la proposition 3.4 du chapitre 5 la forme $\mathrm{rés}_d(I)$ ne peut être différente de 1 que si d contient une telle famille : ceci entraîne que α est de la forme $(a_j + \delta_{i,j})_{1 \leq j \leq q}$ pour un certain i. De plus, pour la même raison, dans ce cas $\mathrm{rés}_d(I)$ est de degré nul en $u_{\mathrm{m}}^{(l)}$ si $d_l \neq \varepsilon_i$. Ceci prouve que $\mathrm{rés}_d(I)$ et $f_{V_i}^{\prod_{j \neq i} d(V_j)}$ ont mêmes degrés : $\prod_{j=1}^{q} d(V_j)$ en les variables correspondant à f_{V_i} (soit les $u_{\mathrm{m}}^{(l)}$ avec $d_l = \varepsilon_i$) et 0 en toutes les autres. Pour conclure à leur proportionnalité, sachant que f_{V_i} est irréductible, il suffit de voir que $\mathrm{élim}_d(I)$ est f_{V_i}. On utilise le théorème de l'élimination en identifiant $\rho : K[d] \to \bar{K}$ à $\bigotimes_{i=1}^{q} \rho_i$ où $\rho_i : K[d_{(i)}] \to \bar{K}$ (voir ci-dessus). Alors

$$\rho(\mathfrak{E}_d(I)) = 0 \iff (I, \rho(U_1), \ldots, \rho(U_r)) \text{ a un zéro non trivial.}$$

Or cet idéal est engendré par les idéaux homogènes

$$(I_i, \rho_i(U_1^{(i)}), \ldots, \rho_i(U_{\alpha_i}^{(i)}))$$

et un zéro multihomogène "non trivial" est un élément de $\prod_{i=1}^{q} \bar{K}^{n_i+1} \setminus 0$ donc correspond à q zéros homogènes non triviaux de chaque $\bar{K}^{n_i+1} \setminus 0$. Ainsi

$$\rho(\mathfrak{E}_d(I)) = 0 \iff \rho_i(\mathfrak{E}_{d_{(i)}}(I_i)) = 0 \text{ pour } 1 \leq i \leq q.$$

Si maintenant d est associé à $\alpha = (a_j + \delta_{i,j})_{1 \leq j \leq q}$ comme ci-dessus on a $\mathfrak{E}_{d_j}(I_j) = 0$ pour $j \neq i$ en vertu du théorème 2.13 du chapitre 5 (car la taille de d_j est $a_j < \dim V_j + 1$). Enfin pour $j = i$ on trouve que $\mathfrak{E}_{d_{(i)}}$ est principal engendré par f_{V_i}. On en déduit que $\sqrt{\mathfrak{E}_d(I)} = (f_{V_i})$. La conclusion découlera alors du fait que $\mathfrak{E}_d(I)$ est radiciel (c'est-à-dire intersection de premiers). D'après le lemme 2.4 du chapitre 5 il suffit de vérifier que I est radiciel : cela signifie que V est réduit ce qui est le cas puisque (à cause de la caractéristique 0) les V_i sont géométriquement réduits. $\square$

On en déduit

COROLLAIRE 2.4. *Soit V_i un sous-schéma fermé intègre de $\mathbf{P}_K^{n_i}$ pour chaque $1 \leq i \leq q$. Les hauteurs multiprojectives de $V_1 \times \cdots \times V_q \hookrightarrow \mathbf{P}_K^{n_1} \times \cdots \times \mathbf{P}_K^{n_q}$ sont nulles à l'exception de*

$$h_{a+\varepsilon_i}(V_1 \times \cdots \times V_q) = \left(\prod_{j \neq i} d(V_j) \right) h(V_i)$$

où $a \in \mathbf{N}^q$ est défini par $a_i = \dim V_i$; de même le seul degré non nul est

$$d_a(V_1 \times \cdots \times V_q) = \prod_{i=1}^{q} d(V_i).$$

DÉMONSTRATION. Conséquence immédiate du lemme (en utilisant encore pour le calcul de la hauteur la possibilité de rajouter des variables muettes).   □

Remarquons que l'on aurait pu aussi démontrer ce corollaire (comme cela était fait dans [**Rem**, lemme III.3.1]) en s'appuyant sur le calcul de [**PPh7**, III, prop. 1] qui explicite la hauteur d'un produit à travers le plongement de Segre (à deux facteurs) puis en identifiant les coefficients de polynômes de la forme de ceux obtenus avant le lemme. Toutefois la proposition citée utilise un calcul qui n'est autre qu'un cas particulier de notre lemme.

2.4. Lien avec la géométrie d'Arakelov. Nous suivons [**BGS**][2] (voir aussi [**Gueb**][3]). Lorsque K est un corps de nombres et S le spectre de son anneau d'entiers, on considère le schéma $\mathbf{P}_S = \mathbf{P}_S^{n_1} \times \cdots \times \mathbf{P}_S^{n_q}$. C'est une variété arithmétique projective au sens de [**BGS**] (*loc. cit.*), 2.1.1 : un schéma plat et projectif sur S dont la fibre générique (ici $\mathbf{P}$) est régulière. La dimension de $\mathbf{P}_S$ est $n + 1 = n_1 + \cdots + n_q + 1$ (tandis que sa dimension relative sur S est égale à n, dimension de sa fibre générique). On dispose sur $\mathbf{P}_S$ de q faisceaux inversibles naturels $\mathcal{L}_i = p_i^* \mathcal{O}(1)$ où $p_i : \mathbf{P}_S \to \mathbf{P}_S^{n_i}$ est la i-ème projection. En munissant $\mathcal{O}(1)$ de sa structure hermitienne standard (métrique de Fubini-Study sur $\mathbf{P}^{n_i}(\mathbf{C})$) on fait de $\mathcal{L}_i$ un faisceau inversible hermitien $\overline{\mathcal{L}_i}$.

Le paragraphe 3.1.3 de [**BGS**] (*loc. cit.*) attache alors à tout cycle Z de $\mathbf{P}_S$ des multihauteurs par la formule

$$h_{\overline{\mathcal{L}_1}, \ldots, \overline{\mathcal{L}_q}}^{\alpha_1, \ldots, \alpha_q}(Z) = \widehat{\deg}(\widehat{c_1}(\overline{\mathcal{L}_1})^{\alpha_1} \cdots \widehat{c_1}(\overline{\mathcal{L}_q})^{\alpha_q} \mid Z)$$

où $\alpha \in \mathbf{N}^q$ donc $\alpha_1 + \cdots + \alpha_q = \dim Z$.

Pour établir une comparaison, soit V un sous-schéma fermé intègre de $\mathbf{P}$ de dimension $r - 1$. L'image fermée de $V \to \mathbf{P}_S$ est un sous-schéma fermé intègre de $\mathbf{P}_S$ de dimension r qui correspond à un cycle noté Z.

PROPOSITION 2.5. *Avec les notations précédentes, si $\alpha_1 + \cdots + \alpha_q = r$, on a*

$$h_{\overline{\mathcal{L}_1}, \ldots, \overline{\mathcal{L}_q}}^{\alpha_1, \ldots, \alpha_q}(Z) = h_\alpha(V).$$

DÉMONSTRATION. Il suffit de prouver que pour tous $(\delta_1, \ldots, \delta_q) \in (\mathbf{N} \setminus 0)^q$ on a

$$\sum_{|\alpha| = r} \binom{r}{\alpha} h_{\overline{\mathcal{L}_1}, \ldots, \overline{\mathcal{L}_q}}^{\alpha_1, \ldots, \alpha_q}(Z) \delta_1^{\alpha_1} \cdots \delta_q^{\alpha_q} = \sum_{|\alpha| = r} \binom{r}{\alpha} h_\alpha(V) \delta_1^{\alpha_1} \cdots \delta_q^{\alpha_q}.$$

[2][**BGS**] J.-B. Bost, H. Gillet, C. Soulé. Heights of projective varieties and positive Green forms, *J. Amer. Math. Soc.* 7, (1994), 903–1027.

[3][**Gueb**] W. Gübler. Höhentheorie, *Math. Ann.* 298, (1994), 427–455.

Or le membre de gauche vaut par multilinéarité (voir [**BGS**] (*loc. cit.*), 2.3.1)

$$\widehat{\deg}\left(\sum_{|\alpha|=r}\binom{r}{\alpha}[\delta_1\widehat{c_1}(\overline{\mathcal{L}_1})]^{\alpha_1}\cdots[\delta_q\widehat{c_1}(\overline{\mathcal{L}_q})]^{\alpha_q}\;\middle|\;Z\right)$$

soit

$$\widehat{\deg}\left(\widehat{c_1}(\overline{\mathcal{L}_1}^{\otimes\delta_1}\otimes\cdots\otimes\overline{\mathcal{L}_q}^{\otimes\delta_q})^r\;\middle|\;Z\right)$$

tandis que l'on a vu que le membre de droite valait $h_s(V)$ où

$$s\colon \mathbf{P}\hookrightarrow\mathbf{P}^{\binom{n_1+\delta_1}{n_1}\cdots\binom{n_q+\delta_q}{n_q}-1}$$

est le plongement de Segre-Veronese (remodelé) associé à $\delta_1,\ldots,\delta_q$. Pour conclure il suffit de noter (comportement des métriques de Fubini-Study par les plongements de Segre et de Veronese remodelés) que

$$\overline{\mathcal{L}_1}^{\otimes\delta_1}\otimes\cdots\otimes\overline{\mathcal{L}_q}^{\otimes\delta_q} = s^*\overline{\mathcal{O}(1)}.$$

Le résultat suit alors du fait que la hauteur $h_s(V)$ ("de Philippon") est égale à $[K:\mathbf{Q}]^{-1}$ fois la hauteur "de Faltings" :

$$\widehat{\deg}(\widehat{c_1}(s^*\overline{\mathcal{O}(1)})\mid Z)$$

(voir [**BGS**] (*loc. cit.*, p. 105), th. 4.3.2, [**Sou**, th. 3], [**PPh7**] : la constante de [**Sou**] disparaît comme il est dit dans [**PPh7**, III, page 346] en raison de la modification par rapport à [**PPh2**]).   □

Signalons que la hauteur définie comme ci-dessus (avec $\widehat{c_1}$) est dite "de Faltings" dans [**BGS**] (*loc. cit.*, p. 105) par opposition à la hauteur "projective" (voir [**BGS**] (*loc. cit.*, p. 105), 4.1.1) qui se définit en utilisant des classes de Chern $\widehat{c_p}$ d'indice supérieur. Cette dernière semble plus normalisée au sens où, positive, elle s'annule sur les sous-espaces linéaires "standards". Toutefois elle ne diffère de la hauteur de Faltings que par un multiple explicite du degré (voir [**BGS**] (*loc. cit.*, p. 105) ou [**PPh7**, III, p. 353]).

3. Une formule d'intersection

Le but de ce paragraphe est de relier la hauteur de $V.Z$ à celle de V lorsque V est un sous-schéma fermé intègre de $\mathbf{P}_K^{n_1}\times\cdots\times\mathbf{P}_K^{n_q}$ et Z une "hypersurface" de $\mathbf{P}_K^{n_1}\times\cdots\times\mathbf{P}_K^{n_q}$ (l'idéal correspondant de $K[X]$ est principal) ne contenant pas V. On va ainsi généraliser au cas multiprojectif le résultat de [**PPh7**, III, prop. 4]. L'outil algébrique est la proposition de spécialisation 3.6 du chapitre 5. Pour le passage aux hauteurs on utilise [**PPh7**, II, par. 1].

3.1. Notations. Soient donc $V\hookrightarrow\mathbf{P}_K^{n_1}\times\cdots\times\mathbf{P}_K^{n_q}$ intègre (K est un corps de nombres) et $\alpha\in\mathbf{N}^q$ tel que $\alpha_1+\cdots+\alpha_q=\dim V$ ainsi que Z une hypersurface définie par un polynôme $P\in K[X]$ de multidegré $\delta\in\mathbf{N}^q\setminus 0$. On considère alors l'indice $d=(\delta,\varepsilon_1,\ldots,\varepsilon_1,\ldots,\varepsilon_q,\ldots,\varepsilon_q)\in(\mathbf{N}^q)^{\dim V+1}$ constitué de δ suivi de α_1 termes ε_1 suivis de α_2 termes ε_2 et ainsi de suite jusqu'à α_q termes ε_q et f_V une forme résultante d'indice d de V. On définit aussi d' comme l'indice de $(\mathbf{N}^q)^{\dim V}$ associé à α c'est-à-dire tel que $d'_l=d_{l+1}$ pour $1\le l\le\dim V$ et on introduit le morphisme de spécialisation $\rho\colon K[d]\to K[d']$ caractérisé par $\rho(U_1)=P$. Dans le même esprit que [**PPh7**, III] on utilise une hauteur de P (ou de Z) relative à V. Nous posons pour cela :

$$h_{V,\alpha}(P) = \sum_v \frac{[K_v : \mathbf{Q}_v]}{[K : \mathbf{Q}]} \mu_{V,v,\alpha}(P)$$

où la somme est prise sur toutes les places de K et où les $\mu_{V,v,\alpha}$ sont définies à présent. Si $d_\alpha(V) = 0$ on pose $\mu_{V,v,\alpha}(P) = 0$ pour toute place v. Dans le cas contraire, on pose si v est une place finie

$$\mu_{V,v,\alpha}(P) = \frac{1}{d_\alpha(V)} \log \frac{M_v(\rho(f_V))}{M_v(f_V)}$$

tandis que si v est infinie on définit $\mu_{V,v,\alpha}(P)$ comme

$$\frac{1}{d_\alpha(V)} \int_{\sigma_v(V)} \log \left(\frac{|P(x)|_v}{\|x^{(1)}\|_v^{\delta_1} \dots \|x^{(q)}\|_v^{\delta_q}} \right) \Omega_1(x^{(1)})^{\wedge \alpha_1} \wedge \dots \wedge \Omega_q(x^{(q)})^{\wedge \alpha_q}$$

formule dans laquelle

- $\sigma_v(V)$ représente la variété analytique complexe des points de V à valeurs dans $\mathbf{C}_v$, partie de $\mathbf{P}^{n_1}(\mathbf{C}) \times \dots \times \mathbf{P}^{n_q}(\mathbf{C})$;
- $(x^{(1)}, \dots, x^{(q)})$ est, pour un point de $\mathbf{P}^{n_1}(\mathbf{C}) \times \dots \times \mathbf{P}^{n_q}(\mathbf{C})$, un de ses antécédents dans $(\mathbf{C}^{n_1+1} \setminus 0) \times \dots \times (\mathbf{C}^{n_q+1} \setminus 0)$ et $\| \cdot \|_v$ désigne la norme hermitienne c'est-à-dire $\|z\| = \left(\sum_{i=0}^n x_i \bar{x}_i \right)^{1/2}$;
- Ω_i est la forme de Fubini-Study sur $\mathbf{P}^{n_i}(\mathbf{C})$ c'est-à-dire, avec les notations précédentes, $\Omega_i(x^{(i)}) = \frac{1}{-i\pi} \partial \bar{\partial} \log \|x^{(i)}\|$.

Notons que le choix de l'antécédent x d'un point de $\sigma_v(V)$ n'influe pas sur le résultat car P est multihomogène de multidegré δ.

3.2. Préparatifs. La manipulation des intégrales ci-dessus appelle un résultat généralisant le théorème de Wirtinger. On déduit effectivement du cas projectif $(q = 1)$ le

LEMME 3.1. *Pour* $V \hookrightarrow \mathbf{P}_{\mathbf{C}}^{n_1} \times \dots \times \mathbf{P}_{\mathbf{C}}^{n_q}$ *et* $\alpha_1 + \dots + \alpha_q = \dim V$ *on a*

$$\int_V \Omega_1(x^{(1)})^{\wedge \alpha_1} \wedge \dots \wedge \Omega_q(x^{(q)})^{\wedge \alpha_q} = d_\alpha(V).$$

DÉMONSTRATION. On considère comme au paragraphe précédent $\delta \in (\mathbf{N} \setminus 0)^q$, $v_1, \dots, v_q$ et s les plongements de Veronese remodelés et de Segre associés à δ. Dans la composée

$$\mathbf{P}_{\mathbf{C}}^{n_1} \times \dots \times \mathbf{P}_{\mathbf{C}}^{n_q} \xrightarrow{s \circ (v_1 \times \dots \times v_q)} \mathbf{P}_{\mathbf{C}}^{\binom{n_1+\delta_1}{n_1} \dots \binom{n_q+\delta_q}{n_q} - 1}$$

si l'on note z l'image de x on a $\|z\| = \|x^{(1)}\|^{\delta_1} \dots \|x^{(q)}\|^{\delta_q}$. Par suite

$$\begin{aligned}
\Omega(z) &= -\frac{1}{i\pi} \partial \bar{\partial} \log \|z\| \\
&= -\frac{1}{i\pi} \partial \bar{\partial} \left[\sum_{i=1}^q \delta_i \log \|x^{(i)}\| \right] \\
&= \sum_{i=1}^q \delta_i \Omega_i(x^{(i)})
\end{aligned}$$

puis

$$\Omega(z)^{\wedge \dim V} = \sum_{|\alpha|=\dim V} \binom{\dim V}{\alpha} \delta_1^{\alpha_1} \dots \delta_q^{\alpha_q} \Omega_1(x^{(1)})^{\wedge \alpha_1} \wedge \dots \wedge \Omega_q(x^{(q)})^{\wedge \alpha_q}.$$

En appliquant le théorème de Wirtinger (voir [**PPh7**, I, Introduction]) au plongement $s \circ (v_1 \times \dots \times v_q) \circ \varphi$ on a

$$d_{s \circ (v_1 \times \dots \times v_q) \circ \varphi}(V) =$$

$$\sum_{|\alpha|=\dim V} \binom{\dim V}{\alpha} \left[\int_V \Omega_1(x^{(1)})^{\wedge \alpha_1} \wedge \dots \wedge \Omega_q(x^{(q)})^{\wedge \alpha_q} \right] \delta_1^{\alpha_1} \dots \delta_q^{\alpha_q}.$$

Pour obtenir le résultat il suffit d'identifier le polynôme du membre de droite (la
formule est vraie pour tout $\delta \in (\mathbf{N} \setminus 0)^q$) avec celui déjà obtenu pour calculer le
membre de gauche à savoir

$$d_{s \circ (v_1 \times \dots \times v_q) \circ \varphi}(V) = \sum_{|\alpha|=r-1} \binom{r-1}{\alpha} d_{\varphi,\alpha}(V) \delta_1^{\alpha_1} \dots \delta_q^{\alpha_q}.$$

$\square$

Nous aurons ensuite besoin de préciser la valeur de $h_{V,\alpha}(P)$ dans le cas particulier
où $V = \mathbf{P} = \mathbf{P}_K^{n_1} \times \dots \times \mathbf{P}_K^{n_q}$. On a en fait l'égalité suivante

LEMME 3.2. *En voyant P comme forme multihomogène on a*

$$h(P) = h_{\mathbf{P},(n_1,\dots,n_q)}(P) + \sum_{i=1}^q \delta_i \sum_{j=1}^{n_i} \frac{1}{2j}.$$

DÉMONSTRATION. En utilisant la formule

$$\sum_{v \text{ infinie}} \frac{[K_v : \mathbf{Q}_v]}{[K : \mathbf{Q}]} = 1$$

on constate qu'il suffit de montrer que

$$\log M_v(P) = \mu_{\mathbf{P},v,(n_1,\dots,n_q)}(P)$$

pour une place v finie et

$$\log M_v(P) = \mu_{\mathbf{P}_K^{n_1} \times \dots \times \mathbf{P}_K^{n_q},v,(n_1,\dots,n_q)}(P) + \sum_{i=1}^q \delta_i \sum_{j=1}^{n_i} \frac{1}{2j}$$

pour une place v infinie. Cette dernière formule se réécrit

$$\int_{S_{n_1+1}(1) \times \dots \times S_{n_q+1}(1)} \log \left(\frac{|P(x)|_v}{\|x^{(1)}\|_v^{\delta_1} \dots \|x^{(q)}\|_v^{\delta_q}} \right) \sigma_{n_1+1}(x^{(1)}) \wedge \dots \wedge \sigma_{n_q+1}(x^{(q)})$$

$$= \int_{\mathbf{P}_\mathbf{C}^{n_1} \times \dots \times \mathbf{P}_\mathbf{C}^{n_q}} \log \left(\frac{|P(x)|_v}{\|x^{(1)}\|_v^{\delta_1} \dots \|x^{(q)}\|_v^{\delta_q}} \right) \Omega_1(x^{(1)})^{\wedge n_1} \wedge \dots \wedge \Omega_q(x^{(q)})^{\wedge n_q}$$

en utilisant que $\|\cdot\|$ vaut 1 sur la sphère et $d_{(n_1,\dots,n_q)}(\mathbf{P}_\mathbf{C}^{n_1} \times \dots \times \mathbf{P}_\mathbf{C}^{n_q}) = 1$. Pour
démontrer cette égalité il suffit de voir que si f est une fonction $S_{n+1}(1) \to \mathbf{R}$ se
factorisant à travers $\mathbf{P}_\mathbf{C}^n$ on a

$$\int_{S_{n+1}(1)} f \sigma_{n+1} = \int_{\mathbf{P}_\mathbf{C}^n} f \Omega^{\wedge n}.$$

On applique en effet cette formule d'abord $(n = n_1)$ à

$$f(x^{(1)}) = \log \frac{|P(x)|}{\|x^{(1)}\|^{\delta_1} \dots \|x^{(q)}\|^{\delta_q}},$$

où $x^{(2)}, \dots, x^{(q)}$ sont fixés, puis à

$$f(x^{(2)}) = \int_{S_{n_1+1}(1)} \log \frac{|P(x)|}{\|x^{(1)}\|^{\delta_1} \dots \|x^{(q)}\|^{\delta_q}} \sigma_{n_1+1}(x^{(1)})$$

avec $n = n_2$ où $x^{(3)}, \dots, x^{(q)}$ sont fixés et ainsi de suite. Pour établir notre relation notons que sur $\mathbf{C}^{n+1} \setminus 0$ on a l'expression (qui suit des définitions de [**PPh7**, I])

$$\sigma_{n+1}(x) = \left(-\frac{1}{2i\pi}\right)^n \left(\frac{\partial\bar{\partial}\|x\|^2}{\|x\|^2}\right)^{\wedge n} \wedge \frac{\partial\|x\|^2 - \bar{\partial}\|x\|^2}{4i\pi\|x\|^2}.$$

Par ailleurs

$$\begin{aligned}
\Omega(x) &= -\frac{1}{i\pi} \partial\bar{\partial} \log \|x\| \\
&= -\frac{1}{2i\pi} \partial\bar{\partial} \log \|x\|^2 \\
&= -\frac{1}{2i\pi} \partial \left(\frac{\bar{\partial}\|x\|^2}{\|x\|^2}\right) \\
&= -\frac{1}{2i\pi} \left(\frac{\partial\bar{\partial}\|x\|^2}{\|x\|^2} - \frac{\partial\|x\|^2 \wedge \bar{\partial}\|x\|^2}{\|x\|^4}\right)
\end{aligned}$$

puis

$$\Omega(x)^{\wedge n} = \left(-\frac{1}{2i\pi}\right)^n \left[\left(\frac{\partial\bar{\partial}\|x\|^2}{\|x\|^2}\right)^{\wedge n} - n \left(\frac{\partial\bar{\partial}\|x\|^2}{\|x\|^2}\right)^{\wedge(n-1)} \wedge \frac{\partial\|x\|^2 \wedge \bar{\partial}\|x\|^2}{\|x\|^4}\right]$$

qui entraîne, en vertu de $\partial\|x\|^2 \wedge \partial\|x\|^2 = 0$ et $\bar{\partial}\|x\|^2 \wedge \bar{\partial}\|x\|^2 = 0$, que l'on peut écrire sur $\mathbf{C}^{n+1} \setminus 0$

$$\sigma_{n+1}(x) = \Omega(x)^{\wedge n} \wedge \frac{\partial\|x\|^2 - \bar{\partial}\|x\|^2}{4i\pi\|x\|^2}.$$

On veut maintenant paramétrer $S_{n+1}(1)$ par $\mathbf{P}^n_{\mathbf{C}} \times S_1(1)$. Pour l'écrire commodément on remplace $S_{n+1}(1)$ par $\mathcal{S} = \{x \in S_{n+1}(1) \mid x_0 \neq 0\}$ pour lequel

$$\begin{aligned}
\mathbf{C}^n \times S_1(1) &\longrightarrow \quad \mathcal{S} \\
(y, \lambda) &\longmapsto \left(\frac{\lambda}{\sqrt{1+\|y\|^2}}, \frac{\lambda y_1}{\sqrt{1+\|y\|^2}}, \dots, \frac{\lambda y_n}{\sqrt{1+\|y\|^2}}\right)
\end{aligned}$$

est un paramétrage. Comme l'image de (y, λ) est le point $(1, y) \in \mathbf{P}^n_{\mathbf{C}}$, $\Omega(x)$ sur $\mathcal{S}$ se lit bien sur $\mathbf{C}^n$ comme la restriction de Ω de $\mathbf{P}^n_{\mathbf{C}}$ à $\mathbf{C}^n$. Par ailleurs si l'on évalue $\frac{\partial\|x\|^2 - \bar{\partial}\|x\|^2}{4i\pi\|x\|^2}$ sur $\mathbf{C}^n \times S_1(1)$ on trouve (avec $x_i = \lambda\frac{y_i}{\sqrt{1+\|y\|^2}}$ et en posant pour simplifier $y_0 = 1$)

$$\frac{1}{4i\pi} \sum_{i=0}^n \bar{x}_i dx_i - x_i d\bar{x}_i = \frac{1}{4i\pi}(\bar{\lambda}d\lambda - \lambda d\bar{\lambda})$$

$$+ \frac{1}{4i\pi\sqrt{1+\|y\|^2}} \sum_{i=0}^n \bar{y}_i d\left(\frac{y_i}{\sqrt{1+\|y\|^2}}\right) - y_i d\left(\frac{\bar{y}_i}{\sqrt{1+\|y\|^2}}\right).$$

Comme le second terme est une combinaison linéaire des dy_i et $d\bar{y}_i$, il est certainement nul lorsqu'on lui applique $\Omega(x)^{\wedge n}\wedge$. Par ailleurs le premier terme $\frac{1}{4i\pi}(\bar{\lambda}d\lambda - \lambda d\bar{\lambda})$ n'est autre que $\sigma_1(\lambda)$ (faire $n = 0$ dans l'expression citée de σ_{n+1}). Ainsi $\sigma_{n+1}(x)$ se lit sur $\mathbf{C}^n \times S_1(1)$ *via* le paramétrage comme le produit de la mesure $\Omega^{\wedge n}$ et de σ_1. Finalement

$$
\begin{aligned}
\int_{S_{n+1}(1)} f\sigma_{n+1} &= \int_{\mathcal{S}} f\sigma_{n+1} \\
&= \int_{\mathbf{C}^n \times S_1(1)} f\Omega^{\wedge n} \wedge \sigma_1 \\
&= \int_{\mathbf{C}^n} \left(\int_{S_1(1)} \sigma_1 \right) f\Omega^{\wedge n} \\
&= \int_{\mathbf{C}^n} f\Omega^{\wedge n} \\
&= \int_{\mathbf{P}^n_{\mathbf{C}}} f\Omega^{\wedge n}
\end{aligned}
$$

les passages de $S_{n+1}(1)$ à $\mathcal{S}$ et de $\mathbf{P}^n_{\mathbf{C}}$ à $\mathbf{C}^n$ étant justifiés parce que l'on écarte à chaque fois un ensemble de mesure nulle.

Pour une place v finie on doit donc vérifier que

$$
M_v(P) = \frac{M_v(\rho(f_{\mathbf{P}}))}{M_v(f_{\mathbf{P}})}
$$

où ρ est la spécialisation correspondant à P. D'après le lemme 3.7 du chapitre 5 et les notations qui s'y rapportent on a $f_{\mathbf{P}} = U_{n+1}(\Delta)$ et donc $\rho(f_{\mathbf{P}}) = P(\Delta)$. En considérant les variables $u_{\mathfrak{m}}^{(n+1)}$ on trouve (forme linéaire)

$$
\begin{aligned}
M_v(U_{n+1}(\Delta)) &= \max_{\mathfrak{m}\in\mathcal{M}_\delta} M_v(\mathfrak{m}(\Delta)) \\
&= \max_{\mathfrak{m}\in\mathcal{M}_\delta} \mathfrak{m}(M_v(\Delta)) \\
&= 1
\end{aligned}
$$

car, vu les définitions, on a évidemment $M_v(\Delta_l) = 1$ pour tout $0 \le l \le n+1$. Il reste à voir $M_v(P) = M_v(P(\Delta))$. Puisque $M_v(P) = \text{moy}|P|_v$, on peut trouver $x \in (\mathbf{C}_v^{n_1+1} \setminus 0) \times \cdots \times (\mathbf{C}_v^{n_q+1} \setminus 0)$ de sorte que $|x_j^{(i)}|_v = 1$ pour $1 \le i \le q$ et $0 \le j \le n_i$ et $M_v(P) = |P(x)|_v$. Considérons alors la spécialisation ρ' suivante

$$
\begin{aligned}
u_{X_0^{(i)}}^{(n_i'+l)} &\longmapsto -x_l^{(i)} & \text{si } & 1 \le l \le n_i \\
u_{X_l^{(i)}}^{(n_i'+l)} &\longmapsto x_0^{(i)} & \text{si } & 1 \le l \le n_i \\
u_{X_k^{(i)}}^{(n_i'+l)} &\longmapsto 0 & \text{si } & 1 \le l, k \le n_i \text{ et } l \ne k.
\end{aligned}
$$

Un calcul direct de déterminant montre que $\rho'(\Delta_{i,j}) = (x_0^{(i)})^{n_i-1}x_j^{(i)}$ pour $1 \le i \le q$ et $0 \le j \le n_i$. Aussi

$$
\rho'(P(\Delta)) = (x_0^{(1)})^{\delta_1(n_1-1)} \ldots (x_0^{(q)})^{\delta_q(n_q-1)} P(x)
$$

d'où $|\rho'(P(\Delta))|_v = M_v(P)$. Par ailleurs

$$
|\rho'(P(\Delta))|_v \le M_v(P(\Delta)) \max_{l,\mathfrak{m}} |\rho'(u_{\mathfrak{m}}^{(l)})|_v = M_v(P(\Delta)).
$$

Ainsi $M_v(P) \leq M_v(P(\Delta))$ et, l'inégalité $M_v(P(\Delta)) \leq M_v(P)$ étant évidente, on a la conclusion souhaitée. □

Au cours de nos calculs apparaîtra encore une autre hauteur du polynôme P que nous introduisons maintenant. Si P s'écrit

$$P = \sum_{m \in \mathcal{M}_\delta} p_m m$$

on note

$$h_m(P) = \sum_{v \text{ finie}} \frac{[K_v : \mathbf{Q}_v]}{[K : \mathbf{Q}]} \log \max_{m \in \mathcal{M}_\delta} |p_m|_v + \sum_{v \text{ infinie}} \frac{[K_v : \mathbf{Q}_v]}{[K : \mathbf{Q}]} \log \|P\|_{v,2}$$

avec pour v infinie

$$\|P\|_{v,2} = \sqrt{\sum_{m \in \mathcal{M}_\delta} \binom{\delta}{m}^{-1} |p_m|^2}.$$

Le lemme suivant fournit la comparaison de ces hauteurs que nous emploierons.

LEMME 3.3. *Pour un polynôme P de multidegré δ comme précédemment on a*

$$h_m(P) \leq h(P) + \frac{\delta_1}{2} \log(n_1 + 1) + \cdots + \frac{\delta_q}{2} \log(n_q + 1).$$

DÉMONSTRATION. On travaille à nouveau place par place. De manière analogue au lemme précédent, il nous suffit d'établir que pour une place v infinie on a

$$\log \sqrt{\sum_{m \in \mathcal{M}_\delta} \binom{\delta}{m}^{-1} |p_m|^2} \leq \log M_v(P) + \frac{\delta_1}{2} \log(n_1 + 1) + \cdots + \frac{\delta_q}{2} \log(n_q + 1).$$

La comparaison de ces deux quantités transite par la mesure de Mahler de P notée $m(P)$ (voir [**PPh2**] et [**Lel2**][4]). On a d'abord par multihomogénéité de P la formule, pour tout m :

$$|p_m| \leq \binom{\delta}{m} m(P).$$

Ceci montre que

$$\sqrt{\sum_{m \in \mathcal{M}_\delta} \binom{\delta}{m}^{-1} |p_m|^2} \leq m(P) \sqrt{\sum_{m \in \mathcal{M}_\delta} \binom{\delta}{m}}$$

$$\leq m(P) \sqrt{(n_1 + 1)^{\delta_1} \dots (n_q + 1)^{\delta_q}}.$$

En prenant le logarithme, on voit qu'il suffit pour prouver notre résultat de majorer $\log m(P)$ par $\log M(P)$. On utilise pour cela le résultat de [**Lel2**] (*loc. cit.*) qui donne pour un polynôme Q en p variables la majoration

$$\log m(Q) \leq \int_{S_p(1)} \log |Q| \sigma_p + \deg(Q) C_p$$

---

[4][**Lel2**] P. Lelong. Mesure de Mahler des polynômes et majorations par convexité. *C. R. Acad. Sci. Paris* 315, (1992), 139–142; Mesure de Mahler et calcul de constantes universelles pour les polynômes de N variables. *Math. Ann.* 299, (1994), 673–695.

où la constante C_p vaut $\sum_{i=1}^{p-1} \frac{1}{2p}$. On applique q fois cette inégalité (à diverses spécialisations de P)

$$
\begin{aligned}
&\log m(P) \\
&= \int_{S_1(1)^{n_1+1} \times \cdots \times S_1(1)^{n_q+1}} \log |P| \sigma_1^{\wedge n_1+1} \wedge \ldots \wedge \sigma_1^{\wedge n_q+1} \\
&= \int_{S_1(1)^{n_1+1} \times \cdots \times S_1(1)^{n_{q-1}+1}} \\
&\qquad \left(\int_{S_1(1)^{n_q+1}} \log |P(x^{(1)}, \ldots, x^{(q-1)}, \cdot)| \sigma_1^{\wedge n_q+1} \right) \sigma_1^{\wedge n_1+1} \wedge \ldots \wedge \sigma_1^{\wedge n_{q-1}+1} \\
&\leq \int_{S_1(1)^{n_1+1} \times \cdots \times S_1(1)^{n_{q-1}+1}} \left(\int_{S_{n_q+1}(1)} \log |P(x^{(1)}, \ldots, x^{(q-1)}, \cdot)| \sigma_{n_q+1} \right. \\
&\qquad \left. + \delta_q C_{n_q+1} \right) \sigma_1^{\wedge n_1+1} \wedge \ldots \wedge \sigma_1^{\wedge n_{q-1}+1} \\
&\leq \cdots \\
&\leq \int_{S_{n_1+1}(1) \times \cdots \times S_{n_q+1}(1)} \log |P| \sigma_{n_1+1} \wedge \ldots \wedge \sigma_{n_q+1} + \sum_{i=1}^{q} \delta_i C_{n_i+1} \\
&\leq \log M(P)
\end{aligned}
$$

et ceci prouve le lemme. $\qquad\qquad\qquad\qquad\qquad\qquad\qquad\qquad\qquad\qquad\qquad$ $\square$

3.3. Résultat. On considère le produit d'intersection $V.Z$ que l'on peut par exemple définir en disant que si $V = Y(I)$ alors $V.Z = Y((I, P))$ (notations de la proposition 2.17 du chapitre 5). On rappelle que l'on suppose $V \not\subset Z$ soit $P \notin I$. Le théorème ci-dessous exprime donc (les degrés et) les hauteurs de $V.Z$ en fonction de V et de P. C'est une formule exacte (et non une inégalité); cependant le terme $h_{V,\alpha}(P)$ n'est pas très aisé à calculer aussi faut-il souvent le majorer dans les applications. Nous donnerons en corollaire de telles estimations. Le théorème fournit aussi, comme conséquence du résultat principal, le lien entre la hauteur (d'indice $(n_1, \ldots, n_q)$) de Z et la hauteur de la forme P.

THÉORÈME 3.4. *Avec les hypothèses et notations ci-dessus et* $\alpha'_1 + \cdots + \alpha'_q = \dim V$ *on a*

$$
d_{\alpha'}(V.Z) = \sum_{i=1}^{q} \delta_i d_{\alpha'+\varepsilon_i}(V),
$$

$$
h_\alpha(V.Z) = \sum_{i=1}^{q} \delta_i h_{\alpha+\varepsilon_i}(V) + d_\alpha(V) h_{V,\alpha}(P)
$$

et

$$
h_{(n_1, \ldots, n_q)}(Z) = h(P) + \sum_{i=1}^{q} \delta_i \sum_{j=1}^{n_i-1} \sum_{l=1}^{j} \frac{1}{2l}.
$$

DÉMONSTRATION. La dernière formule pour $h_{(n_1, \ldots, n_q)}(Z)$ se déduit de celle pour $h(V.Z)$ appliquée avec $V = \mathbf{P} = \mathbf{P}_K^{n_1} \times \cdots \times \mathbf{P}_K^{n_q}$ et $\alpha = (n_1, \ldots, n_q)$; on trouve en effet

$$h_{\mathbf{P},(n_1,\ldots,n_q)}(P) = h_{(n_1,\ldots,n_q)}(Z) - \sum_{i=1}^{q} \delta_i h_{(n_1,\ldots,n_q)+\varepsilon_i}(\mathbf{P})$$

et par le corollaire 2.4 on peut calculer

$$h_{(n_1,\ldots,n_q)+\varepsilon_i}(\mathbf{P}) = h(\mathbf{P}_K^{n_i}) = \sum_{j=1}^{n_i} \sum_{l=1}^{j} \frac{1}{2l}$$

(pour la hauteur de $\mathbf{P}^{n_i}$ voir [**PPh7**, III, p. 346]). D'autre part le lemme 3.2 montre que le membre de gauche est égal à

$$h(P) - \sum_{i=1}^{q} \delta_i \sum_{j=1}^{n_i} \frac{1}{2j}$$

ce qui donne la formule de l'énoncé. Le résultat sur les degrés se borne à exprimer que $\deg(I,P) = \deg(I) * \delta$ (voir lemme 2.11 du chapitre 5). Pour démontrer l'égalité principale, on note tout d'abord que le théorème 2.2 s'énonce dans notre cas

$$h(\alpha_d(f_V)) = \sum_{i=1}^{q} \delta_i h_{\alpha+\varepsilon_i}(V).$$

où f_V est une forme résultante d'indice $d = (\delta, \varepsilon_1, \ldots, \varepsilon_1, \ldots, \varepsilon_q, \ldots, \varepsilon_q)$ de V avec ε_i répété α_i fois. Par ailleurs la proposition 3.6 du chapitre 5 nous donne que $\rho(f_V)$ est une forme résultante d'indice d' ($d'_\ell = d_{\ell+1}$, $1 \le \ell \le \dim V$) de $V.Z$ et donc que $h_\alpha(V.Z) = h(\rho(f_V))$. Ainsi il nous suffira de montrer que

$$h(\rho(f_V)) - h(\alpha_d(f_V)) = d_\alpha(V) h_{V,\alpha}(P).$$

Le premier membre s'écrit

$$\sum_v \frac{[K_v : \mathbf{Q}_v]}{[K : \mathbf{Q}]} \log \frac{M_v(\rho(f_V))}{M_v(\alpha_d(f_V))}$$

tandis que le second est

$$\sum_v \frac{[K_v : \mathbf{Q}_v]}{[K : \mathbf{Q}]} d_\alpha(V) \mu_{V,v,\alpha}(P).$$

La conclusion du théorème découle donc du lemme ci-dessous.   □

LEMME 3.5. *Pour toute place v on a*

$$d_\alpha(V) \mu_{V,v,\alpha}(P) = \log \frac{M_v(\rho(f_V))}{M_v(\alpha_d(f_V))}.$$

DÉMONSTRATION. Pour une place finie v il est suffisant d'établir que les quantités $M_v(\alpha_d(f_V))$ et $M_v(f_V)$ sont égales. Or leur rapport vaut (en vertu du lemme 2.1)

$$\operatorname*{moy}_{\rho'\,:\,u^{(2)},\ldots,u^{(r)}} \frac{M_v(\rho'(f_V))}{M_v(\rho'(\alpha_d(f_V)))}.$$

En utilisant $d_{l,i} \le 1$ pour $2 \le l \le r$ et $1 \le i \le q$ on trouve $\rho' \circ \alpha_d = \alpha_{(\delta)} \circ \rho'$ et sachant que $\rho'(f_v)$ est de la forme

$$\lambda \prod_{i=1}^{t} U_1(z_i)$$

il nous suffit de vérifier que $M_v(\alpha_{(\delta)}(U_1)(z)) = M_v(U_1(z))$ soit

$$\max_{\mathfrak{m} \in \mathcal{M}_\delta} \left| \binom{\delta}{\mathfrak{m}}^{1/2} \mathfrak{m}(z) \right|_v = \max_{\mathfrak{m} \in \mathcal{M}_\delta} |\mathfrak{m}(z)|_v$$

qui est clair (voir aussi démonstration du théorème 2.2).

La formule pour une place infinie v est une généralisation du corollaire 4 de [**PPh7**, I] (voir aussi le lemme 1 de [**PPh7**, II]). Ce dernier repose sur la proposition 5 qui le précède et dont nous allons imiter la preuve. Nous en utilisons les notations $\mathcal{E}_i$, $\Phi(x) \ldots$ Lorsque l'on évalue le membre de droite (pour v infinie), que l'on notera J, il vient (lemme 2.1)

$$J = \log \operatorname*{moy}_{\rho' \,:\, u^{(2)},\ldots,u^{(r)}} \frac{|\rho' \circ \rho(f_V)|}{M(\rho'(\alpha_d(f_V)))}.$$

Grâce aux remarques de [**PPh7**, I] (proposition 1 et *nota bene* attenant) on peut remplacer dans l'écriture de la moyenne les sphères par des boules. Plus précisément si $\alpha_1 + \cdots + \alpha_{i-1} + 2 \le l \le \alpha_1 + \cdots + \alpha_i + 1$ on a $S_{|\mathcal{M}_{d_l}|}(1) = S_{n_i+1}(1)$ et l'on remplace $\int_{S_{n_i+1}(1)} \cdots \sigma_{n_i+1}(u^{(l)})$ par

$$\frac{(n_i+1)!}{\pi^{n_i+1}} \int_{B_{n_i+1}(1)} \cdots \mu_{n_i+1}(u^{(l)})$$

où $B_{n_i+1} = \{x \in \mathbf{C}^{n_i+1} \mid \|x\| \le 1\}$ et μ_{n_i+1} est la mesure de Lebesgue sur $\mathbf{C}^{n_i+1} \simeq \mathbf{R}^{2n_i+2}$. On a donc

$$J = C_\alpha \int_{B_{n_1+1}(1)^{\alpha_1} \times \cdots \times B_{n_q+1}(1)^{\alpha_q}} \log \frac{|\rho' \circ \rho(f_V)|}{M(\rho'(\alpha_d(f_V)))} \mu_{n_1+1}^{\wedge \alpha_1} \wedge \ldots \wedge \mu_{n_q+1}^{\wedge \alpha_q}$$

avec la notation suivante

$$C_\alpha = \left(\frac{(n_1+1)!}{\pi^{n_1+1}} \right)^{\alpha_1} \cdots \left(\frac{(n_q+1)!}{\pi^{n_q+1}} \right)^{\alpha_q}.$$

Comme on l'a déjà vu on peut écrire $\rho'(f_V) = \lambda \prod_{i=1}^{t} U_1(z_i)$ où, si $\rho'(f_V) \ne 0$, les $z_i = (x^{(i,1)}, \ldots, x^{(i,q)})$ sont les points d'intersection de V avec $\{U_2 = \cdots = U_r = 0\}$ dont l'ensemble sera noté $\mathcal{X}_u$. On a encore $\rho' \circ \alpha_d = \alpha_{(d_1)} \circ \rho$ et $\rho' \circ \rho = \rho \circ \rho'$ donc

$$\begin{aligned}
\frac{|\rho' \circ \rho(f_V)|}{M(\rho'(\alpha_d(f_V)))} &= \frac{|\rho(\lambda \prod_{i=1}^{t} U_1(z_i))|}{M(\lambda \prod_{i=1}^{t} \alpha_{(\delta)}(U_1)(z_i))} \\
&= \prod_{i=1}^{t} \frac{|P(z_i)|}{\sqrt{\sum_{\mathfrak{m} \in \mathcal{M}_\delta} \binom{\delta}{\mathfrak{m}} |\mathfrak{m}(z_i)|^2}} \\
&= \prod_{i=1}^{t} \frac{|P(z_i)|}{\|x^{(i,1)}\|^{\delta_1} \ldots \|x^{(i,q)}\|^{\delta_q}}.
\end{aligned}$$

Notre expression J s'écrit alors

$$C_\alpha \int_{B_{n_1+1}(1)^{\alpha_1} \times \cdots \times B_{n_q+1}(1)^{\alpha_q}} \sum_{x \in \mathcal{X}_u} \log \frac{|P(x)|}{\|x^{(1)}\|^{\delta_1} \ldots \|x^{(q)}\|^{\delta_q}} \mu_{n_1+1}^{\wedge \alpha_1} \wedge \ldots \wedge \mu_{n_q+1}^{\wedge \alpha_q} \,;$$

on ne se préoccupe pas des points de $B_{n_1+1}(1)^{\alpha_1} \times \cdots \times B_{n_q+1}(1)^{\alpha_q}$ où $\mathcal{X}_u$ est infini (c'est-à-dire $\rho'(f_V) = 0$) car ils forment un ensemble de mesure nulle. On peut alors voir l'intégrale et la somme ci-dessus comme une seule intégrale sur $\mathcal{E}$ partie de $V \times B_{n_1+1}(1)^{\alpha_1} \times \cdots \times B_{n_q+1}(1)^{\alpha_q}$ définie par

$$(x, (u^{(2)}, \ldots, u^{(r)})) \in \mathcal{E} \iff U_l(x^{(i)}) = 0 \text{ si } 2 \le l - (\alpha_1 + \cdots + \alpha_{i-1}) \le \alpha_i + 1$$

par rapport à la mesure image réciproque (par la deuxième projection) de $\mu_{n_1+1}^{\wedge \alpha_1} \wedge \cdots \wedge \mu_{n_q+1}^{\wedge \alpha_q}$ sur $\mathcal{E}$. C'est cet ensemble $\mathcal{E}$ que l'on va paramétrer par sa première projection (sur V) comme [**PPh7**, I, p. 271]. En copiant cette référence on limite l'intégrale à $x \in V$ tel que $x_0^{(1)} \ldots x_0^{(q)} \ne 0$ (si besoin on permute auparavant les indices pour que cela n'affecte pas l'intégrale c'est-à-dire ne retire qu'un ensemble de mesure nulle). Une fois $x \in V$ fixé avec $x_0^{(i)} = 1$ le paramétrage est le même : $u_{X_0^{(i)}}^{(l)} = -\sum_{j=1}^{n_i} u_{X_j^{(i)}}^{(l)} x_j^{(i)}$ pour $\alpha_1 + \cdots + \alpha_{i-1} + 2 \le l \le \alpha_1 + \cdots + \alpha_i + 1$ et u' défini par $u'^{(l)} = (u_{X_1^{(i)}}^{(l)}, \ldots, u_{X_{n_i}^{(i)}}^{(l)})$ pour $\alpha_1 + \cdots + \alpha_{i-1} + 2 \le l \le \alpha_1 + \cdots + \alpha_i + 1$ parcourt $\mathcal{E}_1 \times \cdots \times \mathcal{E}_q$ (notations de [**PPh7**, I] avec n_i pour n dans $\mathcal{E}_i$). Le calcul des mesures est le même (c'est un produit de mesures sur les facteurs $\mathcal{E}_i$) et l'on trouve donc

$$J = \int_V \log \frac{|P(x)|}{\|x^{(1)}\|^{\delta_1} \ldots \|x^{(q)}\|^{\delta_q}} \Phi_1(x^{(1)})^{\wedge \alpha_1} \wedge \ldots \wedge \Phi_q(x^{(q)})^{\wedge \alpha_q}.$$

Avec les calculs de [**PPh7**, I, p. 271-272] on a (moyennant le changement affine-projectif) $\Phi_i = \Omega_i$ pour $1 \le i \le q$. Ainsi la formule ci-dessus est

$$J = \int_{\sigma_v(V)} \log \left(\frac{|P(x)|_v}{\|x^{(1)}\|_v^{\delta_1} \ldots \|x^{(q)}\|_v^{\delta_q}} \right) \Omega_1(x^{(1)})^{\wedge \alpha_1} \wedge \ldots \wedge \Omega_q(x^{(q)})^{\wedge \alpha_q}$$

dans le second membre duquel on reconnaît exactement $d_\alpha(V)\mu_{V,v,\alpha}(P)$ ce qui donne la conclusion. $\qquad\square$

On remarque dans l'énoncé du théorème que les formules pour hauteur et degré présentent des similitudes. On peut essayer de les exprimer simultanément (pour tous α et α') de manière beaucoup plus compacte quitte à introduire de nouvelles notations. Associons à V sous-schéma fermé de $\mathbf{P}$ un élément $t(V) = (t_\beta(V))_{\beta \in \mathbf{N}^{q+1}} \in \mathbf{R}^{\mathbf{N}^{q+1}}$ de la manière suivante :

- $t_{(0,\alpha)}(V) = h_\alpha(V)$ si $\alpha \in \mathbf{N}^q$,
- $t_{(1,\alpha')}(V) = d_{\alpha'}(V)$ si $\alpha' \in \mathbf{N}^q$ et
- $t_\beta = 0$ si $\beta_0 > 1$.

En particulier $t_\beta = 0$ si $\beta_0 + \beta_1 + \cdots + \beta_q \ne \dim V + 1$. On étend naturellement l'opérateur $*$ à cette situation à $q + 1$ indices en $*: \mathbf{R}^{\mathbf{N}^{q+1}} \times \mathbf{N}^{q+1} \to \mathbf{R}^{\mathbf{N}^{q+1}}$. Ces notations sont adaptées à notre problème à ceci près que la dépendance en α du terme $h_{V,\alpha}(P)$ empêche d'écrire une égalité dans la formule ci-dessous. En revanche, si l'on note $h'_V(P) = \max_\alpha h_{V,\alpha}(P)$ (ce qui ne coûte pas beaucoup dans les applications, en particulier si l'on utilise le corollaire ci-après), alors les expressions de $d_{\alpha'}(V.Z)$ et $h_\alpha(V.Z)$ du théorème entraînent

$$t(V.Z) \le t(V) * \begin{pmatrix} h'_V(P) \\ \delta_1 \\ \vdots \\ \delta_q \end{pmatrix}.$$

Terminons par le corollaire suivant qui estime $h_{V,\alpha}(P)$. On y note

$$\|P\|_{V,v} = \max_{x \in \sigma_v(V)} \frac{|P(x)|_v}{\|x^{(1)}\|_v^{\delta_1} \dots \|x^{(q)}\|_v^{\delta_q}}$$

pour une place v. Noter que le terme $\frac{\delta_1 + \dots + \delta_q}{2} \log 2$ ci-dessous n'est pas optimal et peut être vraisemblablement remplacé par zéro. C'est au moins le cas lorsque $n_i \geq 3$ pour tout i.

COROLLAIRE 3.6. *On a les estimations*

$$h_{V,\alpha}(P) \leq h_m(P) \leq h_{(n_1, \dots, n_q)}(Z) + \frac{\delta_1 + \dots + \delta_q}{2} \log 2$$

et pour une place infinie v

$$\mu_{V,v,\alpha}(P) \leq \log \|P\|_{V,v} \leq \log \|P\|_{v,2}.$$

DÉMONSTRATION. Nous travaillons place par place pour l'établissement des deux premières majorations au cours duquel apparaîtront les estimations à l'infini. On note

$$P = \sum_{m \in \mathcal{M}_\delta} p_m m.$$

Supposons en premier lieu v finie. Les coefficients de $\rho(f_V)$ sont des polynômes de degré $d_\alpha(V)$ en les p_m dont les coefficients figurent parmi ceux de f_V. Ainsi

$$M_v(\rho(f_V)) \leq \max_{m \in \mathcal{M}_\delta} |p_m|_v^{d_\alpha(V)} M_v(f_V)$$

puis

$$\mu_{V,v,\alpha}(P) \leq \log \max_{m \in \mathcal{M}_\delta} |p_m|_v.$$

Si maintenant v est infinie on a par le lemme 3.1

$$\int_{\sigma_v(V)} \Omega_1(x^{(1)})^{\wedge \alpha_1} \wedge \dots \wedge \Omega_q(x^{(q)})^{\wedge \alpha_q} = d_\alpha(V).$$

Ceci montre immédiatement que

$$\mu_{V,v,\alpha}(P)$$
$$= \frac{1}{d_\alpha(V)} \int_{\sigma_v(V)} \log \left(\frac{|P(x)|_v}{\|x^{(1)}\|_v^{\delta_1} \dots \|x^{(q)}\|_v^{\delta_q}} \right) \Omega_1(x^{(1)})^{\wedge \alpha_1} \wedge \dots \wedge \Omega_q(x^{(q)})^{\wedge \alpha_q}$$
$$\leq \log \|P\|_{V,v}.$$

Par la suite on majore simplement $\|P\|_{V,v}$ par $\|P\|_{\mathbf{P},v}$. En écrivant maintenant $P(x)$ comme produit scalaire des vecteurs

$$\left(p_m \binom{\delta}{m}^{-1/2} \right) \quad \text{et} \quad \left(\binom{\delta}{m}^{1/2} m(x) \right)$$

indicés par $m \in \mathcal{M}_\delta$ on a (Cauchy-Schwarz)

$$|P(x)|_v \leq \sqrt{\sum_{m \in \mathcal{M}_\delta} \binom{\delta}{m}^{-1} |p_m|^2} \sqrt{\sum_{m \in \mathcal{M}_\delta} \binom{\delta}{m} |m(x)|^2}.$$

La formule multinomiale montre immédiatement que le deuxième facteur vaut $\|x^{(1)}\|_v^{\delta_1} \cdots \|x^{(q)}\|_v^{\delta_q}$. On a donc

$$\|P\|_{\mathbf{P},v} \leq \sqrt{\sum_{\mathfrak{m} \in \mathcal{M}_\delta} \binom{\delta}{\mathfrak{m}}^{-1} |p_\mathfrak{m}|^2}.$$

Ainsi en prenant le logarithme et en sommant sur toutes les places on constate bien que $h_{V,\alpha}(P) \leq h_m(P)$. En utilisant le lemme 3.3 puis la troisième formule du théorème on a

$$h_m(P) \leq h(P) + \sum_{i=1}^q \frac{\delta_i}{2} \log(n_i + 1)$$

$$\leq h_{(n_1,\dots,n_q)}(Z) + \sum_{i=1}^q \frac{\delta_i}{2} \left(\log(n_i + 1) - \sum_{j=1}^{n_i-1} \sum_{l=1}^j \frac{1}{l} \right)$$

Enfin il est aisé de constater que $v_n = \log(n+1) - \sum_{j=1}^{n-1} \sum_{l=1}^j \frac{1}{l}$ $(n \geq 1)$ est une suite décroissante et que $v_1 = \log 2$. Ainsi $v_n \leq \log 2$ $(n \geq 1)$ et ceci termine la démonstration. On voit également que $v_2 = \log 3 - 1$ et $v_n < 0$ si $n \geq 3$ (car $v_3 = \log 4 - \frac{5}{2} \leq \log 4 - 2 < 0$) et ceci justifie la remarque précédant l'énoncé. $\square$

4. Distances

4.1. Préliminaires. Contrairement à la définition des hauteurs, la théorie des distances est locale. On travaillera essentiellement en une seule place. Soient donc v une place d'un corps de nombres et $K = \mathbf{C}_v$. On applique à ce corps les résultats de notre travail sur l'élimination multihomogène. En particulier on note $B = \mathbf{C}_v[X]$ et d est un élément de $(\mathbf{N}^q)^r$. On utilise aussi des anneaux analogues à $\mathfrak{S}B[d]$ introduit plus haut. En fait, ayant choisi un ordre total arbitraire $\preceq$ sur $\mathcal{M}_k$ pour tout $k \in \mathbf{N}^q$, on peut attacher à tout anneau C l'anneau $\mathfrak{S}C[d]$ des polynômes à coefficients dans C en les variables $s_{\mathfrak{m},\mathfrak{m}'}^{(l)}$ pour $1 \leq l \leq r$ et $\mathfrak{m} \prec \mathfrak{m}'$ dans $\mathcal{M}_{d_l}$. On étend encore la notation $s_{\mathfrak{m},\mathfrak{m}'}^{(l)}$ à $\mathfrak{m} \succeq \mathfrak{m}'$.

Un ingrédient crucial est ici le morphisme (introduit pour démontrer le lemme 2.3 du chapitre 5)

$$\mathfrak{d} \colon B[d] \longrightarrow \mathfrak{S}B[d]$$
$$u_{\mathfrak{m}}^{(l)} \longmapsto \sum_{\mathfrak{m}' \in \mathcal{M}_{d_l}} s_{\mathfrak{m},\mathfrak{m}'}^{(l)} \mathfrak{m}'.$$

Voici un premier fait qui suggère l'utilité de $\mathfrak{d}$ dans la définition d'une notion de distance. Lorsque $x \in (\mathbf{C}_v^{n_1+1} \setminus 0) \times \cdots \times (\mathbf{C}_v^{n_q+1} \setminus 0)$ on dispose d'une évaluation

$$e_x \colon B \longrightarrow \mathbf{C}_v$$
$$\mathfrak{m} \longmapsto \mathfrak{m}(x)$$

que l'on étend en $\mathfrak{S}B[d] \to \mathfrak{S}\mathbf{C}_v[d]$ encore noté e_x (par $e_x(s_{\mathfrak{m},\mathfrak{m}'}^{(l)}) = s_{\mathfrak{m},\mathfrak{m}'}^{(l)}$ bien sûr). Soient V un sous-schéma fermé de $\mathbf{P} = \mathbf{P}_{\mathbf{C}_v}^{n_1} \times \cdots \times \mathbf{P}_{\mathbf{C}_v}^{n_q}$, I l'idéal de $K[X]$ associé et $\hat{x}$ le point fermé de $\mathbf{P}$ correspondant à x. Alors on a

$$\hat{x} \in V \Longrightarrow e_x \circ \mathfrak{d}(\mathfrak{E}_d(I)) = 0$$

(en effet $\hat{x} \in V$ se traduit par $I \mathfrak{S}B[d] \subset \mathrm{Ker} e_x$ et le lemme 2.3 du chapitre 5 montre que si $f \in \mathfrak{E}_d(I)$ alors $\mathfrak{d}(f)\mathcal{M}_k \subset I \mathfrak{S}B[d]$; on conclut en utilisant qu'il existe

$\mathfrak{m} \in \mathcal{M}_k$ tel que $e_x(\mathfrak{m}) = \mathfrak{m}(x) \neq 0$). Dans les cas où $\mathfrak{E}_d(I)$ est principal on trouve que la fonction $x \mapsto e_x \circ \mathfrak{d}(\text{élim}_d(I))$ s'annule sur V. Il reste cependant à passer d'un élément de $\mathfrak{S}\mathbf{C}_v[d]$ à un réel. De la même façon que dans la définitions des hauteurs il avait été commode d'introduire des morphismes α_d nous calculerons ici, si v est infinie, la mesure des polynômes de $\mathfrak{S}\mathbf{C}_v[d]$ après l'application de l'automorphisme

$$\beta_d : \mathfrak{S}\mathbf{C}_v[d] \longrightarrow \mathfrak{S}\mathbf{C}_v[d]$$

$$s^{(l)}_{\mathfrak{m},\mathfrak{m}'} \longmapsto \binom{d_l}{\mathfrak{m}}^{1/2} \binom{d_l}{\mathfrak{m}'}^{1/2} s^{(l)}_{\mathfrak{m},\mathfrak{m}'}$$

(notons que la formule reste valable si $\mathfrak{m} \succeq \mathfrak{m}'$). Comme pour α_d, on constate que si $d_{l,i} \leq 1$ pour $1 \leq l \leq r$ et $0 \leq i \leq q$ alors β_d est l'identité. Lorsque v est finie on note $\beta_d = \text{id}$ et, également, $\alpha_d = \text{id}$.

4.2. Définitions. Avec les notations précédentes on pose pour V sous-schéma fermé de $\mathbf{P}$, $d \in (\mathbf{N}^q)^{\dim V+1}$ et x point fermé de $\mathbf{P}$

$$\text{Dist}_d(x, V) = \frac{M_v(\beta_d \circ e_x \circ \mathfrak{d}(f))}{M_v(\alpha_d(f)) \displaystyle\prod_{l=1}^{r} M_v(\alpha_{(d_l)}(U_l)(x))^{\deg_{u^{(l)}} f}}$$

où

- f est une forme résultante d'indice d de V;
- à droite x désigne un antécédent quelconque de x dans $(\mathbf{C}_v^{n_1+1} \setminus 0) \times \cdots \times (\mathbf{C}_v^{n_q+1} \setminus 0)$;
- la mesure M_v du numérateur s'applique aux formes multihomogènes en les $r = \dim V + 1$ groupes de variables $s^{(l)}$ ($1 \leq l \leq r$).

On vérifie que la formule ne dépend ni du choix de f ni de celui d'un antécédent pour x : comme $\mathfrak{d}(u^{(l)}_{\mathfrak{m}})$ est multihomogène (en X) de degré d_l, $\mathfrak{d}(f)$ est multihomogène de degré $\sum_{l=1}^{r} d_l \deg_{u^{(l)}} f$ tandis que, bien sûr, U_l est multihomogène de degré d_l. Notons que le terme $M_v(\alpha_{(d_l)}(U_l)(x))$ vaut

$$\max_{\mathfrak{m} \in \mathcal{M}_{d_l}} |\mathfrak{m}(x)|_v = \prod_{i=1}^{q} \left(\max_{0 \leq j \leq n_i} |x^{(i)}_j|_v \right)^{d_{l,i}}$$

si v est finie et

$$\sqrt{\sum_{\mathfrak{m} \in \mathcal{M}_{d_l}} \binom{d_l}{\mathfrak{m}} |\mathfrak{m}(x)|_v^2} = \prod_{i=1}^{q} \left(\sum_{j=0}^{n_i} |x^{(i)}_j|_v^2 \right)^{d_{l,i}/2}$$

si v est infinie (voir preuve du théorème 2.2). Comme le choix de l'antécédent de x est libre, on pourra toujours, si on le souhaite, le supposer tel que $M_v(\alpha_{(d_l)}(U_l)(x)) = 1$.

Lorsque $q = 1$ (cas projectif) on omettra l'indice d lorsqu'il est égal à $(1, \dots, 1)$ notant $\text{Dist}(x, V)$ dans ce cas. Remarquons que, pour une place finie, le fait d'avoir remplacé α_d et β_d par les identités ne modifie pas la définition du dénominateur puisque pour une forme résultante $M_v(\alpha_d(f)) = M_v(f)$. En revanche le fait de garder des coefficients multinomiaux dans la définition de β_d aux places finies eût conduit à une définition différente.

Dans les cas inintéressants où $f = 1$ on a $\text{Dist}_d(x, V) = 1$. En revanche si $\sqrt{f} = \text{élim}_d(I)$ engendre $\mathfrak{E}_d(I)$ alors $\text{Dist}_d(x, V) = 0$ pour $x \in V$. Ceci est une des propriétés minimales pour parler de distance. Il serait plus confortable, comme c'est le cas pour $\mathbf{P}^n$ ($q = 1$), d'avoir la réciproque : $\mathrm{Dist}_d(x, V) = 0 \implies x \in V$. Cependant cette propriété ne semble pas compatible avec la situation produit (voir paragraphe suivant pour des exemples montrant qu'elle tombe en défaut).

En réalité nous verrons que les problèmes apparaissent lorsque certaines des composantes $d_{l,i}$ sont nulles. Dans le cas contraire la situation est plus satisfaisante. Commençons par le cas particulier des indices d de la forme $(\delta, \dots, \delta) \in ((\mathbf{N} \setminus 0)^q)^r$. On peut alors écrire

$$\mathrm{Dist}_d(x, V) = \mathrm{Dist}(\iota(x), \iota(V))$$

où ι est le plongement de Segre-Veronese (remodelé lorsque v est infinie) associé à δ. En effet, si $d' = (1, \dots, 1) \in \mathbf{N}^r$, on a déjà vu (voir paragraphe 2.3) que $\mathbf{C}_v[d]$ s'identifie à $\mathbf{C}_v[d']$ (faisant référence au but de ι) de façon à ce que la forme résultante (indice d') de $\iota(V)$ coïncide avec l'image par α_d de celle de V (indice d). De même $\mathfrak{S}\mathbf{C}_v[d]$ s'identifie à $\mathfrak{S}\mathbf{C}_v[d']$ et l'on vérifie alors sans peine par un calcul direct que $\beta_{d'} \circ e_{\iota(x)} \circ \mathfrak{d} \colon \mathfrak{S}\mathbf{C}_v[d'] \to \mathbf{C}_v[d']$ correspond à $\beta_d \circ e_x \circ \mathfrak{d} \circ \alpha_d^{-1}$. En remplaçant dans l'expression de la distance on obtient l'égalité annoncée.

Dans cette situation particulière, il découle clairement du cas projectif que $\mathrm{Dist}_d(x, V) = 0 \iff x \in V$. Bien sûr l'étude multiprojective ne serait guère pertinente si l'on ne pouvait considérer que de tels indices et travailler en fait exclusivement à travers le plongement de Segre-Veronese correspondant. Toutefois l'on va montrer une comparaison utile entre deux distances d'indices d et d' à la condition que $d_{l,i}d'_{l,i} \neq 0$ pour tous l, i.

Cette comparaison ainsi que d'autres propriétés utiles de nos distances se trouve énoncée dans la

PROPOSITION 4.1. *Soient V, d et x comme dans la définition.*

1. $\mathrm{Dist}_d(x, V) \leq 1$.
2. *(Changement d'indices) Si $d' = (d'_1, d_2, \dots, d_r)$ avec $d'_{l,i}d_{l,i} \neq 0$ pour tous $1 \leq l \leq r$ et $1 \leq i \leq q$ alors $\mathrm{Dist}_{d'}(x, V) = \mathrm{Dist}_d(x, V)$ si v est finie tandis que si v est infinie on a en général*

$$\mathrm{Dist}_d(x, V) \leq \mathrm{Dist}_{d'}(x, V) \max\left(1, \frac{d_{1,1}}{d'_{1,1}}, \dots, \frac{d_{1,q}}{d'_{1,q}}\right)^{\frac{1}{2}\deg_{u^{(1)}} f}$$

$$\times \prod_{l=2}^{r}\left(\max_i \gamma_{\varepsilon_i}\gamma_{d_l}\right)^{\deg_{u^{(l)}} f'} \prod_{i=1}^{q} \gamma_{\varepsilon_i}^{d'_{1,i}\deg_{u^{(1)}} f}$$

où $\gamma_k = \exp\left(\sum_{j=1}^{|\mathcal{M}_k|-1}\frac{1}{2j}\right)$ pour $k \in \mathbf{N}^q$ et f' est une forme résultante d'indice d' de V; dans le cas particulier où $d'_{1,i} \leq d_{1,i}$ pour tout $1 \leq i \leq q$ on a

$$\prod_{l=2}^{r}\left(\max_i \gamma_{\varepsilon_i}\gamma_{d_l}\right)^{\deg_{u^{(l)}} f' - \deg_{u^{(l)}} f} \prod_{i=1}^{q} \gamma_{\varepsilon_i}^{(d'_{1,i}-d_{1,i})\deg_{u^{(1)}} f} \leq$$

$$\frac{\mathrm{Dist}_d(x, V)}{\mathrm{Dist}_{d'}(x, V)} \leq \max_{1 \leq i \leq q}\left(\frac{d_{1,i}}{d'_{1,i}}\right)^{\deg_{u^{(1)}} f}.$$

3. *(Lien avec la distance aux points de V) On suppose $d_{l,i} \neq 0$ pour tous l, i. Il existe un point fermé y de V tel que*

$$\mathrm{Dist}_{(d_1)}(x, y) \leq \mathrm{Dist}_d(x, V)^{(\deg_{u^{(1)}} f)^{-1}}$$

si v est finie et

$$\mathrm{Dist}_{(d_1)}(x,y) \leq \mathrm{Dist}_d(x,V)^{(\deg_{u^{(1)}} f)^{-1}} \prod_{l=1}^{r} (\gamma_{d_1} \gamma_{d_l}^{1-\delta_{l,1}})^{\frac{\deg_{u^{(l)}} f}{\deg_{u^{(1)}} f}}$$

si v est infinie (et δ est le symbole de Kronecker).

La démonstration de ces faits nécessitant de nombreux calculs, nous les établirons après l'étude des exemples plus simples où V est un produit $V_1 \times \cdots \times V_q$ ou un point. Dans ce dernier cas on peut donner des formules plus précises et en déduire l'énoncé ci-dessus. La preuve générale utilisera ce cas particulier.

La deuxième assertion permet d'affirmer que si $d_{l,i} \neq 0$ pour tous l,i alors $\mathrm{Dist}_d(\cdot, V)$ ne s'annule que sur V : en effet r applications successives de la comparaison montrent que $\mathrm{Dist}_d(\cdot, V)$ a les mêmes zéros que, par exemple, $\mathrm{Dist}_{d'}(\cdot, V)$ avec $d' = (\varepsilon, \ldots, \varepsilon)$; ce dernier correspondant au plongement de Segre on peut ainsi lui appliquer le résultat précédent.

4.3. Exemples. La première formule que nous donnons concerne le cas d'un produit. Elle montre la séparation des facteurs et permet de se ramener à un calcul projectif.

LEMME 4.2. *Soit V_i un sous-schéma fermé intègre de $\mathbf{P}_{\mathbf{C}_v}^{n_i}$ pour $1 \leq i \leq q$. Notons $d_{(i)}$ l'indice $(\varepsilon_1, \ldots, \varepsilon_1, \ldots, \varepsilon_q, \ldots, \varepsilon_q)$ où ε_j apparaît $\dim V_j + \delta_{i,j}$ fois. Alors pour $1 \leq i \leq q$ on a*

$$\mathrm{Dist}_{d_{(i)}}(x, V_1 \times \cdots \times V_q) = \mathrm{Dist}(x^{(i)}, V_i)^{\prod\limits_{j \neq i} d(V_j)}$$

si x est un point fermé de $\mathbf{P}$.

DÉMONSTRATION. C'est une conséquence immédiate du lemme 2.3. En effet son application donne

$$\mathrm{Dist}_{d_{(i)}}(x, V_1 \times \cdots \times V_q) = \frac{M_v(\beta_d \circ e_x \circ \mathfrak{d}(f'))^D}{M_v(\alpha_d(f'))^D \prod\limits_{l \mid d_l = \varepsilon_i} M_v(\alpha_{(\varepsilon_i)}(U_l)(x))^{D \deg_{u^{(l)}} f'}}$$

où $D = \prod_{j \neq i} d(V_j)$ et f' est une forme résultante de V_i d'indice $d' = (1, \ldots, 1) \in \mathbf{N}^{\dim V_i + 1}$. D'autre part on a

$$\mathrm{Dist}(x^{(i)}, V_i) = \frac{M_v(\beta_{d'} \circ e_{x^{(i)}} \circ \mathfrak{d}(f'))}{M_v(\alpha_{d'}(f')) \prod\limits_{l=1}^{\dim V_i + 1} M_v(\alpha_{d'_l}(U_l)(x^{(i)}))^{\deg_{u^{(l)}} f'}}.$$

On en déduit facilement l'égalité du lemme en remarquant d'une part que tous les morphismes α et β intervenant dans les formules ci-dessus sont des identités et d'autre part que f' ne fait intervenir que les variables $u_{\mathrm{m}}^{(l)}$ avec $d_l = \varepsilon_i$ et de même pour U_l si $d_l = \varepsilon_i$ (ce qui permet de remplacer x par $x^{(i)}$ dans la première relation). $\qquad\square$

Grâce à cette formule on constate que la fonction $\mathrm{Dist}_{d_{(i)}}(\cdot, V_1 \times \cdots \times V_q)$ s'annule non seulement sur les points fermés de $V_1 \times \cdots \times V_q$ mais même sur ceux de $\mathbf{P}_{\mathbf{C}_v}^{n_1} \times \cdots \times V_i \times \cdots \times \mathbf{P}_{\mathbf{C}_v}^{n_q} = \pi_i^{-1}(V_i)$. Ainsi en général l'annulation de $\mathrm{Dist}_d(\cdot, V)$ n'est pas limitée à V, contrairement à ce que pourrait suggérer le terme distance. Il faudrait plutôt considérer cette quantité comme une distance partielle : ainsi dans l'exemple d'un produit $\mathrm{Dist}_{d_{(i)}}$ mesure la distance seulement suivant le i-ème facteur.

Un cas particulier important de produit est celui où V est un point fermé y de $\mathbf{P}$ vu comme produit de points $y^{(i)}$ de $\mathbf{P}_{\mathbf{C}_v}^{n_i}$. Dans ce cas les expressions pour la distance se calculent facilement et les comparaisons entre indices différents sont aisées.

LEMME 4.3. *Soient* $x, y \in \mathbf{P}$ *et* $d \in \mathbf{N}^q = (\mathbf{N}^q)^1$. *Si* v *est une place finie on a*

$$\mathrm{Dist}_d(x,y) = \max_{i|d_i \neq 0} \mathrm{Dist}(x^{(i)}, y^{(i)});$$

ainsi en particulier si $d' \in \mathbf{N}^q$ *vérifie* $d_i d_i' \neq 0$ *pour tout* i *on a* $\mathrm{Dist}_d(x,y) = \mathrm{Dist}_{d'}(x,y)$. *Si* v *est une place infinie on a*

$$\mathrm{Dist}_d(x,y) = \sqrt{1 - \prod_{i=1}^{q} \left(1 - \mathrm{Dist}(x^{(i)}, y^{(i)})^2\right)^{d_i}};$$

par suite

$$\max_{i|d_i \neq 0} \mathrm{Dist}(x^{(i)}, y^{(i)}) \leq \mathrm{Dist}_d(x,y) \leq \sqrt{d_1 + \cdots + d_q} \max_{i|d_i \neq 0} \mathrm{Dist}(x^{(i)}, y^{(i)})$$

pour tout d *tandis que si* d *et* d' *sont tels que* $d_i d_i' \neq 0$ *pour tout* i *on a*

$$\mathrm{Dist}_d(x,y) \leq \sqrt{\max\left(1, \frac{d_1}{d_1'}, \ldots, \frac{d_q}{d_q'}\right)} \mathrm{Dist}_{d'}(x,y).$$

DÉMONSTRATION. Comme une forme résultante d'indice d de y est $U_1(y)$, on calcule

$$e_x \circ \mathfrak{d}(f) = \sum_{\mathfrak{m} \prec \mathfrak{m}' \in \mathcal{M}_d} (\mathfrak{m}(x)\mathfrak{m}'(y) - \mathfrak{m}(y)\mathfrak{m}'(x)) s_{\mathfrak{m}, \mathfrak{m}'}^{(l)}.$$

On en déduit, dans le cas où v est finie, que

$$\mathrm{Dist}_d(x,y) = \max_{\mathfrak{m} \prec \mathfrak{m}'} \left| \mathfrak{m}(x)\mathfrak{m}'(y) - \mathfrak{m}(y)\mathfrak{m}'(x) \right|_v,$$

si l'on choisit des représentants pour x et y tels que $\max_{0 \leq j \leq n_i} |x_j^{(i)}|_v = 1$ et $\max_{0 \leq j \leq n_i} |y_j^{(i)}|_v = 1$ pour tout $1 \leq i \leq q$. Dans ce cas on a aussi

$$\mathrm{Dist}(x^{(i)}, y^{(i)}) = \max_{0 \leq j < k \leq n_i} \left| x_j^{(i)} y_k^{(i)} - x_k^{(i)} y_j^{(i)} \right|_v.$$

On obtient facilement $\mathrm{Dist}_d(x,y) \leq \max_{i|d_i \neq 0} \mathrm{Dist}(x^{(i)}, y^{(i)})$ en écrivant $\mathfrak{m}(x)\mathfrak{m}'(y) - \mathfrak{m}'(x)\mathfrak{m}(y)$ comme somme de termes de la forme

$$\pm \mathfrak{m}_1(x)\mathfrak{m}_1'(y)(x_j^{(i)} y_k^{(i)} - x_k^{(i)} y_j^{(i)})$$

pour $\mathfrak{m}_1, \mathfrak{m}_1' \in \mathcal{M}_{d-\varepsilon_i}$, $0 \leq j < k \leq n_i$ et $1 \leq i \leq q$.

Afin d'aboutir à l'inégalité en sens contraire on considère pour chaque i ($1 \leq i \leq q$) des indices $0 \leq j_i \leq n_i$ et $0 \leq k_i \leq n_i$ tels que $|x_{j_i}^{(i)}|_v = |y_{k_i}^{(i)}|_v = 1$. Distinguons deux cas : si

$$\prod_{i|d_i \neq 0} \left| x_{k_i}^{(i)} y_{j_i}^{(i)} \right|_v \neq 1$$

on a $\max_{i|d_i \neq 0} \mathrm{Dist}(x^{(i)}, y^{(i)}) = \mathrm{Dist}_d(x,y) = 1$. En effet pour le premier on considère i tel que $d_i \neq 0$ et $|x_{k_i}^{(i)} y_{j_i}^{(i)}|_v \neq 1$ ce qui entraîne $|x_{j_i}^{(i)} y_{k_i}^{(i)} - x_{k_i}^{(i)} y_{j_i}^{(i)}|_v = 1$

puis $\text{Dist}(x^{(i)}, y^{(i)}) = 1$ tandis que pour le second on utilise $\mathfrak{m} = \prod_{i=1}^{q} X_{j_i}^{(i)^{d_i}}$ et $\mathfrak{m}' = \prod_{i=1}^{q} X_{k_i}^{(i)^{d_i}}$ (puisque $|\mathfrak{m}(x)\mathfrak{m}'(y)|_v = 1$ et $|\mathfrak{m}'(x)\mathfrak{m}(y)|_v < 1$). Supposons maintenant que le produit ci-dessus vaut 1. Pour i fixé, comme $|x_{j_i}^{(i)}|_v = 1$, il découle de

$$x_j^{(i)} y_k^{(i)} - x_k^{(i)} y_j^{(i)} = \frac{x_j^{(i)}}{x_{j_i}^{(i)}} \left(x_{j_i}^{(i)} y_k^{(i)} - x_k^{(i)} y_{j_i}^{(i)} \right) - \frac{x_k^{(i)}}{x_{j_i}^{(i)}} \left(x_{j_i}^{(i)} y_j^{(i)} - x_j^{(i)} y_{j_i}^{(i)} \right)$$

que $\text{Dist}(x^{(i)}, y^{(i)}) = \max_{0 \le k \le n_i} |x_{j_i}^{(i)} y_k^{(i)} - x_k^{(i)} y_{j_i}^{(i)}|_v$. On considère ensuite $\mathfrak{m} = \prod_{i=1}^{q} X_{j_i}^{(i)^{d_i}}$ et $\mathfrak{m}' = X_k^{(i_0)} \prod_{i=1}^{q} X_{j_i}^{(i)^{d_i - \delta_{i,i_0}}}$ où $0 \le k \le n_{i_0}$ et i_0 est tel que $d_{i_0} \ne 0$. On calcule

$$
\begin{aligned}
\left| \mathfrak{m}(x)\mathfrak{m}'(y) - \mathfrak{m}'(x)\mathfrak{m}(y) \right|_v &= \prod_{i=1}^{q} \left| x_{j_i}^{(i)} y_{j_i}^{(i)} \right|_v^{d_i - \delta_{i,i_0}} \cdot \left| x_{j_{i_0}}^{(i_0)} y_k^{(i_0)} - x_k^{(i_0)} y_{j_{i_0}}^{(i_0)} \right|_v \\
&= \left| x_{j_{i_0}}^{(i_0)} y_k^{(i_0)} - x_k^{(i_0)} y_{j_{i_0}}^{(i_0)} \right|_v
\end{aligned}
$$

puisque par notre hypothèse $|y_{j_i}^{(i)}|_v = 1$. Ainsi on a

$$\text{Dist}_d(x, y) \ge \text{Dist}(x^{(i_0)}, y^{(i_0)})$$

pour tout i_0 tel que $d_{i_0} \ne 0$ ce qui donne bien le résultat.

Traitons à présent le cas où v est infinie. On choisit ici x et y de sorte que $\sum_{j=0}^{n_i} |x_j^{(i)}|_v^2 = \sum_{j=0}^{n_i} |y_j^{(i)}|_v^2 = 1$. On utilise la relation élémentaire suivante pour $z_1, \ldots, z_p$ et $z_1', \ldots, z_p'$ deux familles de nombres complexes

$$\sum_{1 \le i < j \le p} |z_i z_j' - z_j z_i'|^2 + \left| \sum_{i=1}^{p} z_i \overline{z_i'} \right|^2 = \left(\sum_{i=1}^{p} |z_i|^2 \right) \left(\sum_{i=1}^{p} |z_i'|^2 \right).$$

Ainsi on a

$$\text{Dist}(x^{(i)}, y^{(i)})^2 = 1 - \left| \sum_{j=0}^{n_i} x_j^{(i)} \overline{y_j^{(i)}} \right|^2.$$

En considérant les familles $\binom{d}{\mathfrak{m}}^{1/2} \mathfrak{m}(x)$ et $\binom{d}{\mathfrak{m}}^{1/2} \mathfrak{m}(y)$ on trouve que $\text{Dist}_d(x, y)^2$ vaut

$$\left(\sum_{\mathfrak{m} \in \mathcal{M}_d} \binom{d}{\mathfrak{m}} \mathfrak{m}(|x|^2) \right) \left(\sum_{\mathfrak{m} \in \mathcal{M}_d} \binom{d}{\mathfrak{m}} \mathfrak{m}(|y|^2) \right) - \left| \sum_{\mathfrak{m} \in \mathcal{M}_d} \binom{d}{\mathfrak{m}} \mathfrak{m}(x\bar{y}) \right|^2$$

où $|x|^2$, $|y|^2$ et $x\bar{y}$ désignent les éléments de $\mathbf{C}_v^{n_1+1} \times \cdots \times \mathbf{C}_v^{n_q+1}$ de composantes respectives $|x_j^{(i)}|^2$, $|y_j^{(i)}|^2$ et $x_j^{(i)} \overline{y_j^{(i)}}$. Par la formule multinomiale on obtient

$$
\begin{aligned}
\text{Dist}_d(x, y)^2 &= \prod_{i=1}^{q} \left(\sum_{j=0}^{n_i} |x_j^{(i)}|_v^2 \right)^{d_i} \left(\sum_{j=0}^{n_i} |y_j^{(i)}|_v^2 \right)^{d_i} - \prod_{i=1}^{q} \left| \sum_{j=0}^{n_i} x_j^{(i)} \overline{y_j^{(i)}} \right|^{2d_i} \\
&= 1 - \prod_{i=1}^{q} \left(1 - \text{Dist}(x^{(i)}, y^{(i)})^2 \right)^{d_i}.
\end{aligned}
$$

On a ainsi l'égalité de l'énoncé qui entraîne en particulier $\mathrm{Dist}_d(x, y) \leq 1$. En majorant simplement chaque $\mathrm{Dist}(x^{(i)}, y^{(i)})$ tel que $d_i \neq 0$ par leur maximum D, un majorant pour $\mathrm{Dist}_d(x, y)^2$ est

$$1 - \left(1 - D^2\right)^{d_1 + \cdots + d_q} = D^2 \sum_{k=0}^{d_1 + \cdots + d_q - 1} (1 - D^2)^k$$
$$\leq (d_1 + \cdots + d_q)D^2.$$

Au contraire en minorant tous les $\mathrm{Dist}(x^{(i)}, y^{(i)})$ par 0 sauf un on trouve comme minorant (si $d_i \neq 0$)

$$1 - (1 - \mathrm{Dist}(x^{(i)}, y^{(i)})^2)^{d_i} = \mathrm{Dist}(x^{(i)}, y^{(i)})^2 \sum_{k=0}^{d_i - 1} (1 - \mathrm{Dist}(x^{(i)}, y^{(i)})^2)^k$$
$$\geq \mathrm{Dist}(x^{(i)}, y^{(i)})^2.$$

Ceci donne l'encadrement escompté de $\mathrm{Dist}_d(x, y)$. La dernière inégalité s'écrit

$$1 - \prod_{i=1}^{q} z_i^{d_i} \leq \gamma \left(1 - \prod_{i=1}^{q} z_i^{d_i'} \right)$$

si l'on note $z_i = 1 - \mathrm{Dist}(x^{(i)}, y^{(i)})^2$ et $\gamma = \max\left(1, \frac{d_1}{d_1'}, \ldots, \frac{d_q}{d_q'}\right)$, pour simplifier. Comme on a $d_i \leq \gamma d_i'$ et $z_i \in [0, 1]$ pour tout i on déduit $z_i^{d_i} \geq z_i^{\gamma d_i'}$ puis

$$1 - \prod_{i=1}^{q} z_i^{d_i} \leq 1 - \left(\prod_{i=1}^{q} z_i^{d_i'} \right)^{\gamma}.$$

On conclut donc par la formule

$$1 - z^{\gamma} \leq \gamma(1 - z)$$

valable si $z \in [0, 1]$ et $\gamma \in [1, +\infty[$ (par exemple car $1 - z^{\gamma} = \gamma \int_z^1 x^{\gamma - 1} dx \leq \gamma \int_z^1 dx = \gamma(1 - z)$). □

Lorsque v est une place infinie, il résulte des calculs qui précèdent que la distance $\mathrm{Dist}_d(x, y)$ entre deux points est invariante par transformation unitaire (voir [**Jad**, p. 56] pour $q = 1$). Ceci signifie que si $u_i : \mathbf{C}_v^{n_i + 1} \to \mathbf{C}_v^{n_i + 1}$ est un automorphisme unitaire et $u = u_1 \times \cdots \times u_q$ alors $\mathrm{Dist}_d(x, y) = \mathrm{Dist}_d(u(x), u(y))$. Le groupe unitaire agit transitivement sur $S_{n_i + 1}(1)$, on peut par cette remarque supposer pour nos calculs que x est représenté par $((1, 0, \ldots, 0), \ldots, (1, 0, \ldots, 0)) \in \prod_{i=1}^{1} \mathbf{C}_v^{n_i + 1}$. Ceci aura son importance surtout lorsque nous aurons généralisé cette propriété à $\mathrm{Dist}(x, V)$.

Enfin la dernière remarque sur le cas des points que nous ferons est de rappeler la comparaison avec une distance "affine". En effet si x et y sont deux points fermés de $\mathbf{P}_{\mathbf{C}}^n$ de représentants choisis tels que $x_0 = y_0 = 1$ on a

$$\frac{1}{\|x\| \, \|y\|} \leq \frac{\mathrm{Dist}(x, y)}{\|x - y\|} \leq \frac{1}{\|y\|} \leq 1$$

où $\| \cdot \|$ est bien sûr la norme euclidienne sur $\mathbf{C}^{n+1}$; lorsque $x_0 = 1$ mais $y_0 = 0$ on peut écrire

$$\frac{1}{\|x\|} \leq \mathrm{Dist}(x, y) \leq 1$$

(pour ces calculs on utilise une transformation unitaire de $\mathbf{C}^{n+1}$ fixant la première coordonnée pour se réduire au cas où x vérifie $x_i = 0$ si $i > 1$ dans lequel le calcul est direct; voir aussi [**Jad**]).

4.4. Décomposition de la distance. La distance étant définie en utilisant le rapport de la mesure de $\beta_d \circ e_x \circ \mathfrak{d}(f)$ à celle de $\alpha_d(f)$, on va pouvoir la décomposer en introduisant des morphismes intermédiaires entre $\beta_d \circ e_x \circ \mathfrak{d}$ et α_d (cette idée figure dans [**Jad**] par exemple page 70). On définit en effet pour $0 \le i \le r$ un morphisme de $\mathbf{C}_v$-algèbres

$$\mathfrak{d}_{x,d}^{(i)} \colon \mathbf{C}_v[d] \quad\longrightarrow\quad \mathbf{C}_v[u^{(1)}, \ldots, u^{(i)}, s^{(i+1)}, \ldots, s^{(r)}]$$

$$u_{\mathrm{m}}^{(l)} \longmapsto \begin{cases} \alpha_d(u_{\mathrm{m}}^{(l)}) & \text{si} \quad l \le i \\ \beta_d \circ e_x \circ \mathfrak{d}(u_{\mathrm{m}}^{(l)}) & \text{si} \quad l > i. \end{cases}$$

On retrouve $\mathfrak{d}_{x,d}^{(0)} = \beta_d \circ e_x \circ \mathfrak{d}$ et $\mathfrak{d}_{x,d}^{(r)} = \alpha_d$. La décomposition annoncée est

$$\mathrm{Dist}_d(x,V) = \prod_{l=1}^{r} \frac{M_v(\mathfrak{d}_{x,d}^{(l-1)}(f))}{M_v(\mathfrak{d}_{x,d}^{(l)}(f)) M_v(\alpha_{(d_l)}(U_l)(x))^{\deg_{u^{(l)}} f}}.$$

On notera $D_d^{(l)}(x,V)$ le facteur d'indice l (on vérifie qu'il est bien défini indépendamment des choix de f ou du représentant de x qui apparaît pour définir $\mathfrak{d}_{x,d}^{(i)}$). Pour achever de donner un sens à cette écriture il faudrait montrer que $\mathfrak{d}_{x,d}^{(l)}(f) \ne 0$ si $0 < l \le r$. Comme il est clair que $\mathfrak{d}_{x,d}^{(l-1)}(f)$ est une spécialisation de $\mathfrak{d}_{x,d}^{(l)}(f)$, on déduit que si l'un des $\mathfrak{d}_{x,d}^{(l)}(f)$ est nul il en va de même pour tous les indices inférieurs. En particulier il ne peut y avoir de problèmes que lorsque $\mathrm{Dist}_d(x,V) = 0$. On pourrait donc simplement traiter ce cas à part. Toutefois, comme il est effectivement vrai (au moins dès que l'on suppose $d_{l,i} \ne 0$ pour tous l,i) que $\mathfrak{d}_{x,d}^{(l)}(f)$ ne peut s'annuler que si $l = 0$, les facteurs $D_d^{(l)}(x,V)$ sont bien définis et l'annulation de $\mathrm{Dist}_d(\cdot,V)$ est définie par celle de $D_d^{(1)}(\cdot,V)$. Ceci amène à considérer $D_d^{(1)}(x,V)$ comme le terme principal d'autant plus que nous montrerons que, si $2 \le l \le r$, le facteur $D_d^{(l)}(x,V)$ est minoré par un réel strictement positif indépendant de x et V (lemme 4.5). D'ici là les calculs sont faits seulement pour les $D_d^{(l)}(x,V)$ ayant un sens.

La première justification pour l'introduction de notre décomposition de la distance est qu'elle permet d'exprimer cette dernière à l'aide de distances entre points. On calcule en effet $D_d^{(l)}(x,V)$ grâce au lemme 2.1. Son numérateur est la mesure d'une forme de $\mathbf{C}_v[u^{(1)}, \ldots, u^{(l-1)}, s^{(l)}, \ldots, s^{(r)}]$ tandis que (après la normalisation du représentant de x) son dénominateur est la mesure d'une forme de $\mathbf{C}_v[u^{(1)}, \ldots, u^{(l)}, s^{(l+1)}, \ldots, s^{(r)}]$. De plus il est immédiat que $\mathfrak{d}_{x,d}^{(l)}(f)$ et $\mathfrak{d}_{x,d}^{(l-1)}(f)$ ont mêmes degrés en les groupes de variables qui leur sont communs. Par conséquent on a

$$D_d^{(l)}(x,V) = \underset{\rho\colon u^{(1)}, \ldots, u^{(l-1)}, s^{(l+1)}, \ldots, s^{(r)}}{\mathrm{moy}_v} \frac{M_v(\rho(\mathfrak{d}_{x,d}^{(l-1)}(f)))}{M_v(\rho(\mathfrak{d}_{x,d}^{(l)}(f)))}$$

(ici encore pour simplifier l'écriture x est tel que $M_v(\alpha_{(d_l)}(U_l)(x)) = 1$ pour tout l). On associe à présent à un ρ tel qu'il en apparaît dans l'expression ci-dessus un morphisme $\rho' \colon \mathbf{C}_v[d] \to \mathbf{C}_v[u^{(l)}]$ défini par $\rho'(u_{\mathrm{m}}^{(l)}) = u_{\mathrm{m}}^{(l)}$, $\rho'(u_{\mathrm{m}}^{(l')}) = \rho(\alpha_d(u_{\mathrm{m}}^{(l')}))$ si $l' < l$ et $\rho'(u_{\mathrm{m}}^{(l')}) = \rho(\beta_d \circ e_x \circ \mathfrak{d}(u_{\mathrm{m}}^{(l')}))$ si $l' > l$. Bien sûr ceci est fait de telle sorte que $\rho(\mathfrak{d}_{x,d}^{(l-1)}(f)) = \beta_{(d_l)} \circ e_x \circ \mathfrak{d}(\rho'(f))$ et $\rho(\mathfrak{d}_{x,d}^{(l)}(f)) = \alpha_{(d_l)}(\rho'(f))$. De la proposition 2.16 du chapitre 5 (et pour raison de degré) on sait que $\rho'(f)$ s'écrit $\lambda \prod_{h=1}^{\deg_{u^{(l)}} f} U_1(z_{\rho,h})$ ce qui donne finalement

$$D_d^{(l)}(x,V) = \operatorname*{moy}_{\rho}{}_v \prod_{h=1}^{\deg_{u^{(l)}} f} \operatorname{Dist}_{(d_l)}(x, z_{\rho,h})$$

avec le même domaine pour ρ. C'est grâce à cette formule que l'on peut utiliser les résultats du paragraphe précédent sur les points. Par exemple on déduit immédiatement $D_d^{(l)}(x,V) \le 1$.

Signalons aussi que, bien que l'expression ci-dessus nous suffise, elle suggère une écriture faisant apparaître une moyenne de $\operatorname{Dist}_d(x,y)$ pour $y \in V$. Dans le cas d'une place infinie on peut effectivement, par un changement de variables apparenté à celui qui figure dans la démonstration du lemme 3.5, aboutir à

$$D_d^{(l)}(x,V) = \exp \int_V \log \operatorname{Dist}_{(d_l)}(x,y) \Omega_{x,d}(y)$$

pour une certaine mesure $\Omega_{x,d}$ sur V (voir à ce sujet le chapitre 6, lemma 6.1). Plus précisément on montre que la mesure $\Omega_{x,d}(y)$ est proportionnelle à la forme suivante dans laquelle on reconnaît la différence entre les variables $u^{(i)}$ et $s^{(i)}$

$$\bigwedge_{l'<l} \left(\sum_{i=1}^q d_{l',i} \partial \bar\partial \log \|y^{(i)}\| \right) \wedge \bigwedge_{l'>l} \left(\sum_{i=1}^q d_{l',i} \partial \bar\partial \log \|x^{(i)} \wedge y^{(i)}\| \right)$$

(on a noté $x^{(i)} \wedge y^{(i)}$ le vecteur de composantes $x_j^{(i)} y_k^{(i)} - x_k^{(i)} y_j^{(i)}$ pour $0 \le j < k \le n_i$).

Un autre intérêt de faire intervenir une distance entre points se rapporte à l'invariance par transformation unitaire. En effet de même que dans le cas projectif (voir par exemple [**Jad**, page 70]) on déduit de la formule sur les points que si u est une telle transformation on a $D_d^{(l)}(x,V) = D_d^{(l)}(u(x), u(v))$: c'est particulièrement clair sous la forme de l'intégrale précédente puisque tant la fonction (cas des points) que la mesure sont invariantes (en remplaçant x par $u(x)$). Par conséquent on pourra, dans les preuves qui vont suivre, supposer si besoin est que x est de la forme donnée à la fin du paragraphe précédent.

Nos calculs sur $D_d^{(l)}(x,V)$ vont permettre d'établir la proposition restée en suspens. Pour le changement d'indices on distinguera selon si $l > 1$ ou $l = 1$. En effet dans le premier cas on a un comportement multiplicatif tandis que dans le second on trouve une inégalité qui est l'extension du cas des points.

LEMME 4.4. *Dans les notations précédentes on a*

1. *si* $l > 1$

$$D_d^{(l)}(x,V) = \prod_{i=1}^q D_{d^{(i)}}^{(l)}(x,V)^{d_{1,i}}$$

où $d^{(i)} = (\varepsilon_i, d_2, \dots, d_r)$;

2. *si* $d' = (d'_1, d_2, \ldots, d_r)$ *et* $d_{1,i} d'_{1,i} \neq 0$ *pour tout* $1 \leq i \leq q$ *alors* $D_d^{(1)}(x, V) = D_{d'}^{(1)}(x, V)$ *pour une place finie tandis que si* v *est infinie*

$$D_d^{(1)}(x, V) \leq \max\left(1, \frac{d_{1,1}}{d'_{1,1}}, \ldots, \frac{d_{1,q}}{d'_{1,q}}\right)^{\frac{1}{2} \deg_{u^{(1)}} f} D_{d'}^{(1)}(x, V).$$

DÉMONSTRATION. 1. On se reporte à la démonstration du théorème 2.2, où ω et $\tilde{d}$ sont le morphisme de spécialisation et l'indice correspondant introduits au paragraphe 3.4 du chapitre 5. Il est clair que $M_v(\mathfrak{d}_{x,\tilde{d}}^{(l)}(\omega(f))) = \prod_{i=1}^q M_v(\mathfrak{d}_{x,d^{(i)}}^{(l)}(f_i))^{d_{1,i}}$ si $l > 1$ donc notre assertion se ramène à voir que

$$D_d^{(l)}(x, V) = \frac{M_v(\mathfrak{d}_{x,\tilde{d}}^{(l-1)}(\omega(f)))}{M_v(\mathfrak{d}_{x,\tilde{d}}^{(l)}(\omega(f)))}$$

soit

$$\frac{M_v(\mathfrak{d}_{x,\tilde{d}}^{(l)}(\omega(f)))}{M_v(\mathfrak{d}_{x,d}^{(l)}(f))} = \frac{M_v(\mathfrak{d}_{x,\tilde{d}}^{(l-1)}(\omega(f)))}{M_v(\mathfrak{d}_{x,d}^{(l-1)}(f))}.$$

Montrons en fait que les deux termes valent 1. Comme le second correspond à remplacer l par $l-1$ on se limite au premier; celui-ci vaut

$$\operatorname*{moy}_{\substack{v \\ \rho:\, u^{(2)}, \ldots, u^{(l)}, s^{(l+1)}, \ldots, s^{(r)}}} \frac{M_v(\rho(\mathfrak{d}_{x,\tilde{d}}^{(l)}(\omega(f))))}{M_v(\rho(\mathfrak{d}_{x,d}^{(l)}(f)))}$$

par le lemme 2.1. C'est un calcul semblable à celui employé dans la preuve du théorème 2.2 (on y démontrait en fait le cas $l = 0$). On réutilise un ρ' analogue et, $\mathfrak{d}_{x,d}^{(l)}$ étant α_d sur $u^{(1)}$ comme $\mathfrak{d}_{x,\tilde{d}}^{(l)}$ est $\alpha_{\tilde{d}}$ sur $u^{(1)}, u^{(r+1)}, \ldots, u^{(r+q-1)}$, on trouve que l'on fait la moyenne de

$$\frac{M_v(\rho'(\alpha_{\tilde{d}}(\omega(f))))}{M_v(\rho'(\alpha_d(f)))}.$$

Or ce rapport vaut 1 lorsque f est une forme résultante comme on l'a prouvé en établissant le théorème 2.2.

2. Il suffit de voir (en vertu du lemme 4.3) que si les points $(z_{\rho,h})$ pour $\rho: \mathbf{C}_v[s^{(2)}, \ldots, s^{(r)}] \to \mathbf{C}_v$ et $1 \leq h \leq \deg_{u^{(1)}} f$ sont attachés à f (forme résultante d'indice d) alors

$$D_{d'}^{(1)}(x, V) = \operatorname*{moy}_\rho \prod_{h=1}^{\deg_{u^{(1)}} f} \operatorname{Dist}_{(d'_1)}(x, z_{\rho,h}).$$

Or si f' est une forme résultante d'indice d' de V on a d'abord $\deg_{u^{(1)}} f' = \deg_{u^{(1)}} f$ (par la proposition 3.4 du chapitre 5). Ensuite, dans la définition des $(z_{\rho,h})$ attachés à f ou des $(z'_{\rho,h})$ attachés à f', il a été introduit un ρ' ne dépendant que de $d_2, \ldots, d_r$. Ainsi comme les $z_{\rho,h}$ correspondent à l'intersection de V avec les zéros des formes $\rho'(U_2), \ldots, \rho'(U_r)$ on a (à l'ordre près) $z'_{\rho,h} = z_{\rho,h}$ ce qui donne la formule. $\qquad\square$

Le dernier ingrédient dont nous aurons besoin est un minorant de $D_d^{(l)}(x, V)$ si $l \neq 1$. Comme il a été dit plus haut cela achèvera aussi de montrer que ces facteurs (y compris pour $l = 1$) sont bien définis.

LEMME 4.5. *On pose*

$$\tilde{D}_d(x,V) = \frac{\mathrm{Dist}_d(x,V)}{D_d^{(1)}(x,V)} = \prod_{l=2}^{r} D_d^{(l)}(x,V).$$

On suppose $d_{1,i} \le d_{l,i}$ pour tout $1 \le i \le q$ et $1 \le l \le r$. Alors $\tilde{D}_d(x,V) = 1$ si v est finie et

$$\tilde{D}_d(x,V) \ge \prod_{l=1}^{r} \left(\gamma_{d_1} \gamma_{d_l}^{1-\delta_{l,1}} \right)^{-\deg_{u^{(l)}} f}$$

si v est infinie (les symboles sont définis dans l'énoncé de la proposition).

DÉMONSTRATION. Commençons par étendre la remarque 2 de [**PPh2**, p. 16] au cas multiprojectif. Pour $\mathfrak{m} \in \mathcal{M}_{d_l}$ on note

$$\widehat{\mathfrak{m}} = \mathfrak{m} \prod_{i=1}^{q} X_0^{(i)}{}^{d_{1,i} - d_{l,i}}$$

qui est un monôme de multidegré d_1 mais avec des exposants éventuellement négatifs. On considère ensuite des indéterminées $\lambda_2, \ldots, \lambda_r$ et τ le morphisme de K-algèbres donné par

$$
\begin{array}{rcl}
K[d] & \longrightarrow & K[d][\lambda_2, \ldots, \lambda_r] \\
u_{\mathfrak{m}}^{(l)} & \longmapsto & \left\{ \begin{array}{ll} u_{\mathfrak{m}}^{(l)} - \lambda_l u_{\widehat{\mathfrak{m}}}^{(1)} & \text{si } l \ge 2 \text{ et } \widehat{\mathfrak{m}} \in \mathcal{M}_{d_1} \\ u_{\mathfrak{m}}^{(l)} & \text{sinon.} \end{array} \right.
\end{array}
$$

Ce morphisme est tel que son extension à $K[X][d]$ vérifie $\tau(U_1) = U_1$ et $\tau(U_l) = U_l - \lambda_l U_1 \prod_{i=1}^{q} X_0^{(i)}{}^{d_{l,i} - d_{1,i}}$ si $l \ge 2$ (puisque par l'hypothèse $d_l \ge d_1$ tout élément de $\mathcal{M}_{d_1}$ est de la forme $\widehat{\mathfrak{m}}$ pour un certain (aussitôt unique) $\mathfrak{m} \in \mathcal{M}_{d_l}$). L'argument de [**PPh2**] se reproduit alors à l'identique et montre que l'on a $\tau(f) = f$.

On peut alors (utilisant la même technique que dans la démonstration du lemme 4.6 de [**Jad**] que l'on généralise) considérer la spécialisation

$$\sigma \colon \mathbf{C}_v[u^{(1)}, s^{(2)}, \ldots, s^{(r)}] \longrightarrow \mathbf{C}_v[u^{(2)}, \ldots, u^{(r)}](u^{(1)})$$

définie par $\sigma(u_{\mathfrak{m}}^{(1)}) = u_{\mathfrak{m}}^{(1)}$ pour tout $\mathfrak{m} \in \mathcal{M}_{d_1}$ et $(2 \le l \le r)$

$$\sigma(\beta_d(s_{\mathfrak{m},\mathfrak{m}'}^{(l)})) = \alpha_d \left(u_{\mathfrak{m}}^{(l)} u_{\mathfrak{m}'}^{(1)} - u_{\mathfrak{m}'}^{(l)} u_{\widehat{\mathfrak{m}}}^{(1)} \right) \alpha_d \left(\sum_{\mathfrak{m}'' \in \mathcal{M}_{d_l}} u_{\widehat{\mathfrak{m}''}}^{(1)} \mathfrak{m}''(x) \right)^{-1}$$

avec la convention que $u_{\widehat{\mathbf{m}}}^{(1)} = 0$ si $\mathbf{m}$ est tel que $\widehat{\mathbf{m}} \notin \mathcal{M}_{d_1}$. On veut calculer $\sigma(\mathfrak{d}_{x,d}^{(1)}(f))$. Or on a si $2 \leq l \leq r$

$$\sigma(\mathfrak{d}_{x,d}^{(1)}(u_{\mathbf{m}}^{(l)})) = \sigma \circ \beta_d \left(\sum_{\mathbf{m}' \in \mathcal{M}_{d_l}} s_{\mathbf{m},\mathbf{m}'}^{(l)} \mathbf{m}'(x) \right)$$

$$= \frac{\alpha_d \left(\sum_{\mathbf{m}'} u_{\mathbf{m}}^{(l)} u_{\widehat{\mathbf{m}'}}^{(1)} \mathbf{m}'(x) - \sum_{\mathbf{m}'} u_{\mathbf{m}'}^{(l)} \mathbf{m}'(x) u_{\widehat{\mathbf{m}}}^{(1)} \right)}{\alpha_d \left(\sum_{\mathbf{m}'} u_{\widehat{\mathbf{m}'}}^{(1)} \mathbf{m}'(x) \right)}$$

$$= \alpha_d(u_{\mathbf{m}}^{(l)}) - \frac{\alpha_d \left(\sum_{\mathbf{m}'} u_{\mathbf{m}'}^{(l)} \mathbf{m}'(x) \right)}{\alpha_d \left(\sum_{\mathbf{m}'} u_{\widehat{\mathbf{m}'}}^{(1)} \mathbf{m}'(x) \right)} \alpha_d(u_{\widehat{\mathbf{m}}}^{(1)}).$$

Ainsi, d'après ce qui précède, $\sigma(\mathfrak{d}_{x,d}^{(1)}(f)) = \alpha_d(f)$ (car $\alpha_d^{-1} \circ \sigma \circ \mathfrak{d}_{x,d}^{(1)}$ se factorise à travers τ). En particulier $\mathfrak{d}_{x,d}^{(1)}(f)$ est non nul et tous les facteurs $D_d^{(l)}(x,V)$ sont donc bien définis. Notre but étant de majorer $M_v(\alpha_d(f))$ à l'aide de $M_v(\mathfrak{d}_{x,d}^{(1)}(f))$ et au vu de l'écriture

$$M_v(\alpha_d(f)) = \underset{\rho: u^{(1)}, \ldots, u^{(r)}}{\text{moy}_v} |\rho(\alpha_d(f))|_v$$

on souhaite majorer $|\rho(\alpha_d(f))|_v = |\rho \circ \sigma(\mathfrak{d}_{x,d}^{(1)}(f))|_v$ en appliquant la deuxième assertion du lemme 2.1 au polynôme $\mathfrak{d}_{x,d}^{(1)}(f)$ et à la spécialisation $\rho \circ \sigma$. Ceci est loisible car la spécialisation de $\mathfrak{d}_{x,d}^{(1)}(f)$ en tous les groupes de variables sauf un donne bien un produit de formes linéaires (voir par exemple le calcul fait pour écrire $D_d^{(l)}(x,V)$ comme moyenne). On a donc

$$|\rho(\alpha_d(f))|_v \leq M_v(\mathfrak{d}_{x,d}^{(1)}(f)) M_v \left(\sum_{\mathbf{m} \in \mathcal{M}_{d_1}} \rho \circ \sigma(u_{\mathbf{m}}^{(1)}) u_{\mathbf{m}}^{(1)} \right)^{\deg_{u^{(1)}} f}$$

$$\times \prod_{l=2}^r M_v \left(\sum_{\mathbf{m} \prec \mathbf{m}' \in \mathcal{M}_{d_l}} \rho \circ \sigma(s_{\mathbf{m},\mathbf{m}'}^{(l)}) s_{\mathbf{m},\mathbf{m}'}^{(l)} \right)^{\deg_{u^{(l)}} f}.$$

Vu le choix de ρ on a $M_v(\sum_{\mathbf{m}} \rho(u_{\mathbf{m}}^{(l)}) u_{\mathbf{m}}^{(l)}) = 1$ pour tout $1 \leq l \leq r$. Le deuxième facteur du membre de droite disparaît donc. Pour le dernier on majore

$$|\rho \circ \sigma(s_{\mathbf{m},\mathbf{m}'}^{(l)})|_v \leq \left| \sum_{\mathbf{m}'' \in \mathcal{M}_{d_l}} \rho(u_{\widehat{\mathbf{m}''}}^{(1)}) \mathbf{m}''(x) \right|_v^{-1}$$

si v est finie tandis que pour une place infinie on majore $\sum_{\mathfrak{m}\prec\mathfrak{m}'}|\rho\circ\sigma(s_{\mathfrak{m},\mathfrak{m}'}^{(l)})|_v^2$ par

$$\frac{\sum_{\mathfrak{m}\prec\mathfrak{m}'}\left|\binom{d_l}{\mathfrak{m}'}^{-1/2}\binom{d_1}{\widehat{\mathfrak{m}'}}^{1/2}\rho(u_{\mathfrak{m}}^{(l)})\rho(u_{\widehat{\mathfrak{m}'}}^{(1)})-\binom{d_l}{\mathfrak{m}}^{-1/2}\binom{d_1}{\widehat{\mathfrak{m}}}^{1/2}\rho(u_{\mathfrak{m}'}^{(l)})\rho(u_{\widehat{\mathfrak{m}}}^{(1)})\right|_v^2}{\left|\sum_{\mathfrak{m}''\in\mathcal{M}_{d_l}}\rho(\alpha_d(u_{\widehat{\mathfrak{m}''}}^{(1)}))\mathfrak{m}''(x)\right|_v^2}$$

Comme le numérateur est celui du carré de la distance entre les points

$$\left(\rho(u_{\mathfrak{m}}^{(l)})\right)_{\mathfrak{m}\in\mathcal{M}_{d_l}}\qquad\text{et}\qquad\left(\binom{d_l}{\mathfrak{m}}^{-1/2}\binom{d_1}{\widehat{\mathfrak{m}}}^{1/2}\rho(u_{\widehat{\mathfrak{m}}}^{(1)})\right)_{\mathfrak{m}\in\mathcal{M}_{d_l}}$$

on peut le majorer par le produit des carrés des normes de ces points. Or on sait (voir ci-dessus) $\sum_{\mathfrak{m}}|\rho(u_{\mathfrak{m}}^{(l)})|_v^2=1$ et l'on a $\binom{d_1}{\widehat{\mathfrak{m}}}\leq\binom{d_l}{\mathfrak{m}}$ si $\mathfrak{m}\in\mathcal{M}_{d_l}$ (on rappelle l'hypothèse $d_1\leq d_l$). Par suite

$$M_v\left(\sum_{\mathfrak{m}\prec\mathfrak{m}'}\rho\circ\sigma(s_{\mathfrak{m},\mathfrak{m}'}^{(l)})s_{\mathfrak{m},\mathfrak{m}'}^{(l)}\right)$$

est majoré pour toute place par l'inverse de

$$\left|\sum_{\mathfrak{m}''\in\mathcal{M}_{d_l}}\rho(\alpha_d(u_{\widehat{\mathfrak{m}''}}^{(1)}))\mathfrak{m}''(x)\right|_v=|\rho(\alpha_d(U_1))(x)|_v\prod_{i=1}^q|x_0^{(i)}|_v^{d_{1,i}-d_{l,i}}.$$

On peut maintenant supposer que $|x_0^{(i)}|_v=1$ pour tout $1\leq i\leq q$ en utilisant (avant le début des calculs) une permutation d'indices aux places finies et une transformation unitaire aux places infinies. On a donc

$$|\rho(\alpha_d(f))|_v|\rho(\alpha_d(U_1))(x)|_v^{\sum_{l=2}^r\deg_{u^{(l)}}f}\leq M_v(\mathfrak{d}_{x,d}^{(1)}(f)).$$

En calculant moy$_v$ et par multiplicativité on a

$$\text{moy}_v|\alpha_d(f)|_v\left(\text{moy}_v|\alpha_d(U_1(x))|_v\right)^{\sum_{l=2}^r\deg_{u^{(l)}}f}\leq M_v(\mathfrak{d}_{x,d}^{(1)}(f)).$$

Si v est finie moy$_v|\cdot|_v$ est M_v et (avec la normalisation de x) on trouve

$$M_v(\alpha_d(f))\leq M_v(\mathfrak{d}_{x,d}^{(1)}(f))$$

soit $\tilde{D}_d(x,V)\geq 1$ ce qui conclut. Si v est infinie on raisonne de même mais le passage de moy$_v|\cdot|_v$ à M_v fait apparaître des termes de degré; on a au membre de gauche

$$M_v(\alpha_d(f))\prod_{l=1}^r\exp\left(\sum_{j=1}^{|\mathcal{M}_{d_l}|-1}\frac{1}{2j}\right)^{-\deg_{u^{(l)}}f}\exp\left(\sum_{j=1}^{|\mathcal{M}_{d_1}|-1}\frac{1}{2j}\right)^{-\sum_{l=2}^r\deg_{u^{(l)}}f}$$

donc

$$\tilde{D}_d(x,V)\geq\gamma_{d_1}^{-\sum_{l=1}^r\deg_{u^{(l)}}f}\prod_{l=2}^r\gamma_{d_l}^{-\deg_{u^{(l)}}f}$$

qui est le résultat. $\qquad\square$

Ce résultat montre qu'avec la condition de l'énoncé tous les facteurs $D_d^{(l)}(x,V)$ sont bien définis et, si $l > 1$, non nuls. Par le lemme 4.4 on en déduit que cela reste vrai sous la condition plus faible $d_{1,i} \neq 0 \implies d_{l,i} \neq 0$ pour tous l, i (on utilise que $D^{(1)}$ est bien défini dès que $D^{(2)}$ ne s'annule pas). En particulier tous ces facteurs sont définis si $d_{l,i}$ est toujours non nuls (ce que nous supposons pour les assertions 2 et 3 montrées ci-dessous).

DÉMONSTRATION DE LA PROPOSITION. 1. Immédiat à présent que

$$\text{Dist}_d(x,V) = \prod_{l=1}^{r} D_d^{(l)}(x,V)$$

et $D_d^{(l)}(x,V) \leq 1$ pour tout l. Ceci vaut bien sûr seulement lorsque les $D^{(l)}$ sont bien définis mais dans le cas contraire on a $\text{Dist}_d(x,V) = 0$.

2. Pour une place finie on a $\tilde{D}_d(x,V) = \tilde{D}_{d'}(x,V) = 1$ (lemmes 4.5 et 4.4) et $D_d^{(1)}(x,V) = D_{d'}^{(1)}(x,V)$ donc $\text{Dist}_d(x,V) = \text{Dist}_{d'}(x,V)$. Pour v infinie on utilise aussi $\text{Dist}_d(x,V) = \tilde{D}_d(x,V)D_d^{(1)}(x,V)$. On a d'une part

$$D_d^{(1)}(x,V) \leq \max\left(1, \frac{d_{1,1}}{d'_{1,1}}, \dots, \frac{d_{1,q}}{d'_{1,q}}\right)^{\frac{1}{2}\deg_{u^{(1)}} f} D_{d'}^{(1)}(x,V)$$

et d'autre part (lemme 4.4)

$$\frac{\tilde{D}_d(x,V)}{\tilde{D}_{d'}(x,V)} = \prod_{i=1}^{q} \tilde{D}_{d^{(i)}}(x,V)^{d_{1,i}-d'_{1,i}}.$$

Par le lemme 4.5 on a (si $f^{(i)}$ est une forme éliminante d'indice $d^{(i)}$)

$$\tilde{D}_{d^{(i)}}(x,V) \geq \gamma_{\varepsilon_i}^{-\deg_{u^{(1)}} f^{(i)}} \prod_{l=2}^{r} (\max_i(\gamma_{\varepsilon_i})\gamma_{d_l})^{-\deg_{u^{(l)}} f^{(i)}}.$$

Par ailleurs $\deg_{u^{(l)}} f \sum_{i=1}^{q} d_{1,i} \deg_{u^{(l)}} f^{(i)}$ si $l \neq 1$ (en de même par f') et $\deg_{u^{(1)}} f = \deg_{u^{(1)}} f^{(i)}$. Dans le cas général on minore $d_{1,i} - d'_{1,i}$ par $-d'_{1,i}$ et l'on a

$$\frac{\tilde{D}_d(x,V)}{\tilde{D}_{d'}(x,V)} \leq \prod_{i=1}^{q} \gamma_{\varepsilon_i}^{d'_{1,i}\deg_{u^{(1)}} f^{(i)}} \prod_{l=2}^{r} (\max_i(\gamma_{\varepsilon_i})\gamma_{d_l})^{\sum_{i=1}^{q} d'_{1,i}\deg_{u^{(l)}} f^{(i)}}$$

$$\leq \prod_{i=1}^{q} \gamma_{\varepsilon_i}^{d'_{1,i}\deg_{u^{(1)}} f} \prod_{l=2}^{r} (\max_i(\gamma_{\varepsilon_i})\gamma_{d_l})^{\deg_{u^{(l)}} f'}.$$

Ceci donne la première formule. Pour la seconde on obtient la majoration en écrivant

$$\frac{\tilde{D}_d(x,V)}{\tilde{D}_{d'}(x,V)} = \prod_{i=1}^{q} \tilde{D}_{d^{(i)}}(x,V)^{d_{1,i}-d'_{1,i}} \leq 1;$$

pour la minoration on a aisément $D_d^{(1)}(x,V) \geq D_{d'}^{(1)}(x,V)$ puis

$$\frac{\tilde{D}_d(x,V)}{\tilde{D}_{d'}(x,V)} \geq \prod_{i=1}^{q} \left(\prod_{l=2}^{r} (\max_i(\gamma_{\varepsilon_i})\gamma_{d_l})^{-\deg_{u^{(l)}} f^{(i)}} \gamma_{\varepsilon_i}^{-\deg_{u^{(1)}} f^{(i)}} \right)^{d_{1,i}-d'_{1,i}}$$

qui est la formule annoncée.

3. Dans l'écriture

$$D_d^{(l)}(x, V) = \operatorname*{moy}_\rho{}_v \prod_{h=1}^{\deg_{u^{(l)}} f} \operatorname{Dist}_{(d_l)}(x, z_{\rho,h})$$

les points $z_{\rho,h}$ appartiennent à V. Que v soit finie ou infinie la définition de moy_v est telle qu'il existe ρ avec

$$D_d^{(l)}(x, V) \geq \prod_{h=1}^{\deg_{u^{(l)}} f} \operatorname{Dist}_{(d_l)}(x, z_{\rho,h}).$$

On pourrait multiplier entr'elles ces inégalités et obtenir un exposant $1/\deg f$. Pour obtenir une meilleure estimation (au moins — pour v infinie — lorsque $\operatorname{Dist}_d(x, V)$ est petite) on n'emploie l'inégalité que pour $l = 1$ la remplaçant sinon par le lemme 4.5 ainsi pour un certain $z \in V$

$$\operatorname{Dist}_d(x, V) \geq \operatorname{Dist}_{(d_1)}(x, z)^{\deg_{u^{(1)}} f}$$

si v est finie et sinon

$$\operatorname{Dist}_d(x, V) \geq \prod_{l=1}^{r} \left(\gamma_{d_1} \gamma_{d_l}^{1-\delta_{l,1}} \right)^{\deg_{u^{(l)}} f} \operatorname{Dist}_{(d_1)}(x, z)^{\deg_{u^{(1)}} f}$$

ce qui donne la conclusion. $\qquad\square$

CHAPTER 8

Criteria for algebraic independence

In quantitative form criteria for algebraic independence are meant to turn the output of the transcendance method into a measure of algebraic independence. The output of the transcendence method (complemented with a zeros estimate) usually take the following aspect :

(H). Let $x \in \mathbf{P}_n(\mathbf{C})$, $k \in \{0, \ldots, n\}$ and δ, τ, σ, U real numbers satisfying $\delta \geq 1$, $\sigma \geq 1$, $\sigma^{k+1} < \tau < U$. For any S satisfying $\frac{\tau}{\sigma^{k+1}} < S \leq \frac{U}{\sigma^{k+1}}$, there exists a family of forms $Q_1, \ldots, Q_m \in \mathbf{Z}[X_0, \ldots, X_n]$ satisfying for $i = 1, \ldots, m$:
(1) $d^{\circ}Q_i \leq \delta$, $h_1(Q_i) \leq \tau$;
(2) $\frac{\|Q_i(x)\|}{\|Q_i\|} \leq \exp\left(-S\sigma^{k+1}\right)$;
(3) the forms Q_i don't have any common zero in the ball

$$B(x, \exp(-S\sigma^{k+2})) := \left\{ y \in \mathbf{P}_n(\mathbf{C}); \operatorname{Dist}(x, y) \leq \exp(-S\sigma^{k+2}) \right\} \quad .$$

Recall that $\|Q\| = \sqrt{\left(\sum_\alpha \frac{|Q_\alpha|^2}{\binom{d^\circ Q}{\alpha}} \right)^{-1/2}}$, $h_1(Q) = \log \|Q\|$ and $\|Q(x)\| = \frac{|Q(x)|}{\|x\|^{d^\circ_F}}$, for $Q \in \mathbf{Z}[X_0, \ldots, X_n]$.

The following result has been proved by C. Jadot [Jad], improving upon a previous criterium of E.M. Jabbouri [Jab].

MAIN CRITERION FOR ALGEBRAIC INDEPENDENCE (CIA). *Under the hypothesis* (H), *let V be a projective subvariety of $\mathbf{P}_n$ defined over $\mathbf{Q}$ of dimension d such that*

$$(4) \qquad (\delta h(V) + ((k+1)\tau + 3\delta \log(n+1)) d(V)) . \delta^k < \frac{U}{(k+1)\sigma^{k+1}}$$

then, if $d \leq k$ we have $\log \operatorname{Dist}(x, V) \geq -U$.

The proof of this statement goes through a lemma.

LEMMA 0.6. *Under the hypothesis* (H) *let V be a variety defined over $\mathbf{Q}$ of dimension d satisfying* (4) *and $\log \operatorname{Dist}(x, V) < -U$, the following is true. For $h = d + 1, \ldots, \max(0, d - k)$ there exists a projective variety Z_h defined over $\mathbf{Q}$ of dimension $h - 1$, satisfying :*
$d(Z_h) \leq \delta^{d-h+1} d(V)$,
$h(Z_h) \leq (\delta h(V) + (d - h + 1)\tau d(V)) \delta^{d-h}$,
$\log \operatorname{Dist}(x, Z_h) < -h'\sigma^{h'} (\delta h(Z_h) + (h'\tau + 3\delta \log(n+1)) d(Z_h)) \delta^{h'-1}$,
where $h' = h + k - d$.

(*) Chapter's author : Patrice PHILIPPON.

PROOF. Let us first remark that we may assume without loss of generality

$$U = (k+1)\sigma^{k+1}\left(\delta h(V) + ((k+1)\tau + 4\delta \log(n+1))\,d(V)\right)\delta^k \ .$$

We proceed by induction on h, the case $h = d+1$ is clear by setting $Z_{d+1} = V$ since $\log \text{Dist}(x, V) < -U$ with

$$U > (k+1)\sigma^{k+1}.\left(\delta h(V) + ((k+1)\tau + 3\delta \log(n+1))\,d(V)\right)$$

by (4). Assume that we have constructed Z_h and we want to find Z_{h-1} ($h \geq \max(1, d-k+1)$). By the closest point property of chapter 4, § 5, there exists $y \in Z_h(\mathbf{C})$ such that

$$\log \text{Dist}(x, y) \leq \frac{1}{d(Z_h)} \cdot \log \text{Dist}(x, Z_h) + \sum_{i=1}^{n} \frac{h}{i}$$

$$< -h\sigma^{k-d+h}\left(h\tau + 3\delta \log(n+1)\right)\delta^{k-d+h-1} + \sum_{i=1}^{n} \frac{h}{i}$$

$$< -\sigma\tau/\mu \ .$$

Therefore there exists $\frac{\tau}{\sigma^{k+1}} < S' \leq \frac{U}{\sigma^{k+1}}$ such that y lays in the ball $B_{S'} := B(x, \exp(-S'\sigma^{k+2}))$. We take S the upper bound in this interval of the reals S' such that $Z_h(\mathbf{C})$ intersects the ball $B_{S'}$. It follows that one of the corresponding forms Q_i doesnot vanish on Z_h, we consider the cycle $W = Z_h.\mathcal{Z}(Q_i)$ which is equidimensional of dimension $h - 2$, defined over $\mathbf{Q}$ and satisfy

$$d(W) \leq \delta d(Z_h) \leq \delta^{d-h+2}d(V)$$
$$h(W) \leq \delta h(Z_h) + \tau d(Z_h) \leq (\delta h(V) + (d-h+2)\tau d(V))\delta^{d-h+1} \ ,$$

thanks to the geometric and arithmetic Bézout theorems of chapter 4, § 4. In order to bound $\text{Dist}(x, W)$ we distinguish two cases.

First case : $S = U/\sigma^{k+1}$. By (H2) and (4), using the first metric Bézout theorem (FMBT) of chapter 4, § 7, we get

$$\log \text{Dist}(x, W) \leq \max\left(\log \text{Dist}(x, Z_h); -U\right) + \delta h(Z_h) + $$
$$(\tau + 3\delta h \log(n+1))d(Z_h)$$

$$\leq \max\left(-(h'-1)\sigma^{h'}\left(\delta h(Z_h) + (h'\tau + 2\delta \log(n+1))\,d(Z_h)\right)\delta^{h'-1}; \right.$$

$$\left. - k\sigma^{k+1}\left(\delta h(V) + ((k+1)\tau + 3\delta \log(n+1))\,d(V)\right).\delta^k\right)$$

$$\leq -(h'-1)\sigma^{h'-1}\left(\delta h(W) + ((h'-1)\tau + 3\delta \log(n+1))\,d(W)\right)\delta^{h'-2} \ .$$

Second case : $S < U/\sigma^{k+1}$. By the maximality of S we have

$$Z_h \cap B(x, \exp(-S'\sigma^{k+2})) = \emptyset$$

for any $S' > S$ and

$$\frac{1}{\sigma}. \min_{y \in Z_h(\mathbf{C})} \log \text{Dist}(x, y) > -S\sigma^{k+1} \geq \log\left(\frac{\|Q_i(x)\|}{\|Q_i\|}\right) \ .$$

We are thus in position to apply the second metric Bézout theorem (SMBT) of chapter 4, § 8 :

$$\log \mathrm{Dist}(x, W) \le \frac{1}{\sigma} . \log \mathrm{Dist}(x, Z_h) + \delta h(Z_h) + (\tau + 3\delta \log(n+1)) \, d(Z_h)$$

$$< -(h'-1)\sigma^{h'-1} \left(\delta h(Z_h) + (h'\tau + 3\delta \log(n+1)) \, d(Z_h) \right) \delta^{h'-1}$$

$$< -(h'-1)\sigma^{h'-1} \left(\delta h(W) + ((h'-1)\tau + 3\delta \log(n+1)) \, d(W) \right) \delta^{h'-2} ,$$

where $h' = h + k - d$.

In both cases we have checked that W satisfies the assertion in the lemma. Since the functions degree, height and $\log \mathrm{Dist}$ are additive on cycles it results that at least one irreducible component of W also satisfies the same assertions, we set Z_{h-1} one of these components. This concludes the induction and the proof. □

PROOF OF CRITERIA. When $d \le k$ the assertion in the lemma is false since Z_0 must then be the empty cycle (dimension -1) for which $\mathrm{Dist}(x, Z_0) = 1$ (therefore $\log \mathrm{Dist}(x, Z_0)$ cannot be < 0). This proves that the hypothesis made in the lemma are contradictory when $d \le k$ and under (H) we must have $\log \mathrm{Dist}(x, V) \ge -U$ for all variety V defined over $\mathbf{Q}$, of dimension $d \le k$, satisfying (4). □

For $h > 0$ the assertion in the lemma is an approximation property of the type (AP1). More precisely, suppose $V = \mathbf{P}_n$, $k = d = n$, $\sigma = 1$ and let us weaken (4) in

$$(4_\epsilon) \qquad\qquad \epsilon(n+1) \, (\tau + 3\delta \log(n+1)) \, \delta^n \le U$$

where $0 < \epsilon < 1$. Under (H) and this new hypothesis (4_ϵ) the proof of the lemma goes just as well while $h > 1$, assuming $\delta > 1/\epsilon$. We can state

LEMMA 0.7. *Under hypothesis (H) with $\delta > 1/\epsilon$ and assuming (4_ϵ), for $h = n+1, \ldots, 1$ there exists a projective variety Z_h defined over $\mathbf{Q}$ of dimension $h-1$ satisfying :*
$d(Z_h) \le \delta^{n+1-h}$;
$h(Z_h) \le ((n-h+1)\tau + 2\delta n \log(n+1)) \, \delta^{n-h}$;
$\log \mathrm{Dist}(x, Z_h) \le -\epsilon.h \, (\delta h(Z_h) + (h\tau + 3\delta \log(n+1))d(Z_h)) . \delta^{h-1}$.

This shows that (H) and (4_ϵ) imply (AP1), let us see how one can obtain (AP2) in this context. For this we need to assume that hypothesis (H) is valid for the parameters δ, τ, U an also $\epsilon^2\delta, \epsilon^2\tau, \epsilon^{2(k+1)}U$. For convenience, in the next corollary we denote $H(\delta, \tau, \epsilon)$ the hypothesis (H) of the begining of this chapter with $x \in \mathbf{P}_n(\mathbf{C})$, $k = n$, $\sigma = 1$ fixed and $U = \epsilon(n+1)(\tau + 3\delta \log(n+1))\delta^n$.

COROLLARY 0.8. *Under hypothesis $H(\delta, \tau, \epsilon)$ and $H(\epsilon^2\delta, \epsilon^2\tau, \epsilon^{2n+3})$ with $\delta > 1/\epsilon$, there exist $\alpha \in \mathbf{P}_n(\overline{\mathbf{Q}})$ such that :*
$d(\alpha) \le \delta^n . d(V)$;
$d(\alpha)h(\alpha) \le n \, (\tau + 2\delta \log(n+1)) . \delta^{n-1}$;
$\log \mathrm{Dist}(x, \alpha) < -\epsilon^{2n+3}(1-\epsilon).(\delta h(\alpha) + \tau + 3\delta \log(n+1)) . d(\alpha)$.

PROOF. We take α to be one of the points of Z_1 closest to x, $d(\alpha) = d(Z_1)$ and $d(\alpha)h(\alpha) = h(Z_1)$. Suppose

$$\log \mathrm{Dist}(x, \alpha) \ge -\epsilon^{2n+3} \left(\delta h(\alpha) + \tau + 3\delta \log(n+1) \right) . d(\alpha)$$

$$\ge -\epsilon^{2n+3} \left(\delta h(Z_1) + (\tau + 3\delta \log(n+1))d(Z_1) \right)$$

$$\ge -\epsilon^{2n+3}(n+1) \, (\tau + 3\delta \log(n+1)) \, \delta^n .$$

Let S be the upper bound of the reals $\tau < S' \leq U$ such that $Z_1(\mathbf{C})$ intersects $B_{S'}$, we have $S \leq -\epsilon^{2n+3}(n+1)(\tau + 3\delta \log(n+1))\delta^n$. By the maximality of S we have $Z_1 \cap B(x, \exp(-S')) = \emptyset$ for any $S' > S$, and there exists Q_i associated to some $S' > S$ in $H(\epsilon\delta, \tau, \epsilon^{2n+3})$ which doesnot vanish on Z_1 and satisfies

$$\frac{|Q_i(x)|}{\|Q_i\|.\|x\|^{d^\circ Q_i}} \leq \exp(-S') < \min_{y \in Z_1(\mathbf{C})} \log \mathrm{Dist}(x, y) \ .$$

Applying (SMBT) we get

$$\log \mathrm{Dist}(x, Z_0) \leq \log \mathrm{Dist}(x, Z_1) + \epsilon^2 \left(\delta h(Z_1) + (\tau + 3\delta \log(n+1))d(Z_1)\right)$$
$$\leq -\epsilon(1-\epsilon)\left(\delta h(Z_1) + (\tau + 3\delta \log(n+1))d(Z_1)\right)$$
$$< 0$$

which is a contradiction, since Z_0 is the empty cycle. Therefore our assumption on $\log \mathrm{Dist}(x, \alpha)$ is false and this establish the corollary. $\qquad\square$

1. Criteria for algebraic independence

If we can satisfy hypothesis (H) from the previous section for a family of parameters $(\delta, \tau, \sigma, U)$ so that (4) in (CIA) can be verified for any variety V defined over $\mathbf{Q}$ of dimension $\leq k$, we conclude that $\mathrm{Dist}(x, V) \neq 0$ for all these varieties, that is the transcendence degree of the field of definition of x is $> k$. We start with a corollary of (CIA) where such a situation occurs.

COROLLARY 1.1. *Let* $x \in \mathbf{P}_n(\mathbf{C})$, $x_0 \neq 0$ *and* $\delta, \tau, \sigma, U : \mathbf{N} \to \mathbf{R}_{\geq 1}$ *be increasing functions such that* $\tau \geq \delta$, $\sigma^{k+1} < \tau < U$, $\frac{U}{\tau \delta^k \sigma^{k+1}} \to \infty$ *and* $\tau \to \infty$. *Assume that, for all* $j \in \mathbf{N}$ *there exists a family of forms* $Q_{j,1}, \ldots, Q_{j,m} \in \mathbf{Z}[X_0, \ldots, X_n]$ *such that* $d^\circ Q_{j,\ell} \leq \delta(j)$, $h_1(Q_{j,\ell}) \leq \tau(j)$,

$$\frac{|Q_{j,\ell}(x)|}{\|Q_{j,\ell}\|.\|x\|^{d^\circ Q_{j,\ell}}} \leq \exp\left(-U(j)\right)$$

and the forms $Q_{j,\ell}$ *don't have any common zero in* $B\left(x, \exp\left(-U\sigma(j-1)\right)\right)$. *Then, at least* $k+1$ *from the numbers* $\frac{x_1}{x_0}, \ldots, \frac{x_n}{x_0}$ *are algebraically independent over* $\mathbf{Q}$.

PROOF. Let V be a projective subvariety of $\mathbf{P}_n(\mathbf{C})$ defined over $\mathbf{Q}$ of dimension $\leq k$, choose i minimal such that $\tau(i) \geq U(0)$ and

$$(\delta(i)h(V) + ((k+1)\tau(i) + 3\delta(i)\log(n+1))d(V)).\delta(i)^k\sigma(i)^{k+1} < \frac{U(i)}{k+1}$$

(i exists since $\tau \geq \delta$ and $\frac{U}{\tau\delta^k\sigma^{k+1}} \to \infty$). For any $\frac{\tau(i)}{\sigma(i)^{k+1}} < S \leq \frac{U(i)}{\sigma(i)^{k+1}}$ there exists $j \in \{1, \ldots, i\}$ such that

$$\frac{U(j-1)}{\sigma(i)^{k+1}} < S \leq \frac{U(j)}{\sigma(i)^{k+1}} \ ,$$

we set the family of forms attached to S in (H) equal to $(Q_{j,1}, \ldots, Q_{j,m})$. We check $d^\circ Q_\ell \leq \delta(j) \leq \delta(i)$, $h_1(Q_\ell) \leq \tau(j) \leq \tau(i)$,

$$\frac{|Q_\ell(x)|}{\|Q_\ell\|.\|x\|^{d^\circ Q_\ell}} \leq \exp\left(-S\sigma(i)^{k+1}\right)$$

and the Q_ℓ's don't have any common zero in $B(x, \exp(-S\sigma(i)^{k+2}))$. The hypothesis (H) and (4) are fulfilled for V, we deduce from (CIA) that $\log \mathrm{Dist}(x, V) \geq -U(i) > 0$ and $x \notin V(\mathbf{C})$, the corollary is proved. $\qquad\square$

If we are lucky, one form may be enough ($m = 1$) in hypothesis (H), this is in particular the case when the form is not too small at x.

COROLLARY 1.2. *Let* $x \in \mathbf{P}_n(\mathbf{C})$, $x_0 \neq 0$ *and* $\delta, \tau, \sigma, U : \mathbf{N} \to \mathbf{R}_{\geq 1}$ *be increasing functions such that* $\tau \geq \delta$, $\sigma^{k+1} < \tau < U$, $\tau \to \infty$ *and* $\frac{U}{\tau \delta^k \sigma^{k+1}} \to \infty$. *Assume that, for all* $j \in \mathbf{N}$ *there exists a form* $Q_j \in \mathbf{Z}[X_0, \ldots, X_n]$ *such that* $d^\circ Q_j \leq \delta(j)$, $h_1(Q_j) \leq \tau(j)$ *and*

$$(2(n+1))^{2\delta(j)} \cdot \exp\left(-U(j-1)\sigma(j-1)\right) < \frac{|Q_j(x)|}{\|Q_j\| \cdot \|x\|^{d^\circ Q_j}} \leq \exp\left(-U(j)\right) .$$

Then, at least $k+1$ *from the numbers* $\frac{x_1}{x_0}, \ldots, \frac{x_n}{x_0}$ *are algebraically independent over* $\mathbf{Q}$. *We therefore can apply corollary 1.1 with* $m = 1$ *and* $Q_{j,1} = Q_j$.

PROOF. If Q_j vanishes in $B(x, \exp(-U(j-1)\sigma(j-1)))$ then

$$\frac{|Q_j(x)|}{\|Q_j\| \cdot \|x\|^{d^\circ Q_j}} \leq (2(n+1))^{2d^\circ Q_j} \cdot \exp(-U(j-1)\sigma(j-1)) ,$$

contradicting the lower bound in the hypothesis. We can therefore apply corollary 1.1 with $m = 1$ and $Q_{j,1} = Q_j$. $\square$

2. Mixed Segre-Veronese embeddings

In practice transcendence method gives polynomials that we may homogeneize in order to apply (CIA) : $^h P(X_0, \ldots, X_n) = X_0^{d^\circ P} \cdot P\left(\frac{X_1}{X_0}, \ldots, \frac{X_n}{X_0}\right)$. But the degree in the different variables are then mixed, it is sometimes more convenient to multi-homogeneize the polynomials :

$$F(X_1, Y_1; \ldots; X_N, Y_n) = Y_1^{d^\circ P_{X_1}} \ldots Y_n^{d^\circ P_{X_n}} \cdot P\left(\frac{X_1}{Y_1}, \ldots, \frac{X_n}{Y_n}\right) ,$$

keeping track of the possible discrepancies between the degrees. Such a form F defines an hypersurface in the product of projective lines $(\mathbf{P}_1)^n = \mathbf{P}_1 \times \cdots \times \mathbf{P}_1$. In this case, we can attach to F and more generally to subvarieties of $(\mathbf{P}_1)^n$ various degrees and heights as explained in chapter 7, § 2. A shortcut to this setting is to consider the *Segre-Veronese embeddings*, take $\underline{\delta} = (\delta_1, \ldots, \delta_n) \in (\mathbf{N}^*)^n$ and $N + 1 = (\delta_1 + 1) \ldots (\delta_n + 1)$ and define

$$\varphi_{\underline{\delta}} : \qquad (\mathbf{P}_1)^n \qquad \longrightarrow \quad \mathbf{P}_N$$
$$(x_1 : y_1, \ldots, x_n : y_n) \quad \mapsto \quad \left(\cdots : \prod_{i=1}^n x_i^{\alpha_i} y_i^{\delta_i - \alpha_i} : \ldots\right)_{\substack{(\alpha_1, \ldots, \alpha_n) \in \mathbf{N}^n \\ 0 \leq \alpha_i \leq \delta_i}} .$$

Let V be a subvariety of $\mathbf{C}^n$ of dimension $h-1$ and f a Chow form of the Zariski closure of $\varphi_{\underline{\delta}}(V)$ in $(\mathbf{P}_1)^n$ (of index $\underline{1}$), denoting $u_\alpha^{(j,1)}$ the coefficients of generic linear forms on $\mathbf{P}_N$ we may specialize $u_\alpha^{(j,1)}$ to $\prod_{i=1}^n u_{\alpha_i}^{(j,\delta_i)}$ where $\underline{\delta}_i = (0, \ldots, 0, \underset{\substack{\uparrow \\ i\text{th position}}}{\delta_i}$ $, 0, \ldots, 0)$, to the effect that f specializes to $\prod_{i \in \{1, \ldots, n\}^h} f_{\underline{\delta}_{i_1}, \ldots, \underline{\delta}_{i_h}}$ where $f_{\underline{\delta}_{i_1}, \ldots, \underline{\delta}_{i_h}}$ denotes a resulting form of index $(\underline{\delta}_{i_1}, \ldots, \underline{\delta}_{i_h}) \in \mathbf{N}^{nh}$ of $\overline{V}$ (see chapter 5, § 3.2). Simple computations show that the degree and height of f are not changed by the above specialization and therefore we have :

$$d(\varphi_{\underline{\delta}}(V)) = \sum_{i \in \{1, \ldots, n\}^{h-1}} d_{\underline{\delta}_{i_1}, \ldots, \underline{\delta}_{i_{h-1}}}(V)$$

$$h(\varphi_{\underline{\delta}}(V)) = \sum_{i \in \{1, \ldots, n\}^h} h_{\underline{\delta}_{i_1}, \ldots, \underline{\delta}_{i_h}}(V)$$

where $d_{\underline{\delta}_{i_1},\ldots,\underline{\delta}_{i_{h-1}}}(V)$ and $h_{\underline{\delta}_{i_1},\ldots,\underline{\delta}_{i_h}}(V)$ are the partial degrees in U_h and heights of the forms $f_{\underline{\delta}_{i_1},\ldots,\underline{\delta}_{i_h}}$. In particular, viewing a multilinear form on $\mathbf{P}_1^n$ as a special form of degree n on $\mathbf{P}_n$ we get a Chow form of index $\underline{1}$ of $\overline{\varphi_1(V)}$ as a specialization of a Chow form of index $(n,\ldots,n) \in \mathbf{N}^n$ of the Zariski closure of V in $\mathbf{P}_n$ by sending some variables to 0, and we may further write :

$$d(\varphi_{\underline{1}}(V)) = \sum_{i\in\{1,\ldots,n\}^{h-1}} d_{\underline{1}_{i_1},\ldots,\underline{1}_{i_{h-1}}}(V) = n^{h-1}.d_{\mathbf{P}_n}(V)$$

$$h(\varphi_{\underline{1}}(V)) = \sum_{i\in\{1,\ldots,n\}^{h}} h_{\underline{1}_{i_1},\ldots,\underline{1}_{i_h}}(V) \leq n^h.h_{\mathbf{P}_n}(V) \ .$$

Thanks to the multihomogeneous elimination theorem (see chapter 5, thm. 2.1), if two indices i_ℓ are equal and distinct then $f_{\underline{\delta}_{i_1},\ldots,\underline{\delta}_{i_{h-1}}}$ is just a power of a Chow form of $\mathbf{P}_1$ of index $(\delta_{i_\ell},\delta_{i_\ell})$ (that is, according to chapter 5, § 3.5, the resultant of two forms of degree δ_{i_ℓ}), and if more than two indices i_ℓ are equal then $f_{\underline{\delta}_{i_1},\ldots,\underline{\delta}_{i_h}} = 1$.

If the indices i_ℓ are pairwise distinct then

$$d_{\underline{\delta}_{i_1},\ldots,\underline{\delta}_{i_{h-1}}}(V) = \delta_{i_1}\ldots\delta_{i_{h-1}}.d_{\underline{1}_{i_1},\ldots,\underline{1}_{i_{h-1}}}(V)$$

$$h_{\underline{\delta}_{i_1},\ldots,\underline{\delta}_{i_h}}(V) \leq \delta_{i_1}\ldots\delta_{i_h}.h_{\underline{1}_{i_1},\ldots,\underline{1}_{i_h}}(V) \ ,$$

because we can write inductively

$$\frac{f_{\underline{\delta}_{i_1},\ldots,\underline{\delta}_{i_h}}}{(f_{\underline{\delta}_{i_1},\ldots,\underline{\delta}_{i_{h-1}},\underline{1}_{i_h}})^{\delta_{i_h}}} = \prod_{x\in V\cap H_1\cap\cdots\cap H_{h-1}} \frac{U_{h,\underline{\delta}_{i_h}}(x)}{U_{h,\underline{1}_{i_h}}(x)^{\delta_{i_h}}}$$

and the local measure of each factor is ≤ 1. Let us remark that at this point the present setting differs from the one in chapter 5, more precisely the latter is obtained by introducing the coefficients $\prod_{i=1}^n \binom{\delta_i}{\alpha_i}^{1/2}$ in the monomials defining $\varphi_{\underline{\delta}}$ and leads to an equality for heights whereas we only have the above inequality. Now, if exactly two indices i_ℓ and $i_{\ell'}$ are equal then

$$f_{\underline{\delta}_{i_1},\ldots,\underline{\delta}_{i_h}} = \left(g_{\delta_{i_\ell}}\right)^e$$

where $e = d_{\underline{\delta}_{i_1},\ldots,\widehat{\underline{\delta}_{i_\ell}},\ldots,\underline{\delta}_{i_h}}(V)/\delta_{i_\ell}$ and $g_{\delta_{i_\ell}}$ is the resultant of two generic polynomials of degree δ_{i_ℓ}. We have $d_{U_j}^\circ g_{\delta_{i_\ell}} = 2\delta_{i_\ell}$ if $j \in \{\ell,\ell'\}$ and $= 0$ otherwise, and

$$h(g_{\delta_{i_\ell}}) \leq 2\delta_{i_\ell}.(\log(\delta_{i_\ell}+1)+1) \ ,$$

from which follows

$$d_{\underline{\delta}_{i_1},\ldots,\widehat{\underline{\delta}_{i_k}},\ldots,\underline{\delta}_{i_h}}(V) = \begin{cases} \dfrac{\delta_{i_1}\ldots\delta_{i_h}}{\delta_{i_\ell}}.d_{\underline{1}_{i_1},\ldots,\widehat{\underline{1}_{i_\ell}},\ldots,\underline{1}_{i_h}}(V) & \text{if } i_k = i_\ell \\ 0 & \text{otherwise} \end{cases}$$

and

$$h_{\underline{\delta}_{i_1},\ldots,\underline{\delta}_{i_h}}(V) = \frac{d_{\underline{\delta}_{i_1},\ldots,\widehat{\underline{\delta}_{i_\ell}},\ldots,\underline{\delta}_{i_h}}(V)}{\delta_{i_\ell}}.h(g_{\delta_{i_\ell}})$$

$$\leq 2\delta_{i_1}\ldots\delta_{i_h}.\frac{\log(\delta_{i_\ell}+1)+1}{\delta_{i_\ell}}.d_{\underline{1}_{i_1},\ldots,\widehat{\underline{1}_{i_\ell}},\ldots,\underline{1}_{i_h}}(V) \ .$$

Assuming $\delta_1 \geq \cdots \geq \delta_n$, we deduce

(5) $\quad d(\varphi_{\underline{\delta}}(V)) \leq n^{h-1}\delta_1 \ldots \delta_{h-1}.d_{\mathbf{P}_n}(V)$

(6) $\quad h(\varphi_{\underline{\delta}}(V)) \leq n^h \delta_1 \ldots \delta_h.h_{\mathbf{P}_n}(V) + 2n^h \delta_1 \ldots \delta_{h-1}\left(\log(\delta_1+1)+1\right).d_{\mathbf{P}_n}(V)$.

Let now $x, x' \in \mathbf{C}^n$, writing

$$|x_i - x'_i| = \mathrm{Dist}(x_i, x'_i).\sqrt{(1+|x_i|^2)(1+|x'_i|^2)}$$

we check

$$\underline{x}^\alpha \underline{x}'^{\alpha'} - \underline{x}^{\alpha'} \underline{x}'^{\alpha} = \sum_{\beta_1=0}^{\alpha_1} \cdots \sum_{\substack{\beta_n=0 \\ (\beta_1,\ldots,\beta_n)\neq(0,\ldots,0)}}^{\alpha_n} \prod_{i=1}^{n} \binom{\alpha_i}{\beta_i}.x_i^{\alpha'_i+\alpha_i-\beta_i}(x'_i-x_i)^{\beta_i}$$

$$- \sum_{\beta_1=0}^{\alpha'_1} \cdots \sum_{\substack{\beta_n=0 \\ (\beta_1,\ldots,\beta_n)\neq(0,\ldots,0)}}^{\alpha'_n} \prod_{i=1}^{n} \binom{\alpha'_i}{\beta_i}.x_i^{\alpha'_i+\alpha_i-\beta_i}(x'_i-x_i)^{\beta_i} \quad ,$$

and, replacing $|x'_i - x_i|$ by $\mathrm{Dist}(x_i, x'_i).\sqrt{(1+|x_i|^2)(1+|x'_i|^2)}$, one gets

$$\mathrm{Dist}(\varphi_{\underline{\delta}}(x), \varphi_{\underline{\delta}}(x'))^2 = \frac{\sum_{\alpha,\alpha'} \left|\underline{x}^\alpha \underline{x}'^{\alpha'} - \underline{x}^{\alpha'} \underline{x}'^{\alpha}\right|^2}{\sum_\alpha |\underline{x}^\alpha|^2 . \sum_\alpha |\underline{x}'^\alpha|^2}$$

$$\leq 4. \sum_{\alpha,\alpha'} \left(\sum_{\beta_1=0}^{\alpha_1} \cdots \sum_{\substack{\beta_n=0 \\ (\beta_1,\ldots,\beta_n)\neq(0,\ldots,0)}}^{\alpha_n} \prod_{i=1}^{n} \binom{\alpha_i}{\beta_i}.\mathrm{Dist}(x_i, x'_i)^{\beta_i} \right.$$

$$\left. \times \sqrt{\frac{(1+|x_i|^2)^{\beta_i}(1+|x'_i|^2)^{\beta_i}}{(1+|x_i|^{2\beta_i})(1+|x'_i|^{2\beta_i})}} \right)^2 .$$

where the sums run over $\alpha, \alpha' \in \mathbf{N}^n$ subject to the conditions $\alpha_i, \alpha'_i \leq \delta_i$ for $i = 1, \ldots, n$. Since $\sqrt{\frac{(1+|x_i|^2)^{\beta_i}(1+|x'_i|^2)^{\beta_i}}{(1+|x_i|^{2\beta_i})(1+|x'_i|^{2\beta_i})}} \leq 2^{\beta_i}$, we have

$$\mathrm{Dist}(\varphi_{\underline{\delta}}(x), \varphi_{\underline{\delta}}(x')) \leq 2(\delta_1+1)\ldots(\delta_n+1).\left(\prod_{i=1}^{n}(1+2\mathrm{Dist}(x_i,x'_i))^{\delta_i} - 1\right)$$

$$\leq 8(\delta_1+1)\ldots(\delta_n+1).\sum_{i=1}^{n}\delta_i\mathrm{Dist}(x_i,x'_i) \quad ,$$

because the left hand side is ≤ 1. Now $\mathrm{Dist}_{\mathbf{P}_n}(x,x') \geq \frac{\mathrm{Dist}(x_i,x'_i)}{\sqrt{1+\|x\|^2}}$ for all $i = 1,\ldots,n$ and we deduce from the above inequality

(7) $\quad \dfrac{\mathrm{Dist}(\varphi_{\underline{\delta}}(x), \varphi_{\underline{\delta}}(V))}{\mathrm{Dist}_{\mathbf{P}_n}(x,V)} \leq \left(8\sqrt{1+\|x\|^2}.\prod_{i=1}^{n}(\delta_i+1).\sum_{i=1}^{n}\delta_i\right)^{n^h \delta_1 \ldots \delta_{h-1} d_{\mathbf{P}_n}(V)}$

by lemma 5.2 of chapter 6 (see the proof following this lemma).

3. Multi-projective criteria for algebraic independence

In this paragraph we will use the notation $\underline{d}°Q = (d°_{X_1}Q, \ldots, d°_{X_n}Q)$ for $Q \in$ $\mathbf{C}[X_1, \ldots, X_n]$ and, for $x \in \mathbf{C}^n$,

$$\||x\||_{(\delta_1,\ldots,\delta_n)} = \left(\sum_{\alpha_1=0}^{\delta_1} \cdots \sum_{\alpha_n=0}^{\delta_n} |x_1|^{2\alpha_1} \ldots |x_n|^{2\alpha_n} \right)^{1/2} ,$$

we have $1 \leq \||x\||_{(\delta\delta_1,\ldots,\delta\delta_n)} \leq \||x\||^\delta_{(\delta_1,\ldots,\delta_n)}$.

MULTI-PROJECTIVE CRITERIA. *Let* $x \in \mathbf{C}^n$, $\delta, \delta_1, \ldots, \delta_n, \sigma, \tau, U : \mathbf{N} \to \mathbf{R}_{\geq 1}$ *be increasing functions such that* $\delta_1 \geq \cdots \geq \delta_n$, $U > \tau > \sigma^{k+1}$, $\tau \to \infty$, $\frac{U}{(\sigma\delta)^{k+1}\delta_1\ldots\delta_k(\delta_{k+1}+\tau+\log(\delta_1+1))} \to \infty$. *Suppose that for all* $j \in \mathbf{N}$ *there exists a polynomial* $Q_j \in \mathbf{Z}[X_1, \ldots, X_n]$ *of degree* $< (\delta\delta_i)(j)$ *in* X_i, *length* $\leq e^{(\delta\tau)(j)}$ *satisfying*

$$(2 \prod_{i=1}^n (\delta_i(j) + 1))^{3\delta(j)} . e^{-(U\sigma)(j-1)} \leq \frac{|Q_j(x)|}{\|Q_j\| . \||x\||_{\underline{d}°Q_j}} \leq e^{-U(j)} .$$

Then, at least $k + 1$ *of the numbers* $x_1, \ldots, x_n$ *are algebraically independent over* $\mathbf{Q}$.

PROOF. Taking polynomials reciprocal to the Q_j's in the variables X_i such that $|x_i| \geq 1$ we may assume $|x_i| \leq 1$ for all i. Let V be a projective subvariety of $\mathbf{C}^n$ defined over $\mathbf{Q}$ of dimension $\leq k$, choose j_0 minimal such that $\tau(j_0) \geq U(0)$ and

$$\delta_{k+1}(j_0)h_{\mathbf{P}_n}(V) + (\tau(j_0) + 6\log(\delta_1(j_0) + 1))d_{\mathbf{P}_n}(V) \leq \frac{U(j_0)}{n^{k+2}((\sigma\delta)^{k+1}\delta_1\ldots\delta_k)(j_0)} ,$$

according to the estimates (5) and (6) of the previous paragraph and since

$$W = \varphi_{(\delta_1(j_0),\ldots,\delta_n(j_0))}(V) \subset \mathbf{P}_N(\mathbf{C}), \quad N + 1 = \prod_{i=1}^n (\delta_i(j_0) + 1), \quad h = k + 1 ,$$

this ensure

$$(h(W) + ((k+1)\tau(j_0) + 3\log(n+1))d(W)).(\sigma\delta)(j_0)^{k+1} \leq \frac{U(j_0)}{k+1} .$$

For any $\frac{(\tau\delta)(j_0)}{\sigma(j_0)^{k+1}} < S \leq \frac{U(j_0)}{\sigma(j_0)^{k+1}}$ there exists $j \in \{1, \ldots, j_0\}$ such that

$$\frac{U(j-1)}{\sigma(j_0)^{k+1}} < S \leq \frac{U(j)}{\sigma(j_0)^{k+1}} ,$$

we set the family of forms attached to S in (H) equal to Q_j. This polynomial can be viewed as the dehomogeneization of a form $F_S \in \mathbf{Z}[Y_0, \ldots, Y_N]$ of degree $\delta(j) \leq \delta(j_0)$, norm $\|F_S\| = \|Q_j\| \leq e^{\delta\tau(j)} \leq e^{\delta\tau(j_0)}$ satisfying, with $y = \varphi_{(\delta_1(j_0),\ldots,\delta_n(j_0))}(x)$,

$$(N+1)^{-\delta(j)/2} . \frac{|Q_j(x)|}{\||x\||_{\underline{d}°Q_j}} \leq \frac{|F_S(y)|}{\|y\|^{d°F_S}} \leq \frac{|Q_j(x)|}{\||x\||_{\underline{d}°Q_j}} .$$

because $\||x\||_{\underline{d}°Q_j} \leq \|y\|^{\delta(j)} \leq (N+1)^{\delta(j)/2}$, it then follows from the hypothesis

$$\exp\left(-S\sigma(j_0)^{k+2}\right).(2(N+1))^{2\delta(j)} \leq \frac{|F_S(y)|}{\|F_S\|.\|y\|^{d°F_S}} \leq \exp\left(-S\sigma(j_0)^{k+1}\right) .$$

Just as in the proof of corollary 1.2, the lower bound implies that F_S has no zero in $B\left(y, \exp\left(-S\sigma(j_0)^{k+2}\right)\right)$. Therefore the hypothesis (H) and (4) are satisfied for W and it follows from (CIA) $\log \mathrm{Dist}(y, W) > -U(j_0) > 0$, in particular $x \notin V(\mathbf{C})$

and this establishes the result. Furthermore, (7) implies that $\log \mathrm{Dist}_{\mathbf{P}_n}(x, V)$ is bounded below by

$$-U(j_0) - n^{k+1}(\delta_1 + 1) \ldots (\delta_k + 1)(j_0) \cdot \log \left(8n\sqrt{(1 + \|x\|^2)}(\delta_1 + 1)(j_0) \right) . d_{\mathbf{P}_n}(V) \ .$$

$\square$

CHAPTER 9

Upper bounds for (geometric) Hilbert functions

Let $\mathfrak{I}$ be a homogeneous prime (and not maximal) ideal in the ring $\mathbf{C}[x_0, \ldots, x_N]$, and let X be the corresponding (irreducible) algebraic subvariety in the complex projective space $\mathbf{P}_N$. Recall that, for any integer $t \geq 0$, the *Hilbert function* of X (or if you prefer, of $\mathfrak{I}$) provides the maximal number $\mathcal{H}(X, t) := \mathcal{H}(\mathfrak{I}, t)$ of homogeneous polynomials of degree t which are linearly independent over $\mathbf{C}$ modulo $\mathfrak{I}$. For large values of t, $\mathcal{H}(X, t)$ coincides with a polynomial, whose highest term is given by $\deg(X)t^d/d!$, where d denotes the dimension of X, and $\deg(X)$ its degree in $\mathbf{P}_N$. Thus, the asymptotic behaviour of $\mathcal{H}(X, t)$ is well-known. But the bounds needed for application to algebraic independence (see, for instance, Chapter 10) must be valid for all t's. It is this type of bound that we here describe.

In the first section of these Chapter, we give simple geometric proofs of the following upper bounds for $\mathcal{H}(X, t)$.

(32) $$\forall\, t \geq 1, \quad \mathcal{H}(X, t) \leq \deg(X)t^d + d \;;$$

(33) $$\forall\, t \geq 1, \quad \mathcal{H}(X, t) \leq \deg(X)\binom{t + d - 1}{d} + \binom{t + d - 1}{d - 1}.$$

Inequality (32) sharpens Theorem 1 of Nesterenko [Nes3], while (33), which sharpens (32), is due to M. Chardin [Cha][1](see also [Som][2]). Both present proofs were indicated to me by J. Kollár [Kol2][3] in relation with [Ber5] and [Cha] (*loc. cit.*). It has seemed interesting to include them here not only to show the scope of geometric methods – and in the case of (32), their mere elegance –, but also because they readily translate into their own multihomogeneous versions (for applications of this well-known feature to the study of heights , see Chapter 7 – or [Ber6, Proposition 2], in the case of points).

The second section of the Chapter gives another illustration of this feature. In [Nes3] Nesterenko extended his result to the "relative" situation of a prime ideal $\mathfrak{I}$ in the ring $\mathbf{C}(z)[x_0, \ldots, x_N]$, and obtained an upper bound for the maximal number $\mathcal{H}(\mathfrak{I}; (t, s))$ of polynomials in the ring of $\mathbf{C}[z, x_0, \ldots, x_N]$, which are homogeneous of degree t in the x_i's, have degree $\leq s$ in z, and are linearly independent over $\mathbf{C}$ modulo $\mathfrak{I}$. We reprove his result (in a sharpened form) as a corollary of a multihomogeneous version of (32). If X denotes the irreducible variety over $\mathbf{C}(z)$ in $\mathbf{P}_N/\mathbf{C}(z)$ corresponding to $\mathfrak{I}$, with (relative) dimension d, (generic) *degree* $\deg(X)$,

---

(∗) Chapter's author : Daniel BERTRAND.

[1][Cha] M. Chardin. Une majoration de la fonction de Hilbert et ses conséquences pour l'interpolation algébrique. *Bull. Soc. Math. France* 117, (1989), 305–318 (see also Chapter I of his *Thèse de doctorat*, Université de Paris 6, 1990).

[2][Som] M. Sombra. Bounds for the Hilbert functions of polynomial ideals and for the degrees in the Nullstellensatz, *J. Math. Appl. Algebra* 117-118, (1997), 565–599.

[3][Kol2] J. Kollár. Letters to the author (1/8/88 and 18/1/89).

and *height* $ht(X)$, we thus obtain for $\mathcal{H}(X; (t, s)) := \mathcal{H}(\mathfrak{I}; (t, s))$:

(34) $\forall\, t \geq 1, s \geq 1 \quad \mathcal{H}(X; (t, s)) \leq ht(X)t^{d+1} + (d+1)\deg(X)st^d + (d+1)$;

(35)$\forall\, t \geq 0, \forall s \geq 1 \quad \mathcal{H}(X; (t, s)) \leq ht(X)\binom{t+d}{d+1} + \deg(X)(s+1)\binom{t+d}{d}$.

In a nutshell, one may say that the present proofs merely consist in rewriting Hilbert functions as dimensions of linear systems. Hence (and also because they use generic hyperplane sections), (34) and (35) do not readily extend to the situation of a projective scheme over Spec($\mathbf{Z}$) and its (truly) arithmetic Hilbert function, as considered in [**Nes3**, Theorem 3], [**Lau1**], [**PPh7**]. See also [**GS**][4], [**AB**][5], [**Zha**][6], for asymptotic estimates.

Finally, note that the above results easily imply bounds for the Hilbert and relative Hilbert functions of any reduced scheme X of *pure dimension*, i.e. for any reduced ideal $\mathfrak{I}$ all of whose components have the same dimension (the resulting bound is the same for (35), which we shall actually prove in this setting, while a factor $\deg(X)$ must be added in front of the "constant" terms of (32), (33), (34), as in the reduction process of [**Cha**] (*loc. cit.*, p. 143), §1.a) .

1. The absolute case (following Kollár)

1.1. Proof of (32). Let X be an (irreducible) algebraic subvariety of $\mathbf{P}_N/\mathbf{C}$ of dimension d. By the very definition of the Hilbert function, the linear system of hypersurfaces of degree $t \geq 1$ in $\mathbf{P}_N$ defines an embedding ϕ_t of X in a projective space $\mathbf{P}_{N(t)}$ of dimension $N(t) = \mathcal{H}(X, t) - 1$ (in other words, $\mathbf{P}_{N(t)}$ is the projective space generated by the image of X under the t-uple embedding of $\mathbf{P}_N$). If H (resp. H') denotes a generic hyperplane in $\mathbf{P}_N$ (resp. $\mathbf{P}_{N(t)}$), the degree of the image $\phi_t(X)$ is equal to the intersection number

$$\phi_t^* X \cdot H'^d = X \cdot \left(\phi_t^*(H'^d)\right) = X \cdot (\phi_t^* H')^d = X \cdot (tH)^d = t^d(X \cdot H^d) = \deg(X)t^d .$$

Hence, (32) follows from the following "Italian" result, applied to $Y = \phi_t(X)$.

LEMMA 1.1. *Let Y be an irreducible algebraic subvariety in a projective space $\mathbf{P}_M$, not contained in any hyperplane. Then, $M + 1 \leq \deg(Y) + \dim(Y)$.*

This classical lemma can in turn be proven as follows (see also [**EH**][7], Proposition 0 or [**Ful**][8], exercise 8.4.6) : by Bertini's theorem [**Jou**][9], Theorem 6.3.4, the intersection Y' of Y with a general hyperplane $\mathbf{P}_{M'} = \mathbf{P}_{M-1}$ of $\mathbf{P}_M$ remains irreducible as long as $\dim(Y) \geq 2$. Now Y' has the same degree as Y, dimension one less than Y and generates $\mathbf{P}_{M-1}$ (see [**GH**][10], p. 174), so that the expression $\deg(Y) + \dim(Y) - M$ does not change in a general hyperplane section. Therefore,

---

[4][**GS**] H. Gillet, C. Soulé. Amplitude arithmétique. *C.R. Acad. Sci. Paris* 307, (1988), 887–890.

[5][**AB**] M. Abbès, T. Bouche. Sur le théorème de Hilbert "arithmétique". *C.R. Acad. Sci. Paris* 317, (1993), 589–591 and *Ann. Inst. Fourier* 43, (1995), 375–401.

[6][**Zha**] S. Zhang. *J. Amer. Math. Soc.* 8, (1995), 187–221.

[7][**EH**] D. Eisenbud, J. Harris. On varieties of minimal degree (a centennial account). *Proc. Sympos. Pure Math.* 46, (1987), 3–14.

[8][**Ful**] W. Fulton. Intersection theory, Springer, (1984).

[9][**Jou**] J.-P. Jouanolou. Théorèmes de Bertini et applications. *Progr. Math.* 42, Birkhäuser, 1983.

[10][**GH**] P. Griffiths, J. Harris. Principles of algebraic geometry, Wiley, (1978).

it suffices to prove the lemma when Y is a curve, in which case we may proceed in one of the following (equivalent – and again standard –) ways :

- projecting from a point on Y to a hyperplane P_{M-1} lowers both $\deg Y$ and the dimension M of the ambient space by one; we are thus reduced to the case of curves in P_2, where the lemma is trivial.
- any M-tuple of distinct points on the curve Y lie in a hyperplane of $\mathbf{P}_M$, and if the degree of Y was strictly less than M, that hyperplane would have to contain the irreducible curve Y.

REMARK. i) To check the lemma for curves, we could also have argued as follows : $\deg(Y)$ is the degree of the divisor $D = Y \cdot H$ of a hyperplane section of Y, and M is at most equal to the dimension $h^0(Y, D) - 1$ of the complete linear system $|D|$. Since D is effective, $h^1(Y, D) = h^0(Y, \omega(D)) \le p_a$, where ω denotes the dualizing sheaf of the (possibly singular) curve Y and p_a its arithmetic genus, and the Riemann-Roch theorem implies : $M + 1 \le \deg(Y) + 1$, as was to be shown. This (unnecessarily complicated) argument paves the way to the sharpening of (32) provided by inequality (33) below.

ii) We can even do better if Y is a smooth curve : if D is a non-special divisor (i.e. $h^0(Y, \Omega^1(D)) = 0$), cf. [**Har**][11], p. 296, then $h^0(Y, D) - 1 = \deg(Y) - g$; and otherwise, Clifford's theorem asserts that $h^0(Y, D) - 1 \le \deg(Y)/2$, since D is effective. In fact, it is shown in [**EH**] (*loc. cit.*, p. 144) that Lemma 1.1 is optimal for a smooth variety X if and only if X is a quadric hypersurface, the Veronese surface in $\mathbf{P}_5$, or a rational normal scroll.

1.2. Proof of (33). Let again X be an (irreducible) algebraic subvariety of dimension d of $\mathbf{P}_N/\mathbf{C}$, and let $O_X(1)$ be the line bundle attached to this embedding. When X is projectively normal, $\mathcal{H}(X, t)$ is equal to the maximal number $h^0(X, O_X(t))$ of $\mathbf{C}$-linearly independent sections of $O_X(t)$, and in all cases, we certainly have :

$$\mathcal{H}(X, t) \le h^0(X, O_X(t)) \ .$$

We now proceed to bound the latter number by the right-hand side of (33) using induction on $d \ge 1$. (Write

$$\binom{t+a}{d-1} = \binom{t+a+1}{d} - \binom{t+a}{d}$$

to check the last step of the induction – and to deal with the case $d = 0$).

When X is a curve, we already saw by Riemann-Roch that $h^0(X, O_X(t)) \le t \deg(X) + 1$. In higher dimensions, we again apply Bertini to find a general hyperplane H in $\mathbf{P}_N$ such that $X \cap H$ remains irreducible. Multiplying by an equation for H gives for any $t \ge 1$ an exact sequence of sheaves :

$$0 \longrightarrow O_X(t-1) \longrightarrow O_X(t) \longrightarrow O_{X \cap H}(t) \longrightarrow 0 \ ,$$

so that : $h^0(X, O_X(t)) \le h^0(X, O_X(t-1)) + h^0(X \cap H, O_{X \cap H}(t))$. Therefore,

$$h^0(X, O_X(t)) \le 1 + \sum_{i=1,\dots,t} h^0(X \cap H, O_{X \cap H}(i)) \ ,$$

and we conclude by the induction hypothesis, since $X \cap H$ has degree $\deg(X)$, and dimension $d - 1$.

---

[11][**Har**] R. Hartshorne. *Algebraic Geometry*, Springer, Berlin, (1977).

2. The relative case

We now consider a homogeneous prime ideal $\mathfrak{I}$ in the ring $\mathbf{C}(z)[x_0, \ldots, x_N]$ and the corresponding proper integral scheme X in $\mathbf{P}_N/\mathbf{C}(z)$. We denote by d the dimension of $X/\mathbf{C}(z)$, and by $\deg(X)$ its degree. Following [Nes3], we define the *height* $ht(X)$ of $X/\mathbf{C}(z)$ as the height of its Chow form $F\left(\underline{u}_1, \ldots, \underline{u}_{d+1}\right) \in \mathbf{C}(z)\left[\underline{u}_1, \ldots, \underline{u}_{d+1}\right]$. Since the ring $\mathbf{C}[z]$ is principal, we may in turn define this height as the degree in z of a multiple $\Phi\left(z, \underline{u}_1, \ldots, \underline{u}_{d+1}\right) \in C\left[z, \underline{u}_1, \ldots, \underline{u}_{d+1}\right]$ of F whose coefficients (*qua* polynomials in z) are relatively prime.

In the notation $\mathcal{H}(X; (t, s))$ from the introduction, Nesterenko's bound ([Nes3, Theorem 2]) for the characteristic function of the ideal $\mathfrak{I}$ reads as follows :

$$\mathcal{H}(X; (t, s)) \leq 6^d \left(ht(X)t^{d+1} + \deg(X)(s+1)t^d\right)$$

for all $t \geq 1$ and all $s \geq 0$. We shall now deduce two sharpenings of this inequality from the multihomogeneous versions of (32) and of (33) (see $(32)_m$ and $(33)_m$ below).

2.1. Proof of (34).

THEOREM 2.1. *For all integers $t \geq 1$, $s \geq 1$:*

(36) $\mathcal{H}(X; (t, s)) \leq ht(X)t^{d+1} + (d+1)\deg(X)st^d + (d+1)$.

PROOF. Denote by $\pi : \mathbf{P}' = \mathbf{P}_N \times \mathbf{P}_1 \to \mathbf{P}_1$ the projective N-space over the projective line $\mathbf{P}_1/\mathbf{C}$, and by $X' \to \mathbf{P}_1$ the schematic closure of X in $\mathbf{P}'$. In concrete terms, the ideal of definition of $X' - \pi^{-1}(\{\infty\})$ is the intersection of the ideal $\mathfrak{I}$ with the ring $\mathbf{C}[z, x_0, \ldots, x_N]$, and the associated bi-homogeneous ideal in $\mathbf{C}[z_0, z_1; x_0, \ldots, x_N]$ is the ideal of definition $\mathfrak{I}'$ of X' itself. In particular, X' is reduced, cf. [Har] (*loc. cit.*, p. 145), II, ex. 3.11.d, irreducible (because its generic fiber is so) and flat over $\mathbf{P}_1$, and $\mathcal{H}(X; (t, s)) = \mathcal{H}(X'; t, s)$, where we write $\mathcal{H}(X'; t, s) := \mathcal{H}(\mathfrak{I}'; t, s)$ for the bi-homogeneous Hilbert function of the variety $X'/\mathbf{C}$ in $\mathbf{P}'$, i.e. the maximal number of bi-homogeneous polynomials of bi-degrees t in the x_i's, and s in z_j's in the ring of $\mathbf{C}[z_0, z_1; x_0, \ldots, x_N]$, which are linearly independent over $\mathbf{C}$ modulo $\mathfrak{I}'$. ☐

Let $H' = H \times \mathbf{P}_1$ be the closure of a general hyperplane H in $\mathbf{P}_N$, and let $F = \mathbf{P}_N \times \{\text{point}\}$ be a fiber of π. Then, for all positive integers (t, s), the divisor cut out by $tH' + sF$ on $X'/\mathbf{C}$ is very ample, and just as in §1.a, we may interpret $M = \mathcal{H}(X'; t, s) - 1$ as the dimension of the projective embedding $\phi = \phi_{t,s}$ of X' attached to the linear system of hypersurfaces of bi-degrees (t, s) in $\mathbf{P}'$ (in other words, $\mathbf{P}_M$ is the projective space generated by the image of X' under the Segre embedding attached to the t-uple embedding of $\mathbf{P}_N$ and the s-uple embedding of $\mathbf{P}_1$). If H'' denotes a general hyperplane in $\mathbf{P}_M$, we have $f^*H'' = tH' + sF$ in $\text{Pic}(\mathbf{P}')$, so that the degree of the $\phi(X')$ in $\mathbf{P}_M$ is equal to the intersection number $\phi(X') \cdot H''^{d+1} = X' \cdot (f^*H'')^{d+1} = X' \cdot (tH' + sF)^{d+1}$, i.e.

$$\deg(\phi(X')) = t^{d+1}X' \cdot H'^{d+1} + (d+1)st^d X' \cdot F \cdot H'^d$$

(recall that $X'/\mathbf{C}$ has dimension $d+1$ and $F^2 = 0$). Since $\phi(X')/\mathbf{C}$ is an (irreducible) subvariety of $\mathbf{P}_M/\mathbf{C}$, the lemma of §1 now implies :

$(32)_m$ $\mathcal{H}(X'; t, s) \leq t^{d+1}X' \cdot H'^{d+1} + (d+1)st^d X' \cdot F \cdot H^d + d + 1$.

This is the multihomogeneous version of (32) promised above. Of course, it could be extended to the general situation (as considereded in Chapter 7) of a subvariety in a product of projective spaces, but we can here go further by exploiting the flatness of the morphism $X' \to \mathbf{P}_1$. Thus, all the fibres $X' \cdot F$ have the same degree $X' \cdot F \cdot H'^d$ as its generic fiber $X \cdot H^d = \deg(X)$, and Theorem 1 finally follows from

LEMMA 2.2. *The height $ht(X)$ is equal to $X' \cdot H'^{d+1}$.*

PROOF. (see also [Nes3], [BGS][12], 4.3.3, and Chapter 7). The intersection number $X' \cdot H'^{d+1}$ counts the points (with multiplicities) on X' lying on a general "horizontal" linear subspace U of $\mathbf{P}_N \times \mathbf{P}_1$ of codimension $d+1$, and we may assume without loss of generality that none of these points projects to ∞ under π. Their projections to $\mathbf{P}_1 - \{\infty\} = \mathrm{Spec}(\mathbf{C}[z])$ are then precisely given by the roots of the Chow form $\Phi\left(z, \underline{u}_1, \ldots, \underline{u}_{d+1}\right)$, when $\underline{u}_1, \ldots, \underline{u}_{d+1}$ assume the values defining U, so that their number does coincide with the degree $ht(X)$ of Φ in z. $\qquad\square$

2.2. Proof of (35).

THEOREM 2.3. *Let X be a reduced equidimensional subscheme of $\mathbf{P}_N$ over $\mathbf{C}(z)$, of dimension d. For all integers $t \geq 0$, $s \geq 1$:*

$$(37) \qquad \mathcal{H}(X; (t,s)) \leq ht(X) \binom{t+d}{d+1} + \deg(X)(s+1)\binom{t+1}{d} .$$

PROOF. Just as in §2.a, we have to bound $\mathcal{H}(X'; t, s)$, while just as in §1.b, we bound this number from above by $h^0(X', O_{X'}(sF + tH'))$. Since Lemma 2.2 still holds when X is equidimensional (see [BGS] (*loc. cit.*), [Nes3], Chapter 7), we are reduced as in §2.a to showing :

$$\forall\, t \geq 0,\ \forall\, s \geq 1, \quad h^0\left(X', O_{X'}(sF + tH')\right) \leq$$
$$(33)_m \qquad\qquad \left(X' \cdot H'^{d+1}\right)\binom{t+d}{d+1} + \left(X' \cdot F \cdot H'^d\right)(s+1)\binom{t+d}{d} .$$

We prove $(33)_m$ by induction on d. When $d = 0$, $X'/\mathbf{C}$ is a curve and the Riemann-Roch theorem implies

$$h^0\left(X', O_{X'}(sF + tH')\right) \leq (X' \cdot sF + tH') + h^0(X') \leq (X' \cdot H')t + (X' \cdot F)(s+1)$$

for all $t \geq 0$, $s \geq 0$; indeed, the number of connected components $h^0(X')$ is bounded from above by $\deg(X) = X' \cdot F$.

Let now d be ≥ 1. Multiplying by an equation for H' gives for any $t \geq 1$ an exact sequence of sheaves :

$$0 \longrightarrow O_{X'}(sF + (t-1)H') \longrightarrow O_{X'}(sF + tH') \longrightarrow O_{X' \cap H}(sF + tH') \longrightarrow 0 ,$$

hence :

$$\begin{aligned} h^0\left(X, O_{X'}(sF + tH')\right) &\leq h^0\left(X, O_{X'}(sF + (t-1)H')\right) \\ &\quad + h^0\left(X' \cap H, O_{X' \cap H}(sF + tH')\right) , \end{aligned}$$

---

[12][BGS] J.-B. Bost, H. Gillet, C. Soulé. Heights of projective varieties and positive Green forms, *J. Amer. Math. Soc.* 7, (1994), 903–1027.

and

$$h^0\left(X', O_{X'}(sF + tH')\right) \;\leq\; h^0\left(X', O_{X'}(sF)\right)$$
$$+ \sum_{i=1,\dots,t} h^0\left(X' \cap H, O_{X' \cap H}(sF + iH')\right) .$$

Thus, $(33)_m$ will follow from the induction hypothesis on d, once we have established that for any d :

$$h^0\left(X', O_{X'}(sF)\right) \leq \deg(X)(s + 1) .$$

This is clear when $s = 0$, and then follows by induction on s, using the exact sequence

$$0 \longrightarrow O_{X'}((s-1)F) \longrightarrow O_{X'}(sF) \longrightarrow O_{X' \cap F}(sF) \longrightarrow 0 \; :$$

more precisely, the sheaf $O_{X' \cap F}(sF)$ is trivial on the general fiber $X' \cap F$, so that its h^0 is equal to the number of connected components of $X' \cap F$, which is bounded from above by the number of connected components of X over the algebraic closure of $\mathbf{C}(z)$, hence again by $\deg(X)$. $\qquad\square$

REMARK. Let $\partial(X)$ be the number of connected components of $X' \cap F$, i.e. the degree occuring in the Stein factorization of $X' \to \mathbf{P}_1$. If X is irreducible over $\mathbf{C}(z)$, the above proof yields the bound

$$(38) \qquad \mathcal{H}(X; t, s) \leq ht(X)\binom{t+d}{d+1} + \deg(X)s\binom{t+d}{d} + \partial(X)\binom{t+d}{d} .$$

However, I do not know how to exploit this hypothesis to get in this case a sharper bound for $\partial(X)$.

CHAPTER 10

Multiplicity estimates for solutions
of algebraic differential equations

1. Introduction.

Let $f_1(z), \ldots, f_m(z)$ be a set of functions analytic at the point 0 and n, h natural numbers. It's easy to see that there exist a nonzero polynomial $P(z, x_1, \ldots, x_m) \in \mathbf{C}[z, x_1, \ldots, x_m]$, such that $\deg_z P \leq n, \deg_{\underline{x}} P \leq h$ and

$$\operatorname{ord}_{z=0} P(z, f_1(z), \ldots, f_m(z)) \geq (m!)^{-1} n h^m.$$

The upper bounds for this order of zero in terms of n and h depends on individual properties of functions $f_1, \ldots, f_m$. For example, if functions are algebraically dependent over $\mathbf{C}(z)$, and P is a polynomial which realise the dependence we have $\operatorname{ord}_{z=0} P = \infty$. In this article we are interested in upper bounds. Of course, instead of the point $z = 0$ we can ask the same question at any point $z = \xi$.

We will prove that for solutions of algebraic differential equations with a special property (D- property, see Definition 1.2 below) there exists an upper bound for $\operatorname{ord}_{z=0} P$ which differs from the lower bound only by a larger constant multiplier in place of $(m!)^{-1}$.

Let us consider a system of differential equations

$$(39) \qquad y_j' = \frac{A_j(z, y_1, \ldots, y_m)}{A_0(z, y_1, \ldots, y_m)}, \qquad j = 1, \ldots, m,$$

where $A_j(z, y_1, \ldots, y_m) \in \mathbf{C}[z, y_1, \ldots, y_m]$. We assume that these polynomials have no non constant common divisors.

THEOREM 1.1. *Suppose that functions* $\overline{f} = (f_1(z), \ldots, f_m(z))$ *are analytic at the point* $z = 0$ *and form a solution of the system* (39). *If these functions have the D-property at the point* 0, *then there exists a constant* $c_1 > 0$ *depending only on* $\overline{f}$, *such that for any polynomial* $A \in \mathbf{C}[z, x_1, \ldots, x_m]$, $A \neq 0$, *the following inequality holds*

$$\operatorname{ord}_{z=0} A(z, f_1(z), \ldots, f_m(z)) \leq c_1 (\deg_z A + 1)(\deg_{\underline{x}} A + 1)^m.$$

We now define and discuss the D-property, which is used in Theorem 1.1. Define the differential operator

$$(40) \qquad D = A_0(z, x_1, \ldots, x_m) \frac{\partial}{\partial z} + \sum_{j=1}^{m} A_j(z, x_1, \ldots, x_m) \frac{\partial}{\partial x_j},$$

(*) Chapter's author : Yuri V. NESTERENKO.

corresponding to the system (39). If functions $\overline{f}$ form a solution of system (39) and $E \in \mathbf{C}[z, x_1, \ldots, x_m]$ then the following identity holds

$$(41) \qquad\qquad DE(z, \overline{f}) = A_0(z, \overline{f}) \frac{d}{dz} E(z, \overline{f}).$$

DEFINITION 1.2. Set of functions $\overline{f} = (f_1(z), \ldots, f_m(z))$ which are analytic at the point $z = 0$ and form a solution of system (39) has the *D-property at the point* 0, if there exists a constant $c > 0$ depending only on $\overline{f}$ such that for any prime ideal $\mathfrak{a} \subset \mathbf{C}[z, x_1, \ldots, x_m], \mathfrak{a} \neq 0$ satisfying $D\mathfrak{a} \subset \mathfrak{a}$, (such an ideal will be said *stable under D*) one has

$$(42) \qquad\qquad \min_{E \in \mathfrak{a}} \operatorname{ord}_{z=0} E(z, \overline{f}) \leq c.$$

Later on we will use the expression *D-property* and notation ord without reference to the point $z = 0$.

Note that functions having the *D-property* are *algebraically independent* over $\mathbf{C}(z)$. Indeed, if functions $\overline{f}$ are algebraically dependent over $\mathbf{C}(z)$, denote by $\mathfrak{E}$ the prime ideal in the ring $\mathbf{C}[z, \underline{x}] = \mathbf{C}[z, x_1, \ldots, x_m]$ consisting of all polynomials $E \in \mathbf{C}[z, \underline{x}]$ such that $E(z, \overline{f}) = 0$. It follows from (41) that the ideal $\mathfrak{E} \neq (0)$ satisfies $D\mathfrak{E} \subset \mathfrak{E}$ and for any polynomial $E \in \mathfrak{E}$ one has $\operatorname{ord} E(z, \overline{f}) = \infty$.

In the following examples we assume that the functions $\overline{f}$ are analytic at the origin and algebraically independent over $\mathbf{C}(z)$.

Example 1. Let us prove that in case $A_0(0, \overline{f}(0)) \neq 0$ the *D-property* is indeed satisfied. Assume that $\mathfrak{a} \subset \mathbf{C}[z, \underline{x}], \mathfrak{a} \neq (0)$, is a prime ideal stable under the operator D. Choose $E \in \mathfrak{a}$ as a polynomial with minimal $\operatorname{ord} E(z, \overline{f})$ among all polynomials from $\mathfrak{a}$. If $\operatorname{ord} E(z, \overline{f}) \geq 1$ then by (41) the polynomial DE satisfies $\operatorname{ord} DE(z, \overline{f}) < \operatorname{ord} E(z, \overline{f})$ and $DE \in \mathfrak{a}$. This is a contradiction with the minimality in the choise of E, it implies that $\operatorname{ord} E(z, \overline{f}) = 0$, from which we derive the *D-property* with $c = 1$ in (42).

Therefore the assertion of Theorem 1.1 is true whenever the point $(0, f_1(0), \ldots, f_m(0))$ is not a singular point of system (39). For example, this is the case when the right-hand side of the system consists of polynomials ($A_0 = 1$).

Example 2. Let us assume that system (39) is linear over the field $\mathbf{C}(z)$

$$y_j' = \sum_{k=1}^{m} a_{jk}(z) y_k, \quad j = 1, \ldots, m, \quad a_{jk}(z) \in \mathbf{C}(z).$$

In this case the *D-property* is proved in [**Nes1**]. The proof is based on Galois theory for linear differential equations.

Example 3. Ramanujan functions

$$(43) \qquad P(z) = 1 - 24 \sum_{n=1}^{\infty} \sigma_1(n) z^n, \quad Q(z) = 1 + 240 \sum_{n=1}^{\infty} \sigma_3(n) z^n,$$

$$(44) \qquad\qquad R(z) = 1 - 504 \sum_{n=1}^{\infty} \sigma_5(n) z^n,$$

form a solution of the system of differential equations, [Ram][1],

$$z\frac{dP}{dz} = \frac{1}{12}(P^2 - Q), \quad z\frac{dQ}{dz} = \frac{1}{3}(PQ - R),$$

(45)
$$z\frac{dR}{dz} = \frac{1}{2}(PR - Q^2),$$

and are algebraically independent over $\mathbf{C}(z)$ (see [Mah1]). We will prove in section 5 that functions $P(z), Q(z), R(z)$ have the D-property (Proposition 5.1). Therefore we derive the following consequence of Theorem 1.1.

THEOREM 1.3. *Let L_1, L_2 be integers, $L_1 \geq 1$, $L_2 \geq 1$. Then for any polynomial $A(z, x_1, x_2, x_3) \in \mathbf{C}[z, x_1, x_2, x_3]$, $A \not\equiv 0$, $\deg_z A \leq L_1$, $\deg_{x_i} A \leq L_2$, one has*

$$\mathrm{ord}_{z=0}\, A\big(z, P(z), Q(z), R(z)\big) \leq c_2 L_1 L_2^3,$$

where c_2 is an absolute constant.

It is proved in [Nes9] that the bound of Theorem 1.3 is true with $c_2 = 10^{47}$. Theorem 1.3 is an important tool in the proof of algebraic independence of values of Ramanujan functions, Chapter 3.

We finish this section by some historical remarks. It was Siegel, [Sie], who first pointed out that zero bounds for solutions of linear differential equations can be proved with the help of algebraic methods. The methods of this Chapter and the next one are very strong generalizations of his ideas.

Probably the exponential functions form the simplest non-trivial example in this topic. The very sharp bounds for the number of zeros of exponential polynomials were proved in 1970 by R.Tijdeman, [Tij1], by analytic method. Nevertheless we explain here how it is also possible to derive zero bounds in this case by algebraic method, mainly following Siegel [Sie] and Schneider, [Sch1].

PROPOSITION 1.4. *Let $\alpha_0, \ldots, \alpha_m$ be distinct complex numbers and*

$$F(z) = \sum_{i=0}^{m} a_i(z)e^{\alpha_i z},$$

where $a_i(z) \in \mathbf{C}[z], i = 0, \ldots, m$. Then

$$\mathrm{ord}\, F(z) \leq -1 + \sum_{i=0}^{m}(\deg a_i(z) + 1)$$

PROOF. Denote $M = \mathrm{ord}\, F(z)$. Without loss of generality we may assume that $a_i(z) \neq 0, i = 0, \ldots, m$. Denote $a_i(z) = b_i z^{n_i} + \ldots, b_i \neq 0$. It is easy to see that

(46)
$$F^{(k)}(z) = \sum_{i=0}^{m} a_{ki}(z)e^{\alpha_i z}, \quad 0 \leq k \leq m,$$

where $a_{ki}(z) = b_i \alpha_i^k z^{n_i} + \ldots \in \mathbf{C}[z]$. The last equalities imply that for $\Delta(z) = \det \|a_{ki}(z)\|_{0 \leq i,k \leq m}$ the following representation holds

$$\Delta(z) = b_0 \cdots b_m z^{n_0 + \ldots + n_m} \det \|\alpha_i^k\|_{0 \leq i,k \leq m} + \cdots.$$

Therefore we conclude that $\Delta(z) \neq 0$ and $\deg \Delta(z) = n_0 + \ldots + n_m$.

---

[1][Ram] S. Ramanujan. On certain arithmetical functions, *Trans. Cambridge Phil. Soc.* 22/9, (1916), 159–184; Collected papers, Chelsea Publ. Co., (1962), 136–162.

The equalities (46) can be rewritten in the form

$$(47) \qquad \Delta(z)e^{\alpha_i z} = \sum_{k=0}^{m} \Delta_{ik}(z)F^{(k)}(z), \quad 0 \le i \le m,$$

where $\Delta_{ik}(z) \in \mathbf{C}[z]$. Since $\operatorname{ord} F^{(k)}(z) \ge M - k$ we derive from (47) that $\operatorname{ord} \Delta(z) \ge M - m$ and $M \le m + \deg \Delta(z)$. $\qquad \square$

Of course Proposition 1.4 follows easily from the uniqueness theorem for solutions of linear differential equations with constant coefficients. The next example used by Schneider in the proof of Hilbert's 7-th problem exploit the same idea of elimination of exponential functions but in another way of construction of new linear forms in these functions, which is less trivial.

PROPOSITION 1.5. *Let S be a natural number and α, β be complex numbers, $\alpha \ne 0, \beta \notin \mathbf{Q}$, suppose α is not a root of 1 and that the function*

$$F(z) = \sum_{\ell=0}^{m} a_\ell(z)\alpha^{\ell z},$$

where $a_\ell(z) \in \mathbf{C}[z], \ell = 0, \ldots, m$, satisfies

$$F(u + \beta v) = 0, \qquad u, v = 0, \ldots, S - 1,$$

then

$$S^2 \le 2\sum_{\ell=0}^{m} \deg a_\ell(z) + m^2.$$

PROOF. Without loss of generality we may assume that $a_\ell(z) \ne 0, \ell = 0, \ldots, m$. As in the proof of Proposition 1.4 denote $a_\ell(z) = b_\ell z^{n_\ell} + \ldots, b_\ell \ne 0$. It is easy to see that

$$(48) \qquad F(z + k) = \sum_{\ell=0}^{m} a_{k\ell}(z)\alpha^{\ell z}, \quad 0 \le k \le m,$$

where

$$a_{k\ell}(z) = a_\ell(z + k)\alpha^{\ell k} = b_\ell \alpha^{\ell k} z^{n_\ell} + \ldots \in \mathbf{C}[z].$$

The last equalities imply that for $\Delta(z) = \det \|a_{k\ell}(z)\|_{0 \le \ell, k \le m}$ the following representation holds

$$\Delta(z) = b_0 \cdots b_m z^{n_0 + \ldots + n_m} \det \|\alpha^{\ell k}\|_{0 \le \ell, k \le m} + \ldots .$$

Since α is not a root of 1, and therefore $\alpha^{\ell_1} \ne \alpha^{\ell_2}$ for $\ell_1 \ne \ell_2$, we conclude that $\Delta(z) \ne 0$ and $\deg \Delta(z) = n_0 + \ldots + n_m$. The equalities (48) can be rewritten in the form

$$(49) \qquad \Delta(z)\alpha^{\ell z} = \sum_{k=0}^{m} \Delta_{\ell k}(z)F(z + k), \quad 0 \le \ell \le m.$$

Using conditions of the Proposition we derive from (49) that

$$(50) \qquad \Delta(u + \beta v) = 0, \quad 0 \le u < S - m, \quad 0 \le v < S.$$

Since β is irrational, equalities (50) imply that

$$S(S - m) \le \deg \Delta(z) = \sum_{\ell=0}^{m} \deg a_\ell(z).$$

The bound from Proposition 1.5 follows easily from the last inequality.   □

In both cases we proved upper bounds for the number of zeros of the function $F(z)$. The difference in the proof is only the way of construction of linearly independent linear forms

$$L_k(\underline{x}) = \sum_{i=0}^{m} a_{ki}(z) x_i, \qquad 0 \le k \le m,$$

which after substituting the corresponding exponential functions to the x_i's have large set of common zeros. In the case of Proposition 1.4 the differential operator

$$D = \frac{\partial}{\partial z} + \sum_{i=0}^{m} \alpha_i x_i \frac{\partial}{\partial x_i}$$

is used for this purpose. While in the case of Proposition 1.5 we used the transformation

$$\tau : (z, x_0, x_1, \ldots, x_m) \longrightarrow (z+1, x_0, \alpha x_1, \ldots, \alpha^m x_m)$$

which can be thought as a translation on the group equal to the product of the additive group $\mathbb{G}_a$ and the torus $(\mathbb{G}_m)^{m+1}$. In this latter case the structure of the set of zeros is important.

Zero estimate for polynomials on algebraic groups generalizing the proof of Proposition 1.5 were developed by D.Masser, D.Masser and G.Wüstholz, P.Philippon. See next Chapter in this volume.

In 1955 A.B.Shidlovskii, [**Shi**] extended ideas of the proof of Proposition 1.4 to proper set of functions, satisfying a system of linear differential equations over $\mathbf{C}(z)$. This allowed him to prove a general theorem about algebraic independence of values of Siegel's E-functions. In 1977, [**Nes2**], general elimination theory and commutative algebra was used for the proof of zero estimates for polynomials in solutions of linear differential equations over $\mathbf{C}(z)$. Further generalisations and improvements were stated in [**BM1**], [**BM2**], [**Bro1**], [**Nes5**], [**Nes6**]. The proof of Theorem 1.1 is in this line of results.

2. Reduction of Theorem 1.1 to bounds for polynomial ideals.

In the following we use notations, definitions and technical results stated in Section 4 of Chapter 3. The field K is in our case the field of rational functions $\mathbf{C}(z)$. For any $\alpha \in K$ set $|\alpha| = e^{-\operatorname{ord}_{z=0} \alpha}$, this absolute value can be extended to the field of formal series $\mathbf{C}((z))$. With these notations the assertion of Theorem 1.1 can be rewritten in the following form.

THEOREM 2.1. *Suppose that functions* $\overline{f} = (f_1(z), \ldots, f_m(z))$ *are analytic at the point* $z = 0$ *and form a solution of system (39). If these functions have the D-property at the point 0, then there exists a constant* $c_1 > 0$ *depending only on* $\overline{f}$, *such that for any polynomial* $A \in \mathbf{C}[z, x_1, \ldots, x_m]$, $A \ne 0$, *the following inequality holds*

$$\log |A(z, \overline{f})| \ge -c_1 (\deg_z A + 1)(\deg_{\underline{x}} A + 1)^m.$$

In this form Theorem 1.1 looks like Corollary 5.2 from Chapter 3 and the same algebraic method will be used for its proof. We will work with homogeneous polynomials and ideals in the ring $K[x_0, \ldots, x_m]$. Using the set of absolute values on K described in Example 1 of Chapter 3 one can define height $h(I)$ for homogeneous ideals $I \in K[x_0, \dots, x_m]$. The following Theorem 2.2 is analogue to Theorem 5.1 from Chapter 3.

THEOREM 2.2. *Suppose that functions $\overline{f} = (f_1(z), \dots, f_m(z))$ are analytic at the point $z = 0$ and form a solution of system (39). If these functions have the D-property at the point 0, then there exists a constant $\tau > 0$ depending only on $\overline{f}$ such that for any unmixed homogeneous ideal $I \in K[x_0, \dots, x_m] = K[\underline{x}]$, such that $\dim I = r - 1 < m$ the following inequality holds*

$$\log |I(\overline{\omega})| \geq -\tau^{mr}\big(h(I)(\deg I)^{r/(m+1-r)} + (\deg I)^{m/(m+1-r)}\big),$$

where $\overline{\omega} = (1, f_1(z), \dots, f_m(z))$.

Theorem 2.2 is proved in section 4 through induction on r from $r = 1$ till $r = m$. Now we deduce Theorem 2.1 from Theorem 2.2.

PROOF. For the polynomial A from Theorem 2.1 define a homogeneous polynomial

$$P(\underline{x}) = x_0^{\deg_{\underline{x}} A} A\Big(\frac{x_1}{x_0}, \dots, \frac{x_m}{x_0}\Big).$$

Applying Proposition 4.8 from Chapter 3 and Theorem 2.2 to the ideal $I = (P)$, $\dim I = m - 1$, we derive

(51) $$\log \|P\|_{\overline{\omega}} \geq \log |I(\overline{\omega})| \geq -\tau^{m^2}\big(h(P)(\deg P)^m + (\deg P)^m\big).$$

Since

$$\log \|P\|_{\overline{\omega}} \leq \log |P(\overline{\omega})| + \deg_z P,$$

inequality (51) proves Theorem 2.1. $\qquad\qquad\qquad\qquad\qquad\qquad\qquad\square$

3. Auxiliary assertions.

Several assertions necessary to the proof of Theorem 2.2 will be stated in this section. We will use notations introduced in the theorem.

Let $\mathfrak{p}$ be a homogeneous prime ideal in $K[\underline{x}]$. For integers $\nu \geq 1$ and $\mu \geq 0$ let us denote by $\mathcal{L}(\mu, \nu)$, the $\mathbf{C}$ vector space of polynomials $P \in \mathbf{C}[z, \underline{x}]$ which are homogeneous in $x_0, \dots, x_m$ and satisfy $\deg_z P \leq \mu$, $\deg_{\underline{x}} P = \nu$. We denote by $\mathcal{L}_{\mathfrak{p}}(\mu, \nu)$, the $\mathbf{C}$ vector space spanned by the residues modulo $\mathfrak{p}$ of the polynomials from $\mathcal{L}(\mu, \nu)$. We further set $\chi_{\mathfrak{p}}(\mu, \nu) = \dim_{\mathbf{C}} \mathcal{L}_{\mathfrak{p}}(\mu, \nu)$.

LEMMA 3.1. *Suppose that $\mathfrak{p} \subset \mathbf{C}(z)[\underline{x}]$ is a homogeneous prime ideal with $r = \dim \mathfrak{p} + 1 \geq 1$. Then*

$$\chi_{\mathfrak{p}}(\mu, \nu) \leq \gamma_1\Big((\mu + 1)\nu^{r-1} \deg \mathfrak{p} + \nu^r h(\mathfrak{p})\Big),$$

where $\gamma_1 \geq 1$ is a constant depending only on m.

This lemma with the multiple $\gamma_1 = 6^{m-1}$ was proved in [Nes3]. Another proof with $\gamma_1 = m$ is given in Chapter 9.

LEMMA 3.2. *Under the conditions of Lemma 3.1, let ν and μ be integers satisfying the inequalities*

(52) $$\nu^{m-r+1} \geq \gamma_2 \deg \mathfrak{p}, \qquad (\mu + 1)\nu^{m-r} \geq \gamma_2 h(\mathfrak{p}),$$

where $\gamma_2 = 2 \cdot m!\gamma_1$. Then there exists a polynomial $P \in \mathfrak{p} \cap \mathbf{C}[z, \underline{x}]$, which is homogeneous in $\underline{x}$ and such that

$$\deg_{\underline{x}} P = \nu, \qquad \deg_z P \leq \mu.$$

PROOF. Since on one hand

$$\dim \mathcal{L}(\mu, \nu) = (\mu + 1)\binom{\nu + m}{m} > (m!)^{-1}(\mu + 1)\nu^m,$$

and on the other hand by Lemma 3.1

$$\begin{aligned} \chi_{\mathfrak{p}}(\mu, \nu) &\leq \gamma_1((\mu + 1)\nu^{r-1} \deg \mathfrak{p} + \nu^r h(\mathfrak{p})) \\ &\leq (m!)^{-1}(\mu + 1)\nu^m, \end{aligned}$$

we get

(53) $$\chi_{\mathfrak{p}}(\mu, \nu) < \dim \mathcal{L}(\mu, \nu),$$

which proves the Lemma. □

COROLLARY 3.3. *Suppose that $\mathfrak{p} \subset K[\underline{x}]$ is a homogeneous prime ideal with $r = \dim \mathfrak{p} + 1 \geq 1$ and*

(54) $$\nu = 1 + \left[\gamma_2(\deg \mathfrak{p})^{\frac{1}{m-r+1}}\right], \qquad \mu = \left[\gamma_2 h(\mathfrak{p})(\deg \mathfrak{p})^{-\frac{m-r}{m-r+1}}\right].$$

Then there exists a polynomial $P \in \mathfrak{p} \cap \mathbf{C}[z, \underline{x}]$, which is homogeneous in $\underline{x}$ and such that

$$\deg_{\underline{x}} P = \nu, \qquad \deg_z P \leq \mu.$$

PROOF. It is easy to check that choosen parameters ν, μ satisfy (52) □

Let us denote $d = \max\limits_{0 \leq j \leq m} \deg_{\underline{x}} A_j(z, \underline{x})$ and

(55) $$B_j(z, \underline{x}) = x_0^d A_j\left(z, \frac{x_1}{x_0}, \ldots, \frac{x_m}{x_0}\right).$$

Define on the ring $K[\underline{x}]$ a homogeneous analogue of the differential operator D, see (40), by the equality

$$T = B_0(z, \underline{x})\frac{\partial}{\partial z} + \sum_{j=1}^{m} B_j(z, \underline{x})x_0\frac{\partial}{\partial x_j}.$$

Then $T x_0 = 0$.

LEMMA 3.4. *If $\mathfrak{p}$ is a homogeneous prime ideal of the ring $K[\underline{x}]$ with*

$$\log |\mathfrak{p}(\overline{\omega})| < -h(\mathfrak{p}) - cm \deg \mathfrak{p},$$

where c is the constant from (42), then $x_0 \notin \mathfrak{p}$ and there does not exist any homogeneous prime ideal $\mathfrak{q} \subset \mathfrak{p}, \mathfrak{q} \neq (0)$, with $T\mathfrak{q} \subset \mathfrak{q}$.

PROOF. According to Proposition 4.13 of Chapter 3 there exists a zero $\overline{\beta}$ of the ideal $\mathfrak{p}$ such that

$$\deg \mathfrak{p} \cdot \log \|\overline{\omega} - \overline{\beta}\| \leq \frac{1}{r}(\log |\mathfrak{p}(\overline{\omega})| + h(\mathfrak{p})) < -c \deg \mathfrak{p}.$$

This implies that

$$\|\overline{\omega} - \overline{\beta}\| < e^{-c}.$$

With the help of Corollary 4.9 of Chapter 3 we derive that for any homogeneous polynomial $C \in \mathfrak{p}$ the following inequality holds

(56) $$\|C\|_{\overline{\omega}} \leq \|\overline{\omega} - \overline{\beta}\| < e^{-c}.$$

Since $\|x_0\|_{\overline{\omega}} = 1$ on can conclude that $x_0 \notin \mathfrak{p}$.

Assume that there exists a prime homogeneous ideal $\mathfrak{q} \subset \mathfrak{p}, \mathfrak{q} \neq (0)$ such that $T\mathfrak{q} \subset \mathfrak{q}$. Denote by $\mathfrak{a}$ the ideal in $\mathbf{C}[z, x_1, \ldots, x_m]$, generated by all polynomials $A \in \mathbf{C}[z, x_1, \ldots, x_m]$ satisfying

$$x_0^{\deg A} A\left(z, \frac{x_1}{x_0}, \ldots, \frac{x_m}{x_0}\right) \in \mathfrak{q}.$$

Since $\mathfrak{q}$ is prime and $x_0 \notin \mathfrak{q}$ one can conclude that the ideal $\mathfrak{a}$ is prime too. Moreover the ideal $\mathfrak{a}$ is stable under the operator D.

By the D-property, there exists a polynomial $E \in \mathfrak{a}$ such that

$$(57) \qquad\qquad \operatorname{ord} E(z, \overline{f}) \leq c.$$

Without loss of generality we can assume that E is irreducible. It follows from the definition of the ideal $\mathfrak{a}$ that there exists an integer n such that

$$C = x_0^n E\left(\frac{x_1}{x_0}, \ldots, \frac{x_m}{x_m}\right) \in \mathfrak{q} \subset \mathfrak{p}.$$

Since $\|C\|_{\overline{w}} = |C(\overline{w})|$ inequalities (56) and (57) contradict each other. $\qquad \square$

LEMMA 3.5. *Let $I \subset J \subset K[\underline{x}]$ be unmixed homogeneous ideals, $\dim J = \dim I - 1$; let $Q \in K[\underline{x}]$ be a homogeneous polynomial which is not contained in any prime ideal associated to I. If $(I, Q) \subset J$, then*

1. $\deg J \leq \deg I \deg Q$,
2. $h(J) \leq h(I) \deg_{\underline{x}} Q + \deg I \deg_z Q$.

PROOF. The proof is based on the Section 4 of Chapter 3. For details see [**Nes6**, Lemma 4]. $\qquad \square$

The following Proposition is analogous to Lemma 2.2 of Chapter 3. The distance ρ from the point $\overline{w}$ to the variety $V(\mathfrak{p})$ to which it refers, is defined in Chapter 3, (19).

PROPOSITION 3.6. *Let $\mathfrak{p} \subset K[\underline{x}]$ be a prime homogeneous ideal, $r = 1 + \dim I \geq 1$ and*

$$(58) \qquad \log |\mathfrak{p}(\overline{w})| < -\tau\left(h(\mathfrak{p})(\deg \mathfrak{p})^{r/(m+1-r)} + (\deg \mathfrak{p})^{m/(m+1-r)}\right),$$

where τ is a sufficiently large constant depending only on the functions $\overline{f}$. Let ρ be the distance from $\overline{w}$ to the variety $V(\mathfrak{p})$. Then there exists a polynomial $B \in \mathbf{C}[z, \underline{x}]$ homogeneous in the variables $\underline{x}$, $B \notin \mathfrak{p}$, such that

$$(59) \qquad \deg_{\underline{x}} B \leq \sqrt{\tau}(\deg \mathfrak{p})^{1/(m+1-r)},$$

$$(60) \qquad \deg_z B \leq \sqrt{\tau}\left(h(\mathfrak{p})(\deg \mathfrak{p})^{-(m-r)/(m+1-r)} + 1\right),$$

$$(61) \qquad \|B\|_{\overline{w}} \leq \rho.$$

PROOF. Define the parameter λ by the equality $\tau = \lambda^{2^{m+2}}$. Let E be a non-zero polynomial in $\mathfrak{p} \cap \mathbf{C}[z, \underline{x}]$, homogeneous in $\underline{x}$, for which the following expression attains its minimum value:

$$\deg \mathfrak{p} \deg_z E + \left(h(\mathfrak{p}) + 1\right) \deg_{\underline{x}} E.$$

The polynomial E is obviously irreducible. For brevity we set $L = \deg_{\underline{x}} E$, $M = \deg_z E$. We define μ and ν using (54). By Corollary 3.3, if $\lambda \geq \gamma_2$, we find that

$$\begin{aligned} M \deg \mathfrak{p} + L\left(h(\mathfrak{p}) + 1\right) &\leq \mu \deg \mathfrak{p} + \nu\left(h(\mathfrak{p}) + 1\right) \\ &\leq 3\lambda\left(h(\mathfrak{p}) + 1\right)(\deg \mathfrak{p})^{1/(m+1-r)}, \end{aligned}$$

from which it follows that

(62) $$L \leq 3\lambda(\deg \mathfrak{p})^{\frac{1}{m+1-r}}, \quad M \leq 3\lambda\big(h(\mathfrak{p})+1\big)(\deg \mathfrak{p})^{-\frac{m-r}{m+1-r}}.$$

We will prove the existence of a polynomial $A \in \mathfrak{p} \cap \mathbf{C}[z, \underline{x}]$ such that $C = TA \notin \mathfrak{p}$, together with upper bounds for the degrees of C differing from bounds (59), (60) only by a constant multiplier.

Define

$$a_j = \lambda^{2^{j+2}-4}, \quad b_j = \lambda^{2^{j+1}-1}, \quad c_j = \lambda^{2^{j+1}-2}, \quad j = 0, \ldots, m.$$

It is easy to check that for $j = 0, 1, \ldots, m-1$ the following inequalities hold

(63) $$a_{j+1} \geq 2a_j b_{j+1}, \quad b_{j+1} \geq b_j + 2\lambda^2 a_j, \quad c_{j+1} \geq c_j + 2\lambda a_j.$$

For each $i = 0, \ldots, m$ we now denote by J_i the ideal in the ring $K[\underline{x}]$ generated by polynomials $T^j E, 0 \leq j < c_i$. Let n be the largest integer such that $J_n \subset \mathfrak{p}$ and there exists polynomials $E_0, \ldots, E_n \in \mathbf{C}[z, \underline{x}]$ homogeneous in $\underline{x}$ satisfying the conditions

1. $\deg_{\underline{x}} E_j \leq b_j(L+1), \quad \deg_z E_j \leq b_j(M+1), \quad j = 0, \ldots, n,$
2. the ideal $\mathfrak{a}_n = (E_0, \ldots, E_n) \subset K[\underline{x}]$ is contained in J_n,
3. all the primary components of $\mathfrak{a}_n$ contained in $\mathfrak{p}$, have dimension $m - n - 1$, and if $\mathfrak{u}_n$ is the unmixed ideal that is the intersection of these components, then

(64) $$\deg \mathfrak{u}_n \leq a_n(L+1)^{n+1}, \quad h(\mathfrak{u}_n) \leq a_n(M+1)(L+1)^n.$$

Setting $E_0 = E$, by the definition of the numbers L, M and Proposition 4 in Chapter 3 the conditions 1-3 are satisfied with 0 in place of n and we derive $n \geq 0$. On the other hand, the inclusion $\mathfrak{u}_n \subset \mathfrak{p}$ implies that $r - 1 = \dim \mathfrak{p} \leq \dim \mathfrak{u}_n = m - n - 1$ and $n \leq m - r$. We suppose that $J_{n+1} \subset \mathfrak{p}$ and show how this would lead to a contradiction.

Let $\mathfrak{b}$ be a primary component of the ideal $\mathfrak{u}_n$, set $\mathfrak{q} = \sqrt{\mathfrak{b}}$ and let l be the exponent of $\mathfrak{b}$, that is, the smallest natural number such that $\mathfrak{q}^l \subset \mathfrak{b}$. We now prove that

(65) $$l \leq 2\lambda a_n.$$

Suppose that, on the contrary, $l > 2\lambda a_n$. Then according to (64) and Proposition 4.7 in Chapter 3, we have

$$a_n(L+1)^{n+1} \geq \deg \mathfrak{u}_n \geq l \deg \mathfrak{q} \geq l > 2\lambda a_n.$$

This implies that $(L+1)^m \geq (L+1)^{n+1} > 2\lambda$ and $L - 1 > 5m$. Furthermore,

$$\left(\frac{L+1}{L-1}\right)^m = \left(1 + \frac{2}{L-1}\right)^m < \left(1 + \frac{2}{5m}\right)^m < 2,$$

and $(L+1)^n < 2(L-1)^n$. Using Proposition 4.7 in Chapter 3 and (64), from this we see that

$$\lambda h(\mathfrak{q}) \leq \frac{\lambda h(\mathfrak{u}_n)}{l} \leq \frac{h(\mathfrak{u}_n)}{2a_n} \leq \frac{1}{2}(M+1)(L+1)^n < (M+1)(L-1)^n,$$

$$\lambda \deg \mathfrak{q} \leq \frac{\lambda \deg \mathfrak{u}_n}{l} \leq \frac{\deg \mathfrak{u}_n}{2a_n} \leq \frac{1}{2}(L+1)^{n+1} < (L-1)^{n+1}.$$

Thus, for sufficiently large λ the ideal $\mathfrak{q}$ satisfies (52) with $\nu = L - 1$ and $\mu = M$. By Lemma 3.2, there exists a homogeneous polynomial $P \in \mathfrak{q}$, that satisfies the conditions

$$\deg_x P = L - 1, \qquad \deg_z P \leq M.$$

Since $P \in \mathfrak{q} \subset \mathfrak{p}$, these inequalities contradict the definition of the polynomial E. This proves (65).

We next prove that there exist i, j, $0 \leq i \leq n$, $0 \leq j \leq 2\lambda a_n$, such that $T^j E_i \notin \mathfrak{q}$. The ideal $\mathfrak{q}$ is isolated in the set of associated prime ideals of $\mathfrak{a}_n$. Hence, there exists a polynomial $H \notin \mathfrak{q}$, such that $G^l H \in \mathfrak{a}_n$ for any $G \in \mathfrak{q}$. Suppose that there did not exist i, j with the desired properties. Then, since $l \leq 2\lambda a_n$, we should have $T^l(G^l H) \in \mathfrak{q}$. Since $G \in \mathfrak{q}$, this implies that $(TG)^l H \in \mathfrak{q}$; and, since $\mathfrak{q}$ is a prime ideal and $H \notin \mathfrak{q}$, we have $TG \in \mathfrak{q}$. Thus, $T\mathfrak{q} \subset \mathfrak{q}$, which contradicts Lemma 3.4. This proves the existence of indices i, j with the desired properties.

Let $\mathfrak{q}_1, \ldots, \mathfrak{q}_s$ be all of the associated prime ideals of $\mathfrak{u}_n$. By what was proved above for every v with $1 \leq v \leq s$ there exists i_v, j_v with $0 \leq i_v \leq n$, $0 \leq j_v \leq 2\lambda a_n$ such that $T^{j_v} E_{i_v} \notin \mathfrak{q}_v$. We set

$$E_{n+1} = \sum_{v=1}^{s} \eta_v x_0^{r_v} T^{j_v} E_{i_v},$$

where the $\eta_v \in \mathbf{C}$ and $r_v \in \mathbf{Z}$ are chosen so that E_{n+1} is a homogeneous polynomial, $E_{n+1} \notin \mathfrak{q}_v$, $1 \leq v \leq s$. We prove that the polynomials $E_0, \ldots, E_{n+1}$ satisfy the conditions 1-3 with n replaced by $n + 1$. By the assumption $J_{n+1} \subset \mathfrak{p}$ this would lead to a contradiction with the choice of n.

Denote $d_0 = \max_{0 \leq j \leq m}(\deg_z B_j, \deg_x B_j)$, see (55) for the definition of B_j. Using (63), and inequality $d_0 \leq \lambda$ we obtain

$$\deg_x E_{n+1} \leq b_n(L + 1) + 2d_0 \lambda a_n \leq b_{n+1}(L + 1).$$
$$\deg_z E_{n+1} \leq b_n(M + 1) + 2d_0 \lambda a_n \leq b_{n+1}(M + 1).$$

Thus the condition 1 holds with n replaced by $n + 1$. Condition 2 follows from definition of E_{n+1} and (63).

Let $\mathfrak{r}$ be an associated prime ideal of $\mathfrak{a}_{n+1}$ that is contained in $\mathfrak{p}$. Such ideals exist because $\mathfrak{a}_{n+1} \subset J_{n+1} \subset \mathfrak{p}$. Since $\mathfrak{r} \supset \mathfrak{a}_{n+1} = (\mathfrak{a}_n, E_{n+1})$, we have $\mathfrak{p} \supset \mathfrak{r} \supset \mathfrak{a}_n$. If we take into account the fact that the set of all associated prime ideals of $\mathfrak{a}_n$ contained in $\mathfrak{p}$ coincides with $\mathfrak{q}_1, \ldots, \mathfrak{q}_s$, then we conclude that there exists an ideal $\mathfrak{q}_j \subset \mathfrak{r}$. Hence, $\dim \mathfrak{r} \leq \dim \mathfrak{q}_j = m - n - 1$. If $\dim \mathfrak{r} = m - n - 1$, then $\mathfrak{r} = \mathfrak{q}_j$. But this is impossible, since $E_{n+1} \in \mathfrak{r}$ and $E_{n+1} \notin \mathfrak{q}_j$. Thus, $\dim \mathfrak{r} \leq m - n - 2$; since $\mathfrak{a}_{n+1}$ is generated by $n + 2$ polynomials, this means that $\dim \mathfrak{r} = m - n - 2$. We have proved that all the primary components of $\mathfrak{a}_{n+1}$ that are contained in $\mathfrak{p}$ have dimension $m - n - 2$.

Let $\mathfrak{a}_n = \mathfrak{u}_n \cap \mathfrak{a}'$, where $\mathfrak{a}'$ is the intersection of the primary components of $\mathfrak{a}_n$ that do not occur in $\mathfrak{u}_n$. If I were a primary component of $\mathfrak{u}_{n+1}$ and $\mathfrak{r} = \sqrt{I}$, then the inclusion $\mathfrak{r} \supset \mathfrak{a}'$ would imply that $\mathfrak{p} \supset \mathfrak{r} \supset \mathfrak{a}'$, which is impossible. Hence, $\mathfrak{a}'$ is not contained in $\mathfrak{r}$; so from the inclusion $\mathfrak{u}_n \cap \mathfrak{a}' = \mathfrak{a}_n \subset \mathfrak{a}_{n+1} \subset I$ we obtain $\mathfrak{u}_n \subset I$. Thus, $\mathfrak{u}_n \subset \mathfrak{u}_{n+1}$, and since $E_{n+1} \in \mathfrak{a}_{n+1} \subset \mathfrak{u}_{n+1}$, we have

$$(\mathfrak{u}_n, E_{n+1}) \subset \mathfrak{u}_{n+1}.$$

Using Lemma 3.5, (64) and (63), we see that this inclusion implies that

$$\deg \mathfrak{u}_{n+1} \leq \deg \mathfrak{u}_n \deg_x E_{n+1} \leq a_{n+1}(L + 1)^{n+2},$$

$$h(u_{n+1}) \leq h(u_n) \deg_x E_{n+1} + \deg u_n \deg_z E_{n+1} \leq a_{n+1} M (L+1)^{n+1}.$$

We have proved that the polynomials $E_0, \ldots, E_{n+1}$ satisfy all the conditions 1-3 with n replaced by $n+1$. In other words, the assumption that $J_{n+1} \subset \mathfrak{p}$ led us to a contradiction with the definition of n. Thus, J_{n+1} is not contained in $\mathfrak{p}$ and, since $n \leq m - r < m$, the inclusion $J_m \subset \mathfrak{p}$ is impossible. This means that there exists an index i, $0 \leq i < c_m - 1$, such that $A = T^i E \in \mathfrak{p}$, but $C = TA \notin \mathfrak{p}$.

Denote $B = C^2$. Using (62) we find that

$$
\begin{aligned}
\deg_x B &= 2 \deg_x C \leq 2 \deg_x E + 2 c_m d_0 \leq 2L + 2 c_m d_0 \\
&\leq \lambda^{2^{m+1}} (\deg \mathfrak{p})^{1/(m+1-r)}, \\
\deg_z B &= 2 \deg_z C \leq 2 \deg_z E + 2 c_m d_0 \leq 2M + 2 c_m d
\end{aligned}
$$

$$(66) \qquad \qquad \leq \lambda^{2^{m+1}} (h(\mathfrak{p})(\deg \mathfrak{p})^{-(m-r)/(m+1-r)} + 1).$$

This proves (59), (60).

Following is the proof of (61). Since $A \in \mathfrak{p}$ the inequality holds $\|A\|_{\overline{\omega}} \leq \rho$, see Corollary 4.9] in Chapter 3. This implies that

$$\operatorname{ord} A(\overline{\omega}) = -\log|A(\overline{\omega})| \geq -\log\|A\|_{\overline{\omega}} \geq \log \frac{1}{\rho}$$

and

$$(67) \qquad \operatorname{ord} C(\overline{\omega}) \geq \operatorname{ord} A(\overline{\omega}) - 1 \geq \log \frac{1}{\rho} - 1.$$

Using (67) we derive

$$(68) \qquad \log\|C\|_{\overline{\omega}} = -\operatorname{ord} C(\overline{\omega}) - \log|C| \leq \log \rho + 1 + \deg_z C.$$

It follows from Proposition 4.13 in Chapter 3 and (58) that

$$
\begin{aligned}
\deg \mathfrak{p} \log \rho &\leq \tfrac{1}{r} \left(\log|\mathfrak{p}(\overline{\omega})| + h(\mathfrak{p}) \right) \\
&\leq -\tfrac{1}{2r} \tau \left(h(\mathfrak{p})(\deg \mathfrak{p})^{r/(m+1-r)} + (\deg \mathfrak{p})^{m/(m+1-r)} \right).
\end{aligned}
$$

If λ is sufficiently large, then the last inequality and (66) imply

$$
\begin{aligned}
\deg \mathfrak{p}\left(\frac{1}{2} \log \frac{1}{\rho} + 1 + \deg_z C\right) &\leq -\frac{\tau}{4r} \left(h(\mathfrak{p})(\deg \mathfrak{p})^{\frac{r}{m+1-r}} + (\deg \mathfrak{p})^{\frac{m}{m+1-r}} \right) \\
&\quad + \deg \mathfrak{p} + \lambda^{2^m} \left(h(\mathfrak{p})(\deg \mathfrak{p})^{\frac{1}{m+1-r}} + \deg \mathfrak{p} \right) < 0.
\end{aligned}
$$

Hence, by (68) the following inequality holds

$$\log\|C\|_{\overline{\omega}} \leq \frac{1}{2} \log \rho,$$

and this proves (61). $\qquad \qquad \square$

4. End of the proof of Theorem 2.2.

We now proceed directly to the proof of Theorem 2.2. Suppose that there exists ideals that satisfy the hypothesis of the theorem but not its conclusion. Let I be such an ideal having minimal dimension and let $r = 1 + \dim I$. Then

$$(69) \qquad \log|I(\overline{\omega})| < -\tau^{mr} \left(h(I)(\deg I)^{r/(m+1-r)} + (\deg I)^{m/(m+1-r)} \right),$$

Let $I_1, \ldots, I_s$ be all the primary components of I, k_j be their exponents and let $\mathfrak{p}_j = \sqrt{I_j}$ be their radicals. If for all j one has

$$\log|\mathfrak{p}_j(\overline{\omega})| \geq -\tau^{mr} \left(h(\mathfrak{p}_j)(\deg \mathfrak{p}_j)^{\frac{r}{m+1-r}} + (\deg \mathfrak{p}_j)^{\frac{m}{m+1-r}} \right),$$

then from Proposition 4.7 in Chapter 3 we obtain

$$\log |I(\overline{\omega})| = \sum_{j=1}^{s} k_j \log |\mathfrak{p}_j(\overline{\omega})|$$

$$\geq -\tau^{mr} \sum_{j=1}^{s} k_j h(\mathfrak{p}_j)(\deg \mathfrak{p}_j)^{\frac{r}{m+1-r}} - \tau^{mr} \sum_{j=1}^{s} k_j (\deg \mathfrak{p}_j)^{\frac{m}{m+1-r}}$$

$$\geq -\tau^{mr}(\deg I)^{\frac{r}{m+1-r}} \sum_{j=1}^{s} k_j h(\mathfrak{p}_j) - \tau^{mr} \left(\sum_{j=1}^{s} k_j \deg \mathfrak{p}_j \right)^{\frac{m}{m+1-r}}$$

$$= -\tau^{mr}\big(h(I)(\deg I)^{\frac{r}{m+1-r}} + (\deg I)^{\frac{m}{m+1-r}}\big),$$

which contradicts (69). This contradiction means that there exists a prime ideal $\mathfrak{p} \subset K[\underline{x}]$, $\dim \mathfrak{p} = r - 1$ such that

(70) $$\log |\mathfrak{p}(\overline{\omega})| < -\tau^{mr}\big(h(\mathfrak{p})(\deg \mathfrak{p})^{r/(m+1-r)} + (\deg \mathfrak{p})^{m/(m+1-r)}\big),$$

Let B be the polynomial whose existence was proved in Proposition 3.6.

We first consider the case $r = 1$. If we apply Proposition 4.11 in Chapter 3 to the ideal $\mathfrak{p}$ and polynomial B and if we use (59), (60) with $r = 1$, then we find that

$$\log |\mathfrak{p}(\overline{\omega})| \geq -h(\mathfrak{p}) \deg_{\underline{x}} B - \deg \mathfrak{p} \deg_z B$$
$$\geq -2\sqrt{\tau}(h(\mathfrak{p})(\deg \mathfrak{p})^{1/m} + \deg \mathfrak{p}),$$

which contradicts (70) for $r = 1$. Thus this case is impossible.

In the case $r \geq 2$ we consider the ideal J whose existence is ensured by Proposition 4.11 Chapter 3. We apply the Proposition to the ideal $\mathfrak{p}$ and the polynomial B that is constructed in Proposition 3.6. Using (59), (60), according to Proposition 4.11 in Chapter 3 we have the bounds

(71) $$\deg J \leq \deg \mathfrak{p} \deg_{\underline{x}} B \leq \sqrt{\tau}(\deg \mathfrak{p})^{(m+2-r)/(m+1-r)},$$
$$h(J) \leq h(\mathfrak{p}) \deg_{\underline{x}} B + \deg \mathfrak{p} \deg_z B$$
(72) $$\leq 2\sqrt{\tau}(h(\mathfrak{p})(\deg \mathfrak{p})^{1/(m+1-r)} + \deg \mathfrak{p}),$$
$$\log |J(\overline{\omega})| \leq \log |\mathfrak{p}(\overline{\omega})| + h(\mathfrak{p}) \deg_{\underline{x}} B + \deg \mathfrak{p} \deg_z B$$
$$\leq \log |\mathfrak{p}(\overline{\omega})| + 2\sqrt{\tau}(h(\mathfrak{p})(\deg \mathfrak{p})^{\frac{1}{m+1-r}} + \deg \mathfrak{p})$$
(73) $$\leq -\frac{1}{2}\tau^{mr}\big(h(\mathfrak{p})(\deg \mathfrak{p})^{\frac{r}{m+1-r}} + (\deg \mathfrak{p})^{\frac{m}{m+1-r}}\big).$$

The ideal J satisfies the relation $\dim J = \dim \mathfrak{p} - 1 = r - 2 < \dim I$. From the definition of I it follows that

$$\log |J(\overline{\omega})| \geq -\tau^{m(r-1)}\big(h(J)(\deg J)^{\frac{r-1}{m+2-r}} + (\deg J)^{\frac{m}{m+2-r}}\big).$$

Using this inequality and also (71), (72) we find that

$$\log |J(\overline{\omega})| > -\frac{1}{2}\tau^{mr}\big(h(\mathfrak{p})(\deg \mathfrak{p})^{\frac{r}{m+1-r}} + (\deg \mathfrak{p})^{\frac{m}{m+1-r}}\big).$$

This inequality contradicts (73) and thereby completes the proof of Theorem 2.2.

5. D-property for Ramanujan functions.

In this section we prove that the set of Ramanujan functions (43)-(44) has the D-property at the point $z = 0$ with the constant $c = 2$. This assertion follows from the next Proposition.

PROPOSITION 5.1. *If $\mathfrak{p}$ is a prime ideal of $\mathfrak{R} = \mathbf{C}[z, x_1, x_2, x_3]$ with $D\mathfrak{p} \subset \mathfrak{p}$ and having a zero at $(0, 1, 1, 1)$, then either $z \in \mathfrak{p}$ or $\Delta = x_2^3 - x_3^2 \in \mathfrak{p}$.*

For the proof of Proposition 5.1 we need several lemmas.
Let

$$D = z\frac{d}{dz} + \frac{1}{12}(x_1^2 - x_2)\frac{\partial}{\partial x_1} + \frac{1}{3}(x_1 x_2 - x_3)\frac{\partial}{\partial x_2} + \frac{1}{2}(x_1 x_3 - x_2^2)\frac{\partial}{\partial x_3}$$

be the differential operator in the ring $\mathfrak{R}$ corresponding to the system (45) then we have $Dz = z$ and

$$D\Delta = x_2^2(x_1 x_2 - x_3) - x_3(x_1 x_3 - x_2^2) = x_1 \Delta.$$

Thus, the principal ideals generated by z and by Δ in $\mathfrak{R}$ are D-invariant.

LEMMA 5.2. *There exists only two D-invariant principal prime ideals of $\mathfrak{R}$, namely, the ideals generated by z and by Δ.*

PROOF. Suppose that $A \in \mathfrak{R}$ is any irreducible polynomial with the property that $A|DA$. Thus $DA = AB$ where $B \in \mathfrak{R}$.

For any $F \in \mathfrak{R}$ we define the *weight* of F as

$$\varphi(F) = \deg_t F(z, tx_1, t^2 x_2, t^3 x_3).$$

Then φ satisfies the following properties:

1. For any integers k_1, k_2, k_3 all monomials of $D(x_1^{k_1} x_2^{k_2} x_3^{k_3}))$ have the same weight

$$\varphi(D(x_1^{k_1} x_2^{k_2} x_3^{k_3})) = 1 + \varphi(x_1^{k_1} x_2^{k_2} x_3^{k_3}) = k_1 + 2k_2 + 3k_3 + 1.$$

2. For any $F \in \mathfrak{R}$

$$\varphi(DF) \le \varphi(F) + 1.$$

3. For any $F, G \in \mathfrak{R}$

$$\varphi(FG) = \varphi(F) + \varphi(G).$$

These properties follow trivially from the definition of the weight and the definition of D.

Using properties 2 and 3, we find that the relation $DA = AB$ implies that

$$\varphi(A) + \varphi(B) = \varphi(DA) \le \varphi(A) + 1$$

and hence $\varphi(B) \le 1$. Therefore B does not depend on x_2, x_3 and is of degree less or equal to one in x_1. Thus

(74)
$$DA = (ax_1 + b)A$$

with $a, b \in \mathbf{C}[z]$. Also $\deg_z A + \deg_z B = \deg_z DA \le \deg_z A$. Hence $a, b \in \mathbf{C}$.

Let C be the sum of the monomials of A with minimal weight. If we compare the sum of the monomials of weight $\varphi(C)$ on both sides of (74) and use the property 1, then we find that

$$z\frac{\partial C}{\partial z} = bC.$$

In this equality we compare the coefficients of the highest power of z in $C \in \mathfrak{R}$ and derive $b = \deg_z C \in \mathbf{Z}$. Let

$$A(z, P(z), Q(z), R(z)) = c_m z^m + c_{m+1} z^{m+1} + \cdots, \quad c_m \neq 0.$$

Using the identity (74) we find

$$(aP + b)(c_m z^m + \cdots) = DA(z, P(z), Q(z), R(z)) =$$
$$z \frac{d}{dz}(A(z, P(z), Q(z), R(z))) = m c_m z^m + \cdots.$$

Comparing the coefficients of z^m, we get $(a + b)c_m = m c_m$, $a + b = m$ and hence $a \in \mathbf{Z}$. We have thereby shown that a and b are integers.

We observe that

(75)
$$D(\Delta^{-a} z^{-b}) = -(ax_1 + b)\Delta^{-a} z^{-b}.$$

Let

$$S(z, x_1, x_2, x_3) = A \cdot \Delta^{-a} z^{-b} \in \mathbf{C}(z, x_1, x_2, x_3).$$

We easily verify, with the help of (74) and (75), that $DS = 0$. If

$$g(z) = S(z, P(z), Q(z), R(z)),$$

then $\frac{d}{dz} g(z) = 0$. Hence the function $S(z, P(z), Q(z), R(z))$ is a constant.

It has been shown by K.Mahler, [**Mah1**], that Ramanujan functions $P(z)$, $Q(z)$, $R(z)$ are algebraically independent over the field $\mathbf{C}(z)$. Hence $S(z, x_1, x_2, x_3) = c$ for some constant c. Thus $A = c\Delta^a z^b \in \mathfrak{R}$ giving $a \geq 0, b \geq 0$. Since A is an irreducible polynomial, only two cases are possible: either $a = 1, b = 0$ or $a = 0, b = 1$, giving $A = c\Delta$ or $A = cz$, respectively. $\qquad\square$

LEMMA 5.3. *The system of differential equations*

(76)
$$\begin{cases} (x^2 - f)f' = 4(xf - g), \\ (x^2 - f)g' = 6(xg - f^2) \end{cases}$$

has a unique solution in algebraic functions $f(x), g(x)$ *with* $f(1) = 1, g(1) = 1$, *namely,* $f(x) = x^2, g(x) = x^3$.

The proof of this lemma is based on an idea which was first used possibly by Siegel for studying algebraic solutions of the special Riccati differential equations, see [**Sie**], [**Shi**].

PROOF. We observe that $f(x) = x^2$ and $g(x) = x^3$ satisfy the system of differential equations (76). Hence, in what follows we may suppose that $f(x) \neq x^2$.

We set $u(x) = x^2 - f(x)$ and $v(x) = xf(x) - g(x)$. It is easy to verify that functions $u(x), v(x)$ satisfy the system of differential equations

(77)
$$\begin{cases} uu' = 2xu - 4v, \\ 2v' = 10u - 5xu', \end{cases}$$

and $u(1) = 0, v(1) = 0$. Since $u(x)$ and $v(x)$ are algebraic functions, there exists a natural number e and a parametrization (here we are considering parametrizations of branches at $x = 1$)

$$x = 1 + t^e, \quad u = \sum_{k=\lambda}^{\infty} a_k t^k, \quad v = \sum_{k=\mu}^{\infty} b_k t^k$$

where $\lambda \geq 1, \mu \geq 1, a_k, b_k \in \mathbf{C}$ and $a_\lambda b_\mu \neq 0$. We may assume that e has been chosen to be as small as possible.

The expansions of the functions in (77) have the following initial terms:

$$xu = a_\lambda t^\lambda + \cdots, \qquad xu' = \frac{\lambda}{e} a_\lambda t^{\lambda-e} + \cdots,$$

$$uu' = \frac{\lambda}{e} a_\lambda^2 t^{2\lambda-e} + \cdots, \qquad v' = \frac{\mu}{e} b_\mu t^{\mu-e} + \cdots.$$

If we substitute these expansions into the second equation in (77), then we obtain

(78) $$2\frac{\mu}{e} b_\mu t^{\mu-e} + \cdots = 10 a_\lambda t^\lambda + \cdots - 5\frac{\lambda}{e} a_\lambda t^{\lambda-e} + \cdots.$$

Comparing exponents of the smallest powers of t on the left and right, we conclude that $\lambda = \mu$. If we now substitute these expansions into the first equation in (77), then we obtain

(79) $$\frac{\lambda}{e} a_\lambda^2 t^{2\lambda-e} + \cdots = 2 a_\lambda t^\lambda + \cdots - 4 b_\lambda t^\lambda - \cdots.$$

This implies that the inequality $2\lambda - e \geq \lambda$ must hold, and hence $\lambda \geq e$.

We next compare the coefficients of $t^{\lambda-e}$ on the left and right in (78). We obtain

(80) $$2b_\lambda = -5a_\lambda.$$

If $\lambda > e$, then a comparison of the coefficients of t^λ in (79) gives

$$2a_\lambda - 4b_\lambda = 0,$$

which, combined with (80), contradicts the assumption that $a_\lambda \neq 0$. Hence $\lambda = e$, and from (79) we find that

$$a_e^2 = 2a_e - 4b_e.$$

By (80), this means that $a_e^2 = 12a_e$. Since $a_e \neq 0$, we conclude that $a_e = 12$ and $b_e = -30$.

Now suppose that $e \geq 2$. Let r be the smallest number for which the conditions $a_r \neq 0$ and $e \nmid r$ hold. This number exists because e was chosen to be minimal (otherwise, the first equation in (77) would imply that all of the non-zero coefficients b_k have index divisible by e). If we now compare the coefficients of t^r on the left and right of the first equation in (77), then we find that

$$\frac{r+e}{e} a_r a_e = 2a_r - 4b_r$$

or, since $a_e = 12$,

(81) $$12\frac{r+e}{e} a_r = 2a_r - 4b_r.$$

By making the same substitution in the second equation in (77) and comparing the coefficients of t^{r-e}, we find that

$$2\frac{r}{e} b_r = -5\frac{r}{e} a_r$$

or

$$2b_r + 5a_r = 0.$$

The last equation along with (81) give us $ra_r = 0$; since $r > e$, this means that $a_r = 0$, which is impossible. Thus, $e = 1$, the functions $u(x)$ and $v(x)$ are single-valued in a neighbourhood of the point $x = 1$, and

$$(82) \qquad u(1) = v(1) = 0, \qquad u'(1) = 12, \quad v'(1) = -30.$$

We now prove that all the derivatives of $u(x)$ and $v(x)$ at the point $x = 1$ are uniquely determined, that is, there exists a unique solution of the system of differential equations (77) that is analytic in a neighbourhood of $x = 1$ and satisfies the conditions in (82). Let $k \geq 2$. If we differentiate the first equation in (77) k times, then we find that the function

$$uu^{(k+1)} + (k+1)u'u^{(k)} - 2xu^{(k)} + 4v^{(k)}$$

can be expressed as a polynomial in x and $u, u', \ldots, u^{(k-1)}$. Taking (82) into account, we then find that the quantity

$$(6k + 5)u^{(k)}(1) + 2v^{(k)}(1)$$

is uniquely determined by $u(1), u'(1), \ldots, u^{(k-1)}(1)$. In exactly the same way, if we differentiate $k - 1$ times the second equation in (77), we find that the quantity

$$5u^{(k)}(1) + 2v^{(k)}(1)$$

can be expressed uniquely in terms of $u^{(j)}(1), v^{(j)}(1), 0 \leq j \leq k - 1$. But then the derivatives $u^{(k)}(1), v^{(k)}(1)$ are also uniquely determined. This proves the uniqueness of the solution of (77) that is analytic in a neighbourhood of $x = 1$ and satisfies the initial conditions (82).

In some neighbourhood of the point $x = 1$ the equation $x = P(z)$ determines z uniquely as an analytic function of x that vanishes at $x = 1$. We set

$$F(x) = Q(P^{-1}(x)), \qquad G(x) = R(P^{-1}(x)).$$

Then $F(x)$ and $G(x)$ are analytic functions in a neighbourhood of $x = 1$ that satisfy the system of differential equations (76) and the initial conditions $F(1) = 1$, $G(1) = 1$, as we easily verify. In addition,

$$z = -\frac{1}{24}(x - 1) + \cdots,$$
$$F(x) = 1 - 10(x - 1) + \cdots, \qquad G(x) = 1 + 21(x - 1) + \cdots.$$

But then it is easy to see that the functions $U(x) = x^2 - F(x)$, $V(x) = xF(x) - G(x)$, satisfy the system of differential equations (77) and the initial conditions (82). By the uniqueness proved above, we conclude that $u(x) = U(x)$, $v(x) = V(x)$, and hence $f(x) = F(x)$, $g(x) = G(x)$. Thus, $F(x)$ and $G(x)$ are algebraic functions. But if we substitute $x = P(z)$ into some identity $A(x, F(x)) = 0$, $A(x, y) \in \mathbf{C}[x, y]$, then it becomes an algebraic relation $A(P(z), Q(z)) = 0$ between the functions $P(z)$ and $Q(z)$. Since $P(z)$, $Q(z)$ and $R(z)$ are algebraically independent over $\mathbf{C}(z)$, such an algebraic relation is impossible. This contradiction completes the proof of Lemma 5.3. $\qquad \square$

PROOF OF PROPOSITION 5.1. Let $\mathfrak{p}$ be a prime ideal from Proposition 5.1 and let $\mathfrak{q} = \mathfrak{p} \cap \mathbf{C}[x_1, x_2, x_3]$. Then $\mathfrak{q}$ is a prime ideal with $D\mathfrak{q} \subset \mathfrak{q}$.

We note that if $\mathfrak{q} = (0)$, then $\mathfrak{p} = (A)$ with $A \notin \mathbf{C}[x_1, x_2, x_3]$. Using Lemma 5.2 we conclude that in this case $\mathfrak{p} = (z)$.

If $q \cap \mathbf{C}[x_1] \neq 0$, then, since the ideal q is prime, there exists a constant c such that $x_1 - c \in q \subset \mathfrak{p}$. Since $(0,1,1,1)$ is a zero of $\mathfrak{p}$, we get $c = 1$, and $x_1 - 1 \in q$ or $x_1 \equiv 1 \pmod{q}$. The ideal q is D-invariant, therefore $D(x_1 - 1) = \frac{1}{12}(x_1^2 - x_2) \in q$ and $x_2 \equiv x_1^2 \equiv 1 \pmod{q}$. Next $D(x_2 - 1) = \frac{1}{3}(x_1 x_2 - x_3) \in q$ giving $x_3 \equiv 1 \pmod{q}$. Hence, $\Delta = x_2^3 - x_3^2 \equiv 0 \pmod{q}$ and $\Delta \in q$. The Proposition is proved in this case as well.

If $q \neq (0)$ and $q \cap \mathbf{C}[x_1, x_2] = (0)$, then $q = (A)$ with $A \in \mathbf{C}[x_1, x_2, x_3]$, and $A \mid DA$. By Lemma 5.2 we conclude $(A) = (\Delta)$. Hence $\Delta \in q$. The same is true if $q \neq (0)$ and $q \cap \mathbf{C}[x_1, x_3] = (0)$.

The last case is
$$q \cap \mathbf{C}[x_1] = (0), \quad q \cap \mathbf{C}[x_1, x_2] \neq (0), \quad q \cap \mathbf{C}[x_1, x_3] \neq (0).$$
Then there exist irreducible polynomials $A(x_1, x_2), B(x_1, x_3) \in q$. By assumption, we have $A(1,1) = B(1,1) = 0$. Now $\mathbf{C}[x_1, x_2, x_3]/q = \mathbf{C}[\xi_1, \xi_2, \xi_3]$ with ξ_2, ξ_3 algebraic over $\mathbf{C}(\xi_1)$ and ξ_1 is transcendental over $\mathbf{C}$. Further $A(\xi_1, \xi_2) = B(\xi_1, \xi_3) = 0$. Thus there exist algebraic functions $y(x), z(x)$ with $y(1) = z(1) = 1$ and the triple of functions $(x, y(x), z(x))$ is a zero of the ideal q. If we differentiate the equation $A(x, y(x)) \equiv 0$ with respect to x, then we obtain

$$\text{(83)} \qquad \frac{\partial A}{\partial x_1}(x, y(x)) + \frac{\partial A}{\partial x_2}(x, y(x)) y'(x) = 0.$$

If we take into account that the triple $(x, y(x), z(x))$ is a zero of the ideal q and, in particular, of the polynomial
$$DA = \frac{1}{12}(x_1^2 - x_2)\frac{\partial A}{\partial x_1} + \frac{1}{3}(x_1 x_2 - x_3)\frac{\partial A}{\partial x_2} \in q,$$
then we conclude that
$$\frac{1}{12}(x^2 - y(x))\frac{\partial A}{\partial x_1}(x, y(x)) + \frac{1}{3}(xy(x) - z(x))\frac{\partial A}{\partial x_2}(x, y(x)) = 0.$$
Eliminating $\frac{\partial A}{\partial x_1}$ between last equation and (83), we get
$$\frac{\partial A}{\partial x_2}(x, y(x))\left(\frac{1}{12}(x^2 - y(x))y'(x) - \frac{1}{3}(xy(x) - z(x))\right) = 0.$$
Using the fact that
$$\frac{\partial A}{\partial x_2}(x, y(x)) \neq 0,$$
we find that the functions $y(x), z(x)$ satisfy the first differential equation in (76). Applying the above argument to the polynomial B we find that $y(x), z(x)$ also satisfy the second differential equation in (76). By Lemma 5.3 we now know that $y(x) = x^2, z(x) = x^3$, that is, $A(x_1, x_2) = x_1^2 - x_2, B(x_1, x_3) = x_1^3 - x_3$. This means that q contains the polynomial
$$(x_2 - x_1^2)(x_2^2 + x_2 x_1^2 + x_1^4) - (x_3 - x_1^3)(x_3 + x_1^3) = x_2^3 - x_3^2 = \Delta.$$
Proposition 5.1 is proved. $\qquad\square$

This proves the D-property for Ramanujan functions.

CHAPTER 11

Zero Estimates on Commutative Algebraic Groups

1. Introduction

Zero estimates appear as a key stone in the theory of transcendental numbers. The construction of an auxiliary function, by analytic means, typically produces a polynomial vanishing at many points of an algebraic group. The object of a zero estimate is to extract as much information as possible from these data. We shall discuss here a zero estimate due to P. Philippon [**PPh3**], which contains earlier work by D. W. Masser [**Mas2**], by D. W. Masser and G. Wüstholz [**MW1**], [**MW3**], and by G. Wüstholz [**Wus1**], [**Wus2**]. For further study, the reader is encouraged to look at the more recent works of L. Denis [**Den**], M. Nakamaye [**Nak**] and P. Philippon [**PPh9**]. The reader may also look at the paper of J.-C. Moreau [**Mor**] which gives a geometric exposition of the zero estimate of [**MW1**], together with a generalization to multi-projective space. Another exposition of zero estimates including their applications to transcendental number theory can be found in the Bourbaki lecture of D. Bertrand [**Ber5**].

2. Degree of an intersection on an algebraic group

Through this chapter, we work over the field $\mathbf{C}$ of complex numbers, although most of what follows applies over any algebraically closed field. Our references for affine and projective varieties are Chapter I of [**Har**][1] and Chapters I and II of [**Mum**][2].

2.1. Algebraic groups.
Recall that the *Zariski topology* on the projective n-space $\mathbf{P}^n(\mathbf{C})$ is the topology for which a subset E of $\mathbf{P}^n(\mathbf{C})$ is *closed* if it is the zero set $Z(I)$ of a homogeneous ideal I of $\mathbf{C}[\underline{X}] = \mathbf{C}[X_0, \ldots, X_n]$. We say that a subset E of $\mathbf{P}^n(\mathbf{C})$ is *locally closed* if it is the intersection of a closed set and an open set or, equivalently, if E is an open subset of its Zariski closure $\overline{E}$. We say that E is *irreducible* if it cannot be written as a union of two proper closed subsets (in the relative topology). When $\mathbf{x} = (x_0, \ldots, x_n)$ is a nonzero point of $\mathbf{C}^{n+1}$, we denote by $[\mathbf{x}] = (x_0 : \cdots : x_n)$ the corresponding point of $\mathbf{P}^n(\mathbf{C})$.

A *connected commutative algebraic group* is a locally closed irreducible subset G of $\mathbf{P}^n(\mathbf{C})$, for some $n \geq 1$, equipped with a structure of commutative group for

---

(∗) Chapter's author : Damien ROY.

[1][**Har**] R. Hartshorne. *Algebraic Geometry*, Springer, Berlin, (1977).

[2][**Mum**] D. Mumford. *Algebraic Geometry I, Complex Projective Varieties*, Springer, Cambridge (USA), (1976).

which the maps

$$m : \begin{array}{c} G \times G \longrightarrow G \\ (a, b) \longmapsto a + b \end{array} \quad \text{and} \quad i : \begin{array}{c} G \longrightarrow G \\ a \longmapsto -a \end{array}$$

are morphisms of algebraic sets.

An *algebraic subgroup* of G is a subgroup H of G which is closed in the relative Zariski topology of G. It can be shown that the irreducible components of an algebraic subgroup H of G are pairwise disjoint and are translates of the component H_0 which contains the neutral element e of G. This component is also a subgroup of H and therefore is a connected algebraic group.

From now on, we fix a connected commutative algebraic group $G \subseteq \mathbf{P}^n(\mathbf{C})$. The condition that the addition map m is a morphism means that for each point $(a, b) \in G \times G$ there exist a neighborhood $\mathcal{V}$ of (a, b) in $G \times G$ and a collection $\underline{A} = (A_0, \ldots, A_n)$ of bi-homogeneous polynomials in

$$\mathbf{C}[\underline{X}, \underline{Y}] = \mathbf{C}[X_0, \ldots, X_n, Y_0, \ldots, Y_n]$$

of the same bi-degree (i.e. of the same degree in $\underline{X}$ and of the same degree in $\underline{Y}$) such that for any nonzero points $\mathbf{x}, \mathbf{y} \in \mathbf{C}^{n+1}$ with $([\mathbf{x}], [\mathbf{y}]) \in \mathcal{V}$, we have

$$(84) \qquad [\mathbf{x}] + [\mathbf{y}] = \big(A_0(\mathbf{x}, \mathbf{y}) : \cdots : A_n(\mathbf{x}, \mathbf{y})\big) = [\underline{A}(\mathbf{x}, \mathbf{y})].$$

A collection $\underline{A} = (A_0, \ldots, A_n)$ of bi-homogeneous polynomials of the same bi-degree satisfying the condition (84) on a non-empty open subset $\mathcal{V}$ of $G \times G$ is called an *addition law* on $\mathcal{V}$. Hence, there exists an open cover $\{\mathcal{V}_\alpha\}$ of $G \times G$ with an addition law $\underline{A}^{(\alpha)}$ on $\mathcal{V}_\alpha$ for each α. This is called a *complete system of addition laws*. Moreover, since $G \times G$ is quasi-compact, we can always extract from $\{\mathcal{V}_\alpha\}$ a finite cover of $G \times G$. Thus there exist finite complete systems of addition laws.

As in Section 2 of [**MW1**], we define the *associated open set* $\mathcal{U}_{\underline{A}}$ of an addition law $\underline{A} = (A_0, \ldots, A_n)$ to be the complement in $G \times G$ of the set of common zeros of $A_0, \ldots, A_n$. Since $\underline{A}$ defines a regular map from $\mathcal{U}_{\underline{A}}$ to G, and since this map coincides with m on a non-empty open subset $\mathcal{V}$ of $\mathcal{U}_{\underline{A}}$, both maps coincide on the Zariski closure of $\mathcal{V}$ in $\mathcal{U}_{\underline{A}}$. The set G being irreducible, this closure is the whole of $\mathcal{U}_{\underline{A}}$. Hence, $\underline{A}$ defines an addition law on its associated open set $\mathcal{U}_{\underline{A}}$.

In the sequel we denote by $c = c(G)$ the smallest positive integer for which the group G admits a complete system of addition laws, each consisting of families of bi-homogeneous polynomials in $\mathbf{C}[\underline{X}, \underline{Y}]$ having degree $\leq c$ in the set of variables $\underline{X}$. Note that, by a result of H. Lange [**Lange**][3], G is isomorphic to an algebraic group G' with $c(G') \leq 2$, but we shall not use this fact here.

EXAMPLE. The set $G = \{(x_0 : x_1 : \cdots : x_n) \in \mathbf{P}^n(\mathbf{C}) ; x_0 x_1 \cdots x_n \neq 0\}$ is locally closed and irreducible. Define a binary operation $m : G \times G \to G$ by componentwise multiplication:

$$m((x_0 : \cdots : x_n), (y_0 : \cdots : y_n)) = (x_0 y_0 : \cdots : x_n y_n).$$

Then G becomes a commutative algebraic group with $c(G) = 1$. It is isomorphic to $(\mathbf{C}^\times)^n$ via the map which sends a point $(z_1, \ldots, z_n)$ in $(\mathbf{C}^\times)^n$ to $(1 : z_1 : \cdots :$

---

[3][**Lange**] H. Lange. Families of translations of commutative algebraic groups, *J. Algebra* 109, (1987), 260–265.

$z_n) \in G$. Moreover, for each subgroup A of $\mathbf{Z}^n$, the set

$$H_A = \big\{ (x_0 : \cdots : x_n) \in G \,;\, (x_1/x_0)^{a_1} \cdots (x_n/x_0)^{a_n} = 1$$

$$\text{for all } (a_1, \ldots, a_n) \in A \big\}$$

is an algebraic subgroup of G of codimension equal to the rank of A. It can be shown that the map $A \mapsto H_A$ establishes a bijection between the subgroups of $\mathbf{Z}^n$ and the algebraic subgroups of G. The group G is usually denoted $\mathbf{G}_m^n$.

2.2. Degree of an ideal. Let I be a homogeneous ideal of the ring $\mathbf{C}[\underline{X}]$. For each integer $D \in \mathbf{N}$, we denote respectively by $\mathbf{C}[\underline{X}]_D$ and I_D the sets of homogeneous elements of $\mathbf{C}[\underline{X}]$ and I of degree D, and we view them as vector spaces over $\mathbf{C}$. The *Hilbert function of* I is the map from $\mathbf{N}$ to $\mathbf{N}$ given by

$$H(I; D) = \dim_{\mathbf{C}}(\mathbf{C}[\underline{X}]_D / I_D)$$

for any $D \in \mathbf{N}$. It is known from Hilbert that this function is given by a polynomial in D for sufficiently large values of D. If I has no zero in $\mathbf{P}^n(\mathbf{C})$ that is, if I has rank $n + 1$, this polynomial is 0. Otherwise, its degree is the dimension d of the set of zeros $Z(I)$ of I in $\mathbf{P}^n(\mathbf{C})$. Its degree is also given by the formula $d = n - r$ where r denotes the rank of I. Assuming $d \geq 0$, if we write

(85) $$H(I; D) = \frac{c}{d!} D^d + \text{(terms of lower degree)} \qquad \text{for } D \gg 1,$$

then c is a positive integer. This integer is called the *degree of* I and is denoted $\deg(I)$. It is also useful to have a notation for the product by $d!$ of the leading term of the polynomial (85), since it plays a crucial role in this theory. We denote it by

$$\mathcal{H}(I; D) = cD^d.$$

When E is a subset of $\mathbf{P}^n(\mathbf{C})$, we define

$$\mathcal{H}(E; D) = \mathcal{H}(\mathfrak{I}(E); D)$$

where $\mathfrak{I}(E)$ denotes the ideal generated by all homogeneous polynomials vanishing on E. With this definition, we thus have $\mathcal{H}(E; D) = \mathcal{H}(\overline{E}; D)$. We recall some properties of the degree of an ideal and deduce corresponding properties for the function $\mathcal{H}(I; D)$:

(i) When $I = (0)$ is the zero ideal, we have $\deg(I) = 1$ and $\mathcal{H}(I; D) = D^n$.

(ii) Suppose that I has rank $r \leq n$ (so that its set of zeros $Z(I)$ is not empty), and let $\mathfrak{q}_1, \ldots, \mathfrak{q}_t$ be its primary components of rank r. For $i = 1, \ldots, t$, denote by $\mathfrak{p}_i$ the radical of $\mathfrak{q}_i$ and by ℓ_i the length of $\mathfrak{q}_i$. Then, we have

$$\deg(I) = \sum_{i=1}^{t} \deg(\mathfrak{q}_i) \quad \text{and} \quad \deg(\mathfrak{q}_i) = \ell_i \deg(\mathfrak{p}_i), \quad (1 \leq i \leq t).$$

Consequently,

$$\mathcal{H}(I; D) = \sum_{i=1}^{t} \mathcal{H}(\mathfrak{q}_i; D) = \sum_{i=1}^{t} \ell_i \mathcal{H}(\mathfrak{p}_i; D).$$

(iii) (Bézout's Lemma) If I has rank $r < n$ and if P is a homogeneous polynomial which does not belong to any associated prime ideal of I, then the ideal (I, P) has rank $r + 1$ and its degree is $\deg(P) \deg(I)$. Consequently, if D is an integer with $D \geq \deg(P)$, we have

$$\mathcal{H}((I, P); D) \leq \mathcal{H}(I; D).$$

A proof of these assertions in the context of multihomogeneous ideals can be found in [**VdW**][4]. A short proof of (iii) is also given in [**BM2**], Lemma 3, page 281. It is based on the fact that the multiplication by P in the quotient $\mathbf{C}[\underline{X}]/I$ is an injective $\mathbf{C}$-linear map when (and only when) P does not belong to any associated prime ideal of I (see Theorem 11, Chap. IV of [**ZS**][5]).

2.3. Special ideals. Let $\tilde{G}$ be the set of all nonzero points $\mathbf{x} \in \mathbf{C}^{n+1}$ which represent a point $[\mathbf{x}]$ in G. In other words, $\tilde{G}$ is the cone on G in $\mathbf{C}^{n+1}$ without the origin. We denote by $\mathcal{F}$ the set of all maximal ideals M of $\mathbf{C}[\underline{X}]$ defining a point of $\tilde{G}$.

For any $M \in \mathcal{F}$, we denote by $\mathbf{C}[\underline{X}]_M$ the localization of $\mathbf{C}[\underline{X}]$ at M and we view $\mathbf{C}[\underline{X}]$ as a subring of this quotient ring. Then, for an ideal I of $\mathbf{C}[\underline{X}]$, the contracted extension

$$(86) \qquad\qquad \mathbf{C}[\underline{X}] \cap I\mathbf{C}[\underline{X}]_M$$

is the intersection of all primary components of I contained in M (see Theorem 17, Chap. IV of [**ZS**] (*loc. cit.*)). More precisely, if

$$I = \mathfrak{q}_1 \cap \cdots \cap \mathfrak{q}_s$$

is an irredundant primary decomposition of I with the primary ideals $\mathfrak{q}_1, \ldots, \mathfrak{q}_s$ ordered so that the first t of them are contained in M and not the others, then the ideal of $\mathbf{C}[\underline{X}]$ given by (86) is $\mathfrak{q}_1 \cap \cdots \cap \mathfrak{q}_t$, independently of the choice of the primary decomposition.

From this it follows that, for any ideal I of $\mathbf{C}[\underline{X}]$, the ideal

$$(87) \qquad\qquad I^* = \bigcap_{M \in \mathcal{F}} \left(\mathbf{C}[\underline{X}] \cap I\mathbf{C}[\underline{X}]_M \right)$$

is the intersection of all primary components of I contained in at least one ideal M of $\mathcal{F}$, and this is independent of the choice of an irredundant primary representation of I. In more geometrical terms, I^* is the intersection of all primary components of I whose set of zeros in $\mathbf{C}^{n+1}$ meets $\tilde{G}$. In the case where I is homogeneous, its primary components are homogeneous ideals. Then, this definition becomes:

DEFINITION 2.1. For any homogeneous ideal $I \subset \mathbf{C}[\underline{X}]$, we define the ideal I^* to be the intersection of all primary components of I whose set of zeros in $\mathbf{P}^n(\mathbf{C})$ contains at least one point of G. We say that I is *special* if $I = (\mathfrak{G}, I)^*$, where $\mathfrak{G} = \mathfrak{I}(G)$ denotes the ideal of G.

The following exercise provides an alternative definition for I^* in the case where $\mathfrak{G} \subseteq I$.

EXERCISE. Suppose that G is the complement in $\overline{G}$ of the zero set of a homogeneous ideal $\mathfrak{a} \subseteq \mathbf{C}[\underline{X}]$, and let I be a homogeneous ideal of $\mathbf{C}[\underline{X}]$ which contains $\mathfrak{G}$. Then show that $I^* = \cup_{j=1}^{\infty} (I : \mathfrak{a}^j)$ where $(I : \mathfrak{a}^j)$ denotes the ideal consisting of all polynomials P with $P\mathfrak{a}^j \subseteq I$.

In the sequel, we will repeatedly use the following properties of the transformation $I \mapsto I^*$. First of all, we have $I \subseteq I^*$ and also $I\mathbf{C}[\underline{X}]_M = I^*\mathbf{C}[\underline{X}]_M$ for any $M \in \mathcal{F}$, thus $I^* = (I^*)^*$. Moreover, If I_1, I_2 are homogeneous ideals of $\mathbf{C}[\underline{X}]$ with

---

[4][**VdW**] B.V.L. Van der Waerden. On Hilbert's functions, series of composition of ideals and a generalization of a theorem of Bézout, *Proc. Royal Acad. Amsterdam* 31, (1928), 749–770.

[5][**ZS**] O. Zariski, P. Samuel. *Commutative Algebra*, Vol. I & II, Springer, New-York, (1968).

$I_1 \subseteq I_2$, then $I_1 \mathbf{C}[\underline{X}]_M \subseteq I_2 \mathbf{C}[\underline{X}]_M$ for any $M \in \mathcal{F}$ and thus $I_1^* \subseteq I_2^*$. Finally, if I is special, then all its primary components are special as well as its associated prime ideals.

REMARK. When I is a homogeneous ideal (the only case of interest here), the formula (87) also holds if one replaces the set $\mathcal{F}$ by the set of all homogeneous ideals of $\mathbf{C}[\underline{X}]$ of rank n defining a point of G. However, in the proof of theorem 1 below, it is convenient to work with maximal ideals of $\mathbf{C}[\underline{X}]$ corresponding to points of $\tilde{G}$. By contrast, these ideals are not homogeneous, and their rank is $n + 1$.

The following result is due to P. Philippon. More precisely, it is a special case of Proposition 3.3 of [**PPh3**] (see also [**Bro3**]). The general result of P. Philippon deals with multi-homogeneous ideals in a more general ring theoretic context. The proof that we give below uses arguments from [**BM2**] and [**MW1**].

PROPOSITION 2.2. *Put* $\mathfrak{G} = \mathfrak{I}(G)$ *and* $I = (\mathfrak{G}, S)^*$ *where* $S \subseteq \mathbf{C}[\underline{X}]$ *is a family of homogeneous polynomials of degree* $\leq D$. *Assume that* I *admits at least one zero in* G. *Then we have*

$$\mathcal{H}(I; D) \leq \mathcal{H}(G; D).$$

PROOF. Let $r = \operatorname{rank}(\mathfrak{G})$ and $s = \operatorname{rank}(I) - r$. We shall construct recursively homogeneous polynomials $Q_1, \ldots, Q_s \in I$ of degree D such that for $i = 0, \ldots, s$, the ideal $J_i = (\mathfrak{G}, Q_1, \ldots, Q_i)^*$ satisfies the following conditions:

   (a) J_i is unmixed of rank $r + i$;
   (b) $\mathcal{H}(J_i; D) \leq \mathcal{H}(G; D)$.

(Recall that an ideal is said to be *unmixed* if all its associated prime ideals have the same rank.)

If we admit for the moment the existence of such polynomials, then, by (a), J_s has the same rank as I. Since $J_s \subseteq I^* = I$, we also have $H(I; t) \leq H(J_s; t)$ for any integer $t \geq 0$ and so $\mathcal{H}(I; D) \leq \mathcal{H}(J_s; D)$. By (b), we conclude that $\mathcal{H}(I; D) \leq \mathcal{H}(G; D)$, as stated by the proposition.

For $i = 0$, the ideal $J_0 = \mathfrak{G}$ clearly satisfies both conditions. Suppose that, for some integer $i \geq 0$ with $i < s$, we have constructed $Q_1, \ldots, Q_i$ such that $J_i = (J, Q_1, \ldots, Q_i)^*$ satisfies the same conditions. Let $\mathfrak{q}_1, \ldots, \mathfrak{q}_k$ be the primary components of J_i. By hypothesis, they all have the same rank $r + i$, and this is $< \operatorname{rank}(I)$. So, none of the corresponding prime ideals $\mathfrak{p}_1, \ldots, \mathfrak{p}_k$ contains I. Since J_i is special, each of these prime ideals is special and thus, none of $\mathfrak{p}_1, \ldots, \mathfrak{p}_k$ contains the set S. From this, it follows that the ideal generated by S contains a homogeneous polynomial Q_{i+1} of degree D which does not belong to any of $\mathfrak{p}_1, \ldots, \mathfrak{p}_k$ (see Lemma 5 of [**BM2**], page 285). Put $J_{i+1} = (\mathfrak{G}, Q_1, \ldots, Q_{i+1})^*$.

Since Q_{i+1} does not belong to any associated prime ideal of J_i, Bézout's Lemma (see property (iii) in §2.2 above) shows that (J_i, Q_{i+1}) is an ideal of rank $r + i + 1$ with

(88)
$$\mathcal{H}\big((J_i, Q_{i+1}); D\big) \leq \mathcal{H}(J_i; D).$$

On the other hand, for each $M \in \mathcal{F}$, we have

$$(J_i, Q_{i+1})\mathbf{C}[\underline{X}]_M = (J_i \mathbf{C}[\underline{X}]_M, Q_{i+1})$$
$$= (\mathfrak{G}, Q_1, \ldots, Q_{i+1})\mathbf{C}[\underline{X}]_M = J_{i+1}\mathbf{C}[\underline{X}]_M \ ,$$

and thus $(J_i, Q_{i+1})^* = J_{i+1}$. In particular, we have $(J_i, Q_{i+1}) \subseteq J_{i+1}$ and so $\operatorname{rank}(J_{i+1}) \geq r + i + 1$. We will deduce from this that J_{i+1} is unmixed of rank

$r + i + 1$. If we admit this result, then the above inclusion implies $H(J_{i+1}; t) \leq H\big((J_i, Q_{i+1}); t\big)$ for any integer $t \geq 0$. These functions being given by polynomials of the same degree for large values of t, this in turn implies

$$(89) \qquad\qquad \mathcal{H}(J_{i+1}; D) \leq \mathcal{H}\big((J_i, Q_{i+1}); D\big).$$

Combining (88), (89) and the hypotheses on J_i, we get $\mathcal{H}(J_{i+1}; D) \leq \mathcal{H}(G; D)$.

Showing that J_{i+1} is unmixed of rank $r + i + 1$ is equivalent to showing that $J_{i+1}\mathbf{C}[\underline{X}]_M$ is unmixed of rank $r + i + 1$ for any $M \in \mathcal{F}$ with $J_{i+1} \subseteq M$. Since $J_{i+1} \subseteq I$, and since I admits at least one zero in G, such an M exists. Fix one of them and denote by $\mathbf{x}$ the corresponding element of $\tilde{G}$. We already know that the rank of $J_{i+1}\mathbf{C}[\underline{X}]_M$ is $\geq r + i + 1$.

Since G is an algebraic group, it is non-singular. The argument is simple. Any algebraic set admits at least one non-singular point. The translations on G being isomorphisms of the underlying algebraic set, they map a non-singular point to another one. Since they act transitively on the group, it follows that G is non-singular at any point. Since G is non-singular, the projective Jacobian criterion (see [**Har**] (*loc. cit.*, p. 167), page 37) shows that $\tilde{G}$ is non-singular. In particular, $\tilde{G}$ is non-singular at the point $\mathbf{x}$. Since $\mathfrak{G}$ has rank r, this means that there exist polynomials $P_1, \ldots, P_r \in \mathfrak{G}$ whose Taylor series at $\mathbf{x}$ have linearly independent linear terms. By Corollary 1.20 of [**Mum**] (*loc. cit.*, p. 167), since $\mathfrak{G}$ is a prime ideal, this implies

$$\mathfrak{G}\mathbf{C}[\underline{X}]_M = (P_1, \ldots, P_r)\mathbf{C}[\underline{X}]_M.$$

Thus

$$J_{i+1}\mathbf{C}[\underline{X}]_M = (P_1, \ldots, P_r, Q_1, \ldots, Q_{i+1})\mathbf{C}[\underline{X}]_M$$

is generated by $r + i + 1$ elements. By Krull's theorem (Theorem 30 in Chap. IV of [**ZS**] (*loc. cit.*, p. 170)), it follows that $J_{i+1}\mathbf{C}[\underline{X}]_M$ has rank $\leq r + i + 1$. Since we know that its rank is also $\geq r + i + 1$, we have equality. Thus, $J_{i+1}\mathbf{C}[\underline{X}]_M$ has rank $r + i + 1$ and is generated by $r + i + 1$ elements. Since $\mathbf{C}[\underline{X}]_M$ is a regular local ring, this fact implies, by virtue of a theorem of I. S. Cohen (Theorem 21 of [**ISC**][6], page 99), that $J_{i+1}\mathbf{C}[\underline{X}]_M$ is unmixed of rank $r + i + 1$. Thus J_{i+1} satisfies the conditions (a) and (b) with i replaced by $i + 1$. This completes the induction step. $\qquad\qquad\qquad\qquad\qquad\qquad\qquad\qquad\qquad\qquad\qquad\qquad \square$

When A and B are subsets of the group G, we denote by $A + B$ the set consisting of all sums $a + b$ with $a \in A$ and $b \in B$. When H is a subgroup of G and Σ a finite subset of G, the sum $\Sigma + H$ is a finite union of translates of H. We denote by $\mathrm{Card}\big((\Sigma + H)/H\big)$ the number of these translates.

PROPOSITION 2.3. *Let H be an algebraic subgroup of G, let H_0 be its connected component of the identity, and let Σ be a finite subset of G. Then, for the constant $c = c(G)$ defined in §2.1, we have, for any integer $D \geq 0$,*

$$\mathcal{H}(\Sigma + H\,;\, cD) \geq \mathrm{Card}\big((\Sigma + H_0)/H_0\big)\,\mathcal{H}(H_0; D).$$

SKETCH OF PROOF. Since $\Sigma + H$ is a finite union of translates of H, this sum is a finite union of translates of H_0. Thus, it is a closed subset of G of the same

---

[6][**ISC**] I. S. Cohen. On the structure and ideal theory of complete local rings, *Trans. Amer. Math. Soc.* 59, (1946), 54–106.

dimension as H_0 and its irreducible components are the translates of H_0 contained in $\Sigma + H$. Since the number of these is $\geq \text{Card}\left((\Sigma + H_0)/H_0\right)$, it suffices to show

$$(90) \qquad \mathcal{H}(a + H_0\,;\, cD) \geq \mathcal{H}(H_0; D)$$

for any $a \in G$ and any integer $D \geq 0$ (see the property (ii) in §2.2). More precisely, since $a + H_0$ has the same dimension as H_0, it suffices to prove (90) with $D = 1$.

A general linear subvariety L of $\mathbf{P}^n(\mathbf{C})$ of codimension equal to the dimension of H_0 meets H_0 in exactly $\deg(H_0)$ distinct points. Following P. Philippon's argument, if we fix such an L, we get

$$\begin{aligned}
\mathcal{H}(H_0; 1) &= \text{Card}(H_0 \cap L) \\
&= \text{Card}\left((a + H_0) \cap (a + L)\right) \\
&= \mathcal{H}\left((a + H_0) \cap (a + L)\,;\, c\right),
\end{aligned}$$

where the last equality comes from the fact that, for any finite subset E of $\mathbf{P}^n(\mathbf{C})$, the function $\mathcal{H}(E\,;\, t)$ is a constant equal to the cardinality of E. The considerations of § 3 below show that $a + L$ is defined by homogeneous polynomials of degree $\leq c$. On the other hand, Proposition 2.2 applies in fact to any non-singular variety. If we apply it with $a + H_0$ instead of G, it gives

$$\mathcal{H}\left((a + H_0) \cap (a + L)\,;\, c\right) \leq \mathcal{H}(a + H_0\,;\, c). \qquad \square$$

3. Translations and derivations

Since G is non-singular, it is an analytic subvariety of $\mathbf{P}^n(\mathbf{C})$. More precisely, it is, for this structure, a commutative complex Lie group. As such, its admits an exponential map

$$\exp_G \colon T_G \longrightarrow G \subseteq \mathbf{P}^n(\mathbf{C}),$$

where T_G denotes the tangent space of G at the identity. This map is a local isomorphism with the property $\exp_G(u + v) = \exp_G(u) + \exp_G(v)$ for any $u, v \in T_G$. If H is an algebraic subgroup of G, its tangent space T_H at the identity is a subspace of T_G and $\exp_G$ restricts to a local isomorphism from T_H to H.

For the rest of this section, we fix a subspace W of T_G of dimension t and a linear isomorphism $\varphi \colon \mathbf{C}^t \xrightarrow{\sim} W$. We denote by Φ the composite map

$$\Phi \colon \mathbf{C}^t \xrightarrow{\varphi} W \subseteq T_G \xrightarrow{\exp_G} G \subseteq \mathbf{P}^n(\mathbf{C}).$$

This map satisfies

$$(91) \qquad \Phi(\mathbf{z} + \mathbf{w}) = \Phi(\mathbf{z}) + \Phi(\mathbf{w})$$

for any $\mathbf{z}, \mathbf{w} \in \mathbf{C}^t$.

3.1. Derivatives of functions along W. We denote by $\mathcal{O}(\mathcal{U})$ the ring of regular functions on a Zariski open subset $\mathcal{U}$ of G. Recall that an element of this ring is a function $f \colon \mathcal{U} \to \mathbf{C}$ which can be expressed locally as a quotient of homogeneous polynomials of the same degree. More precisely, one requires that, for any point $a \in \mathcal{U}$, there exists an open neighborhood $\mathcal{V}$ of a in $\mathcal{U}$ and two homogeneous polynomials P, Q of the same degree with Q vanishing nowhere on $\mathcal{V}$ such that f coincides with P/Q on $\mathcal{V}$.

For any $a \in G$, we denote by $\tau_a \colon G \to G$ the translation by a in the group G. Since this is an isomorphism, the set $-a + \mathcal{U} = \tau_a^{-1}(\mathcal{U})$ is an open set for any open set $\mathcal{U}$ of G. Moreover, if $f \colon \mathcal{U} \to \mathbf{C}$ is a regular map, then the composite

$f \circ \tau_a \colon (-a + \mathcal{U}) \to \mathbf{C}$ is also regular. We first introduce a family of differential operators ∂^κ on $\mathcal{O}(\mathcal{U})$:

LEMMA 3.1. *Let* $f \colon \mathcal{U} \to \mathbf{C}$ *be a regular map on an open subset* $\mathcal{U}$ *of* G. *For any* $\kappa \in \mathbf{N}^t$, *the map*

$$\partial^\kappa f \colon \mathcal{U} \longrightarrow \mathbf{C}$$

$$a \longmapsto \frac{\partial^\kappa}{\partial \mathbf{z}^\kappa} f(a + \Phi(\mathbf{z}))\Big|_{\mathbf{z}=0}$$

is also a regular map.

PROOF. Let $a \in \mathcal{U}$. Then there exists an open neighborhood $\mathcal{V}$ of a in $\mathcal{U}$ and two homogeneous polynomials P, Q of the same degree with Q vanishing nowhere on $\mathcal{V}$ such that $f = P/Q$ on $\mathcal{V}$. Replacing $\mathcal{V}$ by a smaller open neighborhood of a if needed, we may also assume that there exists an addition law $\underline{T} = (T_0, \ldots, T_n)$ whose associated open set contains $\mathcal{V} \times \{0\}$. Let $\phi_0, \ldots, \phi_n$ be holomorphic maps defined in an open neighborhood of 0 in $\mathbf{C}^t$ such that, on this neighborhood,

$$\Phi(\mathbf{z}) = [\underline{\phi}(\mathbf{z})] \quad \text{where} \quad \underline{\phi}(\mathbf{z}) = (\phi_0(\mathbf{z}), \ldots, \phi_n(\mathbf{z})).$$

Then, for fixed $\mathbf{x} \in \mathbf{C}^{n+1}$ with $[\mathbf{x}] \in \mathcal{V}$ and any $\mathbf{z} \in \mathbf{C}^t$ with norm sufficiently small in function of $\mathbf{x}$, the point $[\mathbf{x}] + \Phi(\mathbf{z})$ belongs to $\mathcal{V}$ and we have

$$f([\mathbf{x}] + \Phi(\mathbf{z})) = \frac{P(\underline{T}(\mathbf{x}, \underline{\phi}(\mathbf{z})))}{Q(\underline{T}(\mathbf{x}, \underline{\phi}(\mathbf{z})))}.$$

Taking derivatives on both sides, we get

$$\partial^\kappa f([\mathbf{x}]) = \frac{\partial^\kappa}{\partial \mathbf{z}^\kappa} f([\mathbf{x}] + \Phi(\mathbf{z}))\Big|_{\mathbf{z}=0} = \frac{A(\mathbf{x})}{B(\mathbf{x})},$$

where $B(\underline{X}) = Q(\underline{T}(\underline{X}, \underline{\phi}(0)))^{|\kappa|+1}$ is a homogeneous polynomial not vanishing on $\mathcal{V}$ and where $A(\underline{X})$ is a homogeneous polynomial of the same degree in the ideal of $\mathbf{C}[\underline{X}]$ generated by the partial derivatives of order $\leq |\kappa|$ of $P(\underline{T}(\underline{X}, \underline{\phi}(\mathbf{z})))$ at the point $\mathbf{z} = 0$. $\qquad \square$

DEFINITION 3.2. A regular function $f \in \mathcal{O}(\mathcal{U})$ is said to *vanish to order* $> T$ *along* W *at a point* $a \in \mathcal{U}$ if $\partial^\kappa f(a) = 0$ for any $\kappa \in \mathbf{N}^t$ with $|\kappa| \leq T$. A homogeneous polynomial $P \in \mathbf{C}[\underline{X}]$ is said to *vanish to order* $> T$ *along* W *at a point* $a \in G$ if there exists a neighborhood $\mathcal{U}$ of a in G and another homogeneous polynomial Q of the same degree not vanishing on $\mathcal{U}$ such that the regular function on $\mathcal{U}$ defined by P/Q vanishes to order $> T$ along W at a.

Note that both definitions are independent of the choice of the linear isomorphism $\varphi \colon \mathbf{C}^t \to W$ giving rise to Φ. So, they really depend only on W.

Observe also that, for any regular function $f \colon \mathcal{U} \to \mathbf{C}$, any $\sigma \in G$ and any $\kappa \in \mathbf{N}^t$, we have

(92) $$\partial^\kappa (f \circ \tau_\sigma) = (\partial^\kappa f) \circ \tau_\sigma.$$

We denote this element of $\mathcal{O}(-\sigma + \mathcal{U})$ by $\partial_\sigma^\kappa f$.

LEMMA 3.3. *For any regular function* $f \colon \mathcal{U} \to \mathbf{C}$, *any* $\gamma, \sigma \in G$ *and any* $\lambda, \kappa \in \mathbf{N}^t$, *we have*

$$\partial_\gamma^\lambda (\partial_\sigma^\kappa (f)) = \partial_{\gamma+\sigma}^{\lambda+\kappa}(f).$$

PROOF. Because of (92), it suffices to show $\partial^\lambda(\partial^\kappa(f)) = \partial^{\lambda+\kappa}(f)$. To prove this, take $a \in \mathcal{U}$. Using (91), we get

$$\partial^\lambda(\partial^\kappa(f))(a) = \frac{\partial^\lambda}{\partial \mathbf{w}^\lambda} \frac{\partial^\kappa}{\partial \mathbf{z}^\kappa} f(a + \Phi(\mathbf{w}) + \Phi(\mathbf{z}))\Big|_{\mathbf{z}=0}\Big|_{\mathbf{w}=0}$$
$$= \frac{\partial^{\lambda+\kappa}}{\partial \mathbf{t}^{\lambda+\kappa}} f(a + \Phi(\mathbf{t}))\Big|_{\mathbf{t}=0}$$
$$= \partial^{\lambda+\kappa} f(a). \quad \square$$

3.2. Operations on ideals.

DEFINITION 3.4. Let I be a homogeneous ideal of $\mathbf{C}[X]$. For each Zariski open set $\mathcal{U} \subseteq G$, define $I(\mathcal{U})$ to be the set of all functions $f : \mathcal{U} \to \mathbf{C}$ with the following property: for any $a \in \mathcal{U}$ there is an open neighborhood $\mathcal{V}$ of a in $\mathcal{U}$ and homogeneous polynomials $P \in I$ and $Q \in \mathbf{C}[X]$ of the same degree, with Q not vanishing on $\mathcal{V}$, such that f is represented by P/Q on $\mathcal{V}$.

Clearly, $I(\mathcal{U})$ is an ideal of $\mathcal{O}(\mathcal{U})$. Its elements are regular functions which can locally be written as quotients of homogeneous polynomials with numerator in I.

DEFINITION 3.5. Let I be as in Definition 3.4, let Σ be any subset of G and let T be any integer ≥ 0. We denote by $\partial_\Sigma^T(I)$ the ideal of $\mathbf{C}[X]$ generated by all homogeneous polynomials P with the following property: for any $a \in G$, there is an open neighborhood $\mathcal{U}$ of a in G and a homogeneous polynomial Q of the same degree as P, not vanishing on $\mathcal{U}$, such that the function $P/Q : \mathcal{U} \to \mathbf{C}$ belongs to the ideal $\mathcal{J}(\mathcal{U})$ of $\mathcal{O}(\mathcal{U})$ generated by all functions $\partial_\sigma^\kappa(f)$ with $\sigma \in \Sigma$, $\kappa \in \mathbf{N}^t$, $|\kappa| \leq T$, and $f \in I(\sigma + \mathcal{U})$.

When $\Sigma = \{\sigma\}$ consists of one point, we write $\partial_\sigma^T(I)$ instead of $\partial_{\{\sigma\}}^T(I)$. In this case, as we will see below, the above definition is equivalent to that of [PPh3]. When the point σ is the neutral element e of G, we simply write $\partial^T(I)$ instead of $\partial_e^T(I)$. The next proposition collects the properties of these transformations that we will need to establish the zero estimate.

PROPOSITION 3.6. Let I be a homogeneous ideal of $\mathbf{C}[X]$, let Γ, Σ be subsets of G, and let S, T be integers ≥ 0. Then, we have:

(i) $\partial^0 I = (\mathfrak{G}, I)^*$;

(ii) $\partial_\Gamma^S(\partial_\Sigma^T(I)) = \partial_{\Gamma+\Sigma}^{S+T}(I)$;

(iii) the set of zeros of $\partial_\Sigma^T(I)$ in G consists of all points $a \in G$ such that, for any homogeneous polynomial $P \in I$ and any point $\sigma \in \Sigma$, P vanishes to order $> T$ along W at $a + \sigma$;

(iv) if I is generated by a set E of homogeneous polynomials of degree $\leq D$, then $\partial_\Sigma^T(I) = (\mathfrak{G}, F)^*$ for a set F of homogeneous polynomials of degree $\leq cD$, where $c = c(G)$ is the constant defined in § 2.1.

The proof of these assertions relies on the following:

LEMMA 3.7. Let I, Σ and T be as in Proposition 3.6. Put $J = \partial_\Sigma^T(I)$ and, for each open subset $\mathcal{U}$ of G, denote by $\mathcal{J}(\mathcal{U})$ the ideal of $\mathcal{O}(\mathcal{U})$ generated by the functions $\partial_\sigma^\kappa(f)$ with $\sigma \in \Sigma$, $\kappa \in \mathbf{N}^t$, $|\kappa| \leq t$ and $f \in I(\sigma + \mathcal{U})$. Then we have

$$\mathcal{J}(\mathcal{U}) \subseteq J(\mathcal{U}).$$

Moreover, for any $g \in J(\mathcal{U})$ and any $a \in \mathcal{U}$, there is an open neighborhood $\mathcal{V}$ of a in $\mathcal{U}$ such that the restriction $g|_{\mathcal{V}}$ of g to $\mathcal{V}$ belongs to $J(\mathcal{V})$.

This can be formulated simply in terms of sheaves (see Chapter II, § 1 of [**Har**] (*loc. cit.*, p. 167)), by saying that the map $\mathcal{U} \mapsto J(\mathcal{U})$ is the sheaf of ideals associated to the pre-sheaf $\mathcal{U} \mapsto J(\mathcal{U})$.

PROOF. We start with the second assertion, which is simpler. So, let $g \in J(\mathcal{U})$ and $a \in \mathcal{U}$. By definition, there exist a neighborhood $\mathcal{U}_0$ of a in $\mathcal{U}$ and homogeneous polynomials P, Q of the same degree, with $P \in J$ and Q not vanishing on $\mathcal{U}_0$, such that g is represented by P/Q on $\mathcal{U}_0$. Since $P \in J$, it can be written as a linear combination $P = A_1 P_1 + \cdots + A_s P_s$ where $A_1, \ldots, A_s$ and $P_1, \ldots, P_s$ are homogeneous polynomials with the property that, for $k = 1, \ldots, s$, there exists an open neighborhood $\mathcal{U}_k$ of a in $\mathcal{U}_0$ and a homogeneous polynomial Q_k of the same degree as P_k, not vanishing on $\mathcal{U}_k$, such that $P_k/Q_k \in J(\mathcal{U}_k)$. Define $\mathcal{V} = \mathcal{U}_1 \cap \cdots \cap \mathcal{U}_s$. Then, $\mathcal{V}$ is an open neighborhood of a in $\mathcal{U}$ and we find

$$g|_{\mathcal{V}} = \frac{P}{Q} = \sum_{k=1}^{s} \left(\frac{A_k Q_k}{Q} \right) \frac{P_k}{Q_k} \in J(\mathcal{V}).$$

To prove the first assertion, fix $g \in \mathcal{J}(\mathcal{U})$ and a point $a \in \mathcal{U}$. We need simply to show that there exists an open neighborhood $\mathcal{V}$ of a in $\mathcal{U}$ such that $g|_{\mathcal{V}} \in J(\mathcal{V})$. In fact, it suffices to show this for a function g of the form $g = \partial_{\sigma}^{\kappa}(f)$ with $\sigma \in \Sigma$, $\kappa \in \mathbf{N}^t$, $|\kappa| \le T$ and $f \in I(\sigma + \mathcal{U})$. Suppose that g is of this form. At the expense of replacing $\mathcal{U}$ by a smaller open neighborhood of a if needed, we may also assume that f is given by a quotient P/Q of two homogeneous polynomials of the same degree with $P \in I$ and Q not vanishing on $\sigma + \mathcal{U}$. Let $\{Q_1, \ldots, Q_s\}$ be a finite set of homogeneous polynomials of the same degree as P which do not vanish simultaneously at any point of G. For $k = 1, \ldots, s$, define $\mathcal{U}_k$ to be the open set for which $\sigma + \mathcal{U}_k = \{Q_k \ne 0\}$. Define

$$h_k = \frac{Q_k}{Q} \in \mathcal{O}(\sigma + \mathcal{U}) \quad \text{and} \quad f_k = \frac{P}{Q_k} \in \mathcal{O}(\sigma + \mathcal{U}_k),$$

so that

$$f = h_k f_k \quad \text{on} \quad \sigma + (\mathcal{U} \cap \mathcal{U}_k).$$

Applying $\partial_{\sigma}^{\kappa}$ on both sides of the above equality, we get

$$g = \sum_{|\lambda| \le T} c_{\lambda,k} \partial_{\sigma}^{\lambda}(f_k) \quad \text{on} \quad \mathcal{U} \cap \mathcal{U}_k,$$

where $c_{\lambda,k}$ are elements of $\mathcal{O}(\mathcal{U})$. Choose an open neighborhood $\mathcal{V}$ of a in $\mathcal{U}$ and a homogeneous polynomial B not vanishing on $\mathcal{V}$ such that g and the functions $c_{\lambda,k}$ can be represented on $\mathcal{V}$ by fractions with denominator B:

$$g = \frac{A}{B} \quad \text{and} \quad c_{\lambda,k} = \frac{C_{\lambda,k}}{B} \quad \text{on} \quad \mathcal{V}.$$

We claim that $A \in J$. To see this, take any point $b \in G$. Since $G \subseteq \bigcup_{k=1}^{s} \mathcal{U}_k$, there exists an index k for which $b \in \mathcal{U}_k$. Choose a homogeneous polynomial C of the same degree as A with $C(b) \ne 0$. By construction, we have

$$\frac{A}{C} = \sum_{|\lambda| \le T} \frac{C_{\lambda,k}}{C} \partial_{\sigma}^{\lambda}(f_k) \quad \text{on} \quad \mathcal{V} \cap \mathcal{U}_k \cap \{C \ne 0\}.$$

Since both sides of this equality are regular functions on the open set $W := \mathcal{U}_k \cap \{C \neq 0\}$, they coincide on all this open set. This shows that $A/C \in \mathcal{J}(W)$. The choice of b being arbitrary, this proves that $A \in J$, and thus $g \in J(\mathcal{V})$. $\qquad\square$

PROOF OF PROPOSITION 3.6. (i) Observe first that $J = \partial^0 I$ and $(\mathfrak{G}, I)^*$ are homogeneous ideals. To show that they coincide, it suffices to show that they have the same homogeneous elements.

First, let P be a homogeneous polynomial in J. Choose a maximal ideal $M \in \mathcal{F}$ defining a point $\mathbf{x} \in \tilde{G}$ (see § 2.3) and put $a = [\mathbf{x}]$. Choose also a homogeneous polynomial Q of the same degree as P with $Q(a) \neq 0$, and let $\mathcal{U} = G \cap \{Q \neq 0\}$. Since $P/Q \in J(\mathcal{U})$, Lemma 3.7 shows that P/Q defines an element of $\mathcal{J}(\mathcal{V}) = I(\mathcal{V})$ on some open neighborhood $\mathcal{V}$ of a in $\mathcal{U}$. So, there exists another pair of homogeneous polynomials P_1, Q_1 of the same degree, with $P_1 \in I$ and Q_1 non vanishing on $\mathcal{V}$, such that P/Q and P_1/Q_1 define the same functions on $\mathcal{V}$. The difference $Q_1 P - Q P_1$ thus vanishes identically on $\mathcal{V}$, and so it belongs to $\mathfrak{G}$. This implies $Q_1 P \in (\mathfrak{G}, I)$ since $P_1 \in I$ and therefore $P \in (\mathfrak{G}, I)\mathbf{C}[X]_M$ since $Q_1 \notin M$. This proves $P \in (\mathfrak{G}, I)^*$ and so $J \subseteq (\mathfrak{G}, I)^*$.

For the reverse inclusion, choose a homogeneous polynomial $P \in (\mathfrak{G}, I)^*$, a point $a \in G$, and another homogeneous polynomial Q of the same degree as P with $Q(a) \neq 0$. Put $\mathcal{U} = G \cap \{Q \neq 0\}$. We need to construct an open neighborhood $\mathcal{V}$ of a in $\mathcal{U}$ such that $P/Q \in I(\mathcal{U})$. By definition, there exists a polynomial $B \in \mathbf{C}[X]$ with $B(a) \neq 0$ and $BP \in (\mathfrak{G}, I)$. Since $(\mathfrak{G}, I)$ is homogeneous, we may take B to be homogeneous. Write $BP = P_1 + P_2$ where $P_1 \in \mathfrak{G}$ and $P_2 \in I$ are homogeneous. Then, on $\mathcal{V} = \mathcal{U} \cap \{B \neq 0\}$, the quotient P/Q represents the same function as $P_2/(BQ)$ and this function belongs to $I(\mathcal{V})$, as required.

(ii) For any open set $\mathcal{U}$ of G, define $\mathcal{J}(\mathcal{U})$ as in Lemma 3.7, and put $J = \partial_{\Sigma}^T(I)$. Similarly, define $\mathcal{K}(\mathcal{U})$ to be the ideal of $\mathcal{O}(\mathcal{U})$ generated by the functions $\partial_\gamma^\lambda(g)$ with $\gamma \in \Gamma$, $\lambda \in \mathbf{N}^t$, $|\lambda| \leq S$, and $g \in J(\gamma + \mathcal{U})$, and put $K = \partial_\Gamma^S(J)$. Finally, define $\mathcal{L}(\mathcal{U})$ to be the ideal of $\mathcal{O}(\mathcal{U})$ generated by the functions $\partial_{\gamma+\sigma}^{\lambda+\kappa}(f)$ with $\gamma \in \Gamma$, $\sigma \in \Sigma$, $\lambda, \kappa \in \mathbf{N}^t$, $|\lambda| \leq S$, $|\kappa| \leq T$ and $f \in I(\gamma + \sigma + \mathcal{U})$, and put $L = \partial_{\Gamma+\Sigma}^{S+T}(I)$. To prove that $K = L$, it suffices to show that, for any open subset $\mathcal{U}$ of G, we have

$$\mathcal{L}(\mathcal{U}) \subseteq \mathcal{K}(\mathcal{U}),$$

and that, for any open set $\mathcal{U}$ of G, any $h \in \mathcal{K}(\mathcal{U})$ and any $a \in \mathcal{U}$, there exists an open neighborhood $\mathcal{V}$ of a in $\mathcal{U}$ such that $h|_{\mathcal{V}} \in \mathcal{L}(\mathcal{V})$.

Observe first that, by virtue of Lemma 3.3, the generators of $\mathcal{L}(\mathcal{U})$ are of the form $\partial_\gamma^\lambda(g)$ with $\gamma \in \Gamma$, $\lambda \in \mathbf{N}^t$, $|\lambda| \leq S$ and $g \in \mathcal{J}(\gamma + \mathcal{U})$. Since Lemma 3.7 gives $\mathcal{J}(\gamma + \mathcal{U}) \subseteq J(\gamma + \mathcal{U})$, these generators belong to $\mathcal{K}(\mathcal{U})$ and so, we have $\mathcal{L}(\mathcal{U}) \subseteq \mathcal{K}(\mathcal{U})$. In the other direction, if $h \in \mathcal{K}(\mathcal{U})$ and $a \in \mathcal{U}$, Lemma 3.7 shows the existence of an open neighborhood $\mathcal{V}$ of a in $\mathcal{U}$ such that $h|_{\mathcal{V}}$ belongs to the ideal generated by the functions $\partial_\gamma^\lambda(g)$ with $\gamma \in \Gamma$, $\lambda \in \mathbf{N}^t$, $|\lambda| \leq S$ and $g \in \mathcal{J}(\gamma + \mathcal{V})$. The formulas for differentiating a product show that the latter ideal is also generated by all functions $\partial_\gamma^\lambda(\partial_\sigma^\kappa(f))$ with $\gamma \in \Gamma$, $\sigma \in \Sigma$, $\lambda, \kappa \in \mathbf{N}^t$, $|\lambda| \leq S$, $|\kappa| \leq T$ and $f \in I(\gamma + \sigma + \mathcal{V})$. Thus, by Lemma 3.3, this ideal coincides with $\mathcal{L}(\mathcal{V})$, and so $h|_{\mathcal{V}} \in \mathcal{L}(\mathcal{V})$.

(iii) Let $J = \partial_\Sigma^T(I)$ and let a be a point of G. Clearly, a is a zero of J if and only if it is a zero of $J(\mathcal{U})$ for any open neighborhood $\mathcal{U}$ of a. In the notation of Lemma 3.7, this is in turn equivalent to the condition that a is a zero of $\mathcal{J}(\mathcal{U})$ for any open neighborhood $\mathcal{U}$ of a. By definition, this happens if and only if, for any $\sigma \in \Sigma$ and any open neighborhood $\mathcal{U}$ of a, the elements of $I(\sigma + \mathcal{U})$ vanish to order $> T$ at the point $a + \sigma$. Finally, this is just asking that any $P \in I$ vanishes to order $> T$ at the points $a + \sigma$ with $\sigma \in \Sigma$.

(iv) Assume that I is generated by a set E of homogeneous polynomials of degree $\leq D$. Choose a complete system of addition laws $\mathcal{A}$ on G and, for each $\sigma \in \Sigma$, choose a holomorphic map ψ_σ, defined in a neighborhood of 0 in $\mathbf{C}^t$, with values in $\mathbf{C}^{n+1}$, such that, for any $\mathbf{z}$ in this neighborhood of 0, we have $[\psi_\sigma(\mathbf{z})] = \sigma + \Phi(\mathbf{z})$. We will show more precisely that $\partial_\Sigma^T(I) = (\mathfrak{G}, F)^*$ where

$$F = \left\{ \frac{\partial^\kappa}{\partial \mathbf{z}^\kappa} P\big(\underline{A}(\underline{X}, \psi_\sigma(\mathbf{z}))\big) \Big|_{\mathbf{z}=0} ; \right.$$
$$\left. P \in E, \ \sigma \in \Sigma, \ \kappa \in \mathbf{N}^t, \ |\kappa| \leq T \text{ and } \underline{A} \in \mathcal{A} \right\} .$$

Since this is precisely how P. Philippon defines $\partial_\Sigma^T(I)$ in [**PPh3**] in the case where Σ is reduced to a point, this result will show that his definition and the one adopted here are compatible. Moreover, if $\mathcal{A}$ is chosen so that the components of each addition law are polynomials in $\mathbf{C}[\underline{X}, \underline{Y}]$ of degree $\leq c$ in $\underline{X}$, then the above set F consists of homogeneous polynomials of degree $\leq cD$. So this will also prove the last assertion of the Proposition.

Put $J = \partial_\Sigma^T(I)$ and define $\mathcal{J}(\mathcal{U})$ as in Lemma 3.7, for any open subset $\mathcal{U}$ of G. Moreover, let $K = (F)$ be the ideal of $\mathbf{C}[\underline{X}]$ generated by F. By (i), we have $\partial^0(K) = (\mathfrak{G}, F)^*$. So, in order to show $J \subseteq (\mathfrak{G}, F)^*$, it suffices to show that, for each open set $\mathcal{U}$ of G, each function $g \in \mathcal{J}(\mathcal{U})$ and each $a \in \mathcal{U}$, there exists an open neighborhood $\mathcal{V}$ of a in $\mathcal{U}$ such that $g|_{\mathcal{V}} \in K(\mathcal{V})$. To this end, we may assume that g is of the form $g = \partial_\sigma^\kappa(f)$, with $\sigma \in \Sigma$, $\kappa \in \mathbf{N}^t$, $|\kappa| \leq T$ and $f \in I(\sigma + \mathcal{U})$. Replacing $\mathcal{U}$ by a smaller neighborhood of a if necessary, we may also assume that f is represented on $\sigma + \mathcal{U}$ by a quotient P/Q where P, Q are homogeneous polynomials of the same degree with $P \in I$ and Q vanishing nowhere on $\sigma + \mathcal{U}$. Finally, we may assume that $P \in E$. Choose an addition law $\underline{A}$ in $\mathcal{A}$ whose associated open set $\mathcal{U}_{\underline{A}}$ contains the point (a, σ). Define $\mathcal{V}$ to be the open set consisting of all $b \in \mathcal{U}$ such that $(b, \sigma) \in \mathcal{U}_{\underline{A}}$. Then, for any point $\mathbf{x} \in \mathbf{C}^{n+1}$ with $[\mathbf{x}] \in \mathcal{V}$, there is a neighborhood of 0 in $\mathbf{C}^t$ such that, for $\mathbf{z}$ in this neighborhood, we have

$$[\mathbf{x}] + \sigma + \Phi(\mathbf{z}) = [\underline{A}(\mathbf{x}, \psi_\sigma(\mathbf{z}))]$$

and thus

$$f\big([\mathbf{x}] + \sigma + \Phi(\mathbf{z})\big) = \frac{P\big(\underline{A}(\mathbf{x}, \psi_\sigma(\mathbf{z}))\big)}{Q\big(\underline{A}(\mathbf{x}, \psi_\sigma(\mathbf{z}))\big)}.$$

The fact that $\partial_\sigma^\kappa(f)|_{\mathcal{V}} \in \mathcal{K}(\mathcal{V})$ then follows by differentiating with respect to $\mathbf{z}$ on both sides of this equality.

By (iii), we also have $\partial^0(J) = J$ so, in order to show the reverse inclusion, it suffices to show $F \subseteq J$. Hence, choose $P \in E$, $\sigma \in \Sigma$, $\kappa \in \mathbf{N}^t$ with $|\kappa| \leq T$ and $\underline{A} \in \mathcal{A}$. Choose also a point $a \in G$, a homogeneous polynomial Q of the same degree as P with $Q(a + \sigma) \neq 0$ and an open neighborhood $\mathcal{U}$ of a in G such that Q vanishes nowhere on $\sigma + \mathcal{U}$. Denote by f the element of $I(\sigma + \mathcal{U})$ represented by P/Q. We have

$$\frac{P(\underline{A}(\mathbf{x}, \mathbf{y}))}{Q(\underline{A}(\mathbf{x}, \mathbf{y}))} = f([\mathbf{x}] + [\mathbf{y}])$$

on some (non-empty) open subset $\mathcal{W}_1$ of $G \times G$ where $Q(\underline{A}(\underline{X}, \underline{Y}))$ does not vanish. Finally, choose a bi-homogeneous polynomial $B(\underline{X}, \underline{Y}) \in \mathbf{C}[\underline{X}, \underline{Y}]$ of the same bi-degree as $Q(\underline{A}(\underline{X}, \underline{Y}))$ such that $B(a, \sigma) \neq 0$, and let $\mathcal{W}$ be the open subset of $G \times G$ where B does not vanish. We have

$$\frac{P(\underline{A}(\mathbf{x}, \mathbf{y}))}{B(\mathbf{x}, \mathbf{y})} = \frac{Q(\underline{A}(\mathbf{x}, \mathbf{y}))}{B(\mathbf{x}, \mathbf{y})} f([\mathbf{x}] + [\mathbf{y}])$$

on $\mathcal{W} \cap \mathcal{W}_1$. On the other hand, both sides of the above equality are regular functions on the open neighborhood $\mathcal{W}$ of (a, σ). Since $G \times G$ is irreducible, they coincide on the whole of $\mathcal{W}$. Let $\mathcal{V}$ be an open neighborhood of a in G such that $\mathcal{V} \times \{\sigma\} \subseteq \mathcal{W}$. We get

$$\frac{\partial^\kappa}{\partial \mathbf{z}^\kappa} \left(\frac{P(\underline{A}(\underline{X}, \psi_\sigma(\mathbf{z})))}{B(\underline{X}, \psi_\sigma(\mathbf{z}))} \right)_{\mathbf{z}=0} \in \mathcal{J}(\mathcal{V}).$$

By induction over $|\kappa|$, we deduce that

$$\frac{(\partial^\kappa / \partial \mathbf{z}^\kappa) P(\underline{A}(\underline{X}, \psi_\sigma(\mathbf{z})))|_{\mathbf{z}=0}}{B(\underline{X}, \psi_\sigma(0))} \in \mathcal{J}(\mathcal{V}).$$

Since the denominator of this fraction does not vanish on the neighborhood $\mathcal{V}$ of a, and since a was taken arbitrarily, this shows

$$\frac{\partial^\kappa}{\partial \mathbf{z}^\kappa} P(\underline{A}(\underline{X}, \psi_\sigma(\mathbf{z})))\Big|_{\mathbf{z}=0} \in J.$$

So, we have $F \subseteq J$ and thus $K \subseteq J$. $\qquad\qquad\qquad\qquad\qquad\qquad\qquad \square$

3.3. Multiplicities. The next result is the last one needed in the proof of the zero estimate. It is a special case of results of G. Wüstholz in §3 of [**Wus2**]. The more general result of Wüstholz concerns a subvariety V of G (ie. a closed irreducible subset of G). Assuming that V is both the set of zeros in G of a homogeneous primary ideal $\mathfrak{q}$ and an irreducible component of the set of zeros of $\partial^T(\mathfrak{q})$ in G for some integer $T \geq 0$, his result gives a lower bound for the length of $\mathfrak{q}$. The approach of P. Philippon requires this lower bound only when V is a translate of an algebraic subgroup of G (see Proposition 4.7 of [**PPh3**]). For a union of translates of an algebraic subgroup, we have:

PROPOSITION 3.8. *Let* $\mathfrak{A} \subseteq \mathbf{C}[\underline{X}]$ *be a special homogeneous ideal,* H *an algebraic subgroup of* G, *and* Σ *a finite subset of* G. *Assume that* $\Sigma + H$ *is both a union of irreducible components of* $Z(\mathfrak{A}) \cap G$ *and a union of irreducible components of* $Z(\partial^T \mathfrak{A}) \cap G$. *Assume further that* $\Sigma + H$ *and* $Z(\mathfrak{A}) \cap G$ *have the same dimension. Then, for any integer* $D \geq 0$, *we have*

$$\mathcal{H}(\mathfrak{A}; D) \geq \binom{T+s}{s} \mathcal{H}(\Sigma + H; D)$$

where $s = \dim_{\mathbf{C}} \left((W + T_H)/T_H \right)$.

The proof of this result is divided in three parts. We start with a reduction.

Step 1. *It suffices to prove Proposition 3.8 in the case where* $\mathfrak{A}$ *is a primary ideal,* H *is irreducible and* $\Sigma = \{\sigma\}$ *consists of one point.*

Let H_0 be the connected component of the neutral element in H. The irreducible components of $\Sigma + H$ are the translates of H_0 contained in $\Sigma + H$. Let $V = v + H_0$ be one of them. By hypothesis, V is an irreducible component of

$Z(\mathfrak{A}) \cap G$. So, there is a primary component $\mathfrak{q}$ of $\mathfrak{A}$ with $V = Z(\mathfrak{q}) \cap G$. We claim that V is also an irreducible component of $Z(\partial^T(\mathfrak{q})) \cap G$.

Let $\mathfrak{p}$ be the radical of $\mathfrak{q}$ and let P be a homogeneous polynomial with $P \notin \mathfrak{p}$ such that $P\mathfrak{q} \subseteq \mathfrak{A}$. Denote by $\mathcal{V}$ the non-empty open subset of V where P is $\neq 0$. To establish the claim, we need simply to show that, for any $a \in \mathcal{V}$ and any sufficiently small open neighborhood $\mathcal{U}$ of a in G, the point a is a zero of the functions $\partial^\kappa f$ with $f \in \mathfrak{q}(\mathcal{U})$, $\kappa \in \mathbf{N}^t$, $|\kappa| \leq T$. So, fix a and $\mathcal{U}$. Since, by hypothesis, V is an irreducible component of $Z(\partial^T(\mathfrak{A})) \cap G$, we already know that a is a zero of the functions $\partial^\kappa h$ with $h \in \mathfrak{A}(\mathcal{U})$, $\kappa \in \mathbf{N}^t$, $|\kappa| \leq T$. To go from $\mathfrak{A}$ to $\mathfrak{q}$, choose a homogeneous polynomial Q of the same degree as P with $Q(a) \neq 0$. We may assume that Q does not vanish on $\mathcal{U}$. Then, $g = P/Q$ is an element of $\mathcal{O}(\mathcal{U})$ with $g(a) \neq 0$, and we have $gf \in \mathfrak{A}(\mathcal{U})$ for any $f \in \mathfrak{q}(\mathcal{U})$. For fixed f, this implies $\partial^\kappa(gf)(a) = 0$ for all $\kappa \in \mathbf{N}^t$, $|\kappa| \leq T$. Since $g(a) \neq 0$, we deduce that $(\partial^\kappa f)(a) = 0$ for the same values of κ and the claim is proved.

If the proposition is true in the case of a primary ideal, this implies

$$(93) \qquad \mathcal{H}(\mathfrak{q}; D) \geq \binom{T+s}{s} \mathcal{H}(V; D).$$

The conclusion of the proposition then follows by summing over all the irreducible components V of $\Sigma + H$. This uses the property 2 of the function $\mathcal{H}$ in §2.2, together with the fact that, since $\mathfrak{A}$ is a special homogeneous ideal, its set of zeros $Z(\mathfrak{A})$ has the same dimension as $Z(\mathfrak{A}) \cap G$.

This reduction step allows us to suppose that H is irreducible, that $\mathfrak{A}$ is a primary ideal $\mathfrak{q}$ whose set of zeros in G is a translate $V = \sigma + H$ of H, and that $\partial^T \mathfrak{q}$ is contained in the radical $\mathfrak{p} = \mathfrak{J}(V)$ of $\mathfrak{q}$. The next step also involves a reduction.

Step 2. Let $e_1, \ldots, e_t$ be the canonical basis of $\mathbf{C}^t$ and let $W_1 \subseteq T_G$ be the image under φ of the subspace of $\mathbf{C}^t$ generated by $e_1, \ldots, e_s$. We may also assume that $W_1 \cap T_H = \{0\}$. Then, there exist an open neighborhood $\mathcal{U}$ of σ in G and functions $g_1, \ldots, g_s \in \mathfrak{p}(\mathcal{U})$ such that $\partial^{e_i} g_i \notin \mathfrak{p}(\mathcal{U})$ for $i = 1, \ldots, s$ and $\partial^{e_i} g_j \in \mathfrak{p}(\mathcal{U})$ when $i \neq j$.

The first affirmation follows from the fact that the definition of $\partial^T \mathfrak{q}$ is independent of the choice of the linear isomorphism $\varphi \colon \mathbf{C}^t \to W$ which, composed with $\exp_G$, gives rise to Φ. Assuming $W_1 \cap T_H = \{0\}$, we claim that there exist an open neighborhood $\mathcal{U}$ of σ in G and functions $f_1, \ldots, f_s \in \mathfrak{p}(\mathcal{U})$ such that $\det\left(\partial^{e_i} f_j(\sigma)\right)_{1 \leq i,j \leq s} \neq 0$.

If this were not the case, there would exist $\lambda_1, \ldots, \lambda_s \in \mathbf{C}$ not all zero such that $(\lambda_1 \partial^{e_1} + \cdots + \lambda_s \partial^{e_s}) f(\sigma) = 0$ for any regular function f defined in an open neighborhood $\mathcal{U}$ of σ in G and vanishing identically on $\mathcal{U} \cap (\sigma + H)$. Since $\partial^{e_i} f(\sigma) = \partial^{e_i}(f \circ \tau_\sigma)(e)$ for any regular function f at σ, this would imply $(\lambda_1 \partial^{e_1} + \cdots + \lambda_s \partial^{e_s}) g(e) = 0$ for any regular function g defined in an open neighborhood $\mathcal{V}$ of e in G with $g|_{\mathcal{V} \cap H} \equiv 0$, in contradiction with the hypothesis $W_1 \cap T_H = \{0\}$.

Put $h = \det\left(\partial^{e_i} f_j\right)_{1 \leq i,j \leq s}$. Then, for each $k = 1, \ldots, s$, there exists $h_{k,1}, \ldots, h_{k,s} \in \mathcal{O}(\mathcal{U})$ such that

$$(94) \qquad h_{k,1} \partial^{e_i} f_1 + \cdots + h_{k,s} \partial^{e_i} f_s = \begin{cases} h & \text{if } i = k \\ 0 & \text{if } i \neq k. \end{cases}$$

The functions $g_k = h_{k,1}f_1 + \cdots + h_{k,s}f_s \in \mathfrak{p}(\mathcal{U})$, $(1 \leq k \leq s)$, have the required properties since, for each i, the left hand side of (94) is congruent to $\partial^{e_i}g_k$ modulo $\mathfrak{p}(\mathcal{U})$.

Step 3. Let the notation be as in step 2. Without loss of generality, we may assume that there exist homogeneous polynomials $A_1, \ldots, A_s$ and B of the same degree d with B vanishing nowhere on $\mathcal{U}$ such that A_k/B represents g_k on $\mathcal{U}$ for each $k = 1, \ldots, s$. We claim that

$$(95) \qquad H(\mathfrak{q}; D + dT) \geq \binom{T + s}{s} H(\mathfrak{p}; D)$$

for each integer $D \geq 1$. If we take this affirmation for granted, then (93) holds for each $D \geq 1$ and the proposition is proved.

To prove the claim, fix an integer $D \geq 1$ and put $N = H(\mathfrak{p}; D)$. Let $P_1, \ldots, P_N$ be homogeneous polynomials of $\mathbf{C}[\underline{X}]$ of degree D whose images in $\mathbf{C}[\underline{X}]/\mathfrak{p}$ are linearly independent over $\mathbf{C}$. Choose also an open neighborhood $\mathcal{V}$ of σ in $\mathcal{U}$ and a homogeneous polynomial Q of degree D with Q vanishing nowhere on $\mathcal{V}$. Then the ratios $P_1/Q, \ldots, P_N/Q$ define functions $f_1, \ldots, f_N \in \mathcal{O}(\mathcal{V})$ whose images in $\mathcal{O}(\mathcal{V})/\mathfrak{p}(\mathcal{V})$ are linearly independent over $\mathbf{C}$. Now, consider the functions

$$(96) \qquad h_{i,\kappa} = f_i g_1^{\kappa_1} \cdots g_s^{\kappa_s} \in \mathcal{O}(\mathcal{V}), \qquad (1 \leq i \leq N, \ \kappa \in \mathbf{N}^s, \ |\kappa| \leq T).$$

If their images in $\mathcal{O}(\mathcal{V})/\mathfrak{q}(\mathcal{V})$ were linearly dependent over $\mathbf{C}$, there would exist complex numbers $c_{i,\kappa}$ not all zero such that

$$(97) \qquad \sum_{\substack{\kappa \in \mathbf{N}^s \\ |\kappa| \leq T}} \sum_{i=1}^{N} c_{i,\kappa} h_{i,\kappa} \equiv 0 \quad \mathrm{mod}\ \mathfrak{q}(\mathcal{V}).$$

Suppose that it is the case and, according to the lexicographical ordering, let λ be the smallest $\kappa \in \mathbf{N}^s$ for which there is at least one index i with $c_{i,\kappa} \neq 0$. Applying ∂^λ to both sides of (97) and using the hypothesis that $\partial^T \mathfrak{q} \subseteq \mathfrak{p}$, we get

$$\left(\sum_{i=1}^{N} c_{i,\lambda}f_i\right)(\partial^{e_1}g_1)^{\lambda_1} \cdots (\partial^{e_s}g_s)^{\lambda_s} \equiv 0 \quad \mathrm{mod}\ \mathfrak{p}(\mathcal{V}).$$

Since none of the functions $\partial^{e_1}g_1, \ldots, \partial^{e_s}g_s$ belong to $\mathfrak{p}(\mathcal{V})$, this implies that $\sum_{i=1}^{N} c_{i,\lambda}f_i$ belongs to $\mathfrak{p}(\mathcal{V})$. This is a contradiction since not all the numbers $c_{1,\lambda}, \ldots, c_{N,\lambda}$ are zero. Thus, the images of the functions (96) in $\mathcal{O}(\mathcal{V})/\mathfrak{q}(\mathcal{V})$ are linearly independent over $\mathbf{C}$. On the other hand, each of these functions $h_{i,\kappa}$ can be written as a quotient $P_{i,\kappa}/(QB^{|\kappa|})$ where $P_{i,\kappa}$ is a homogeneous polynomial of degree $D + d|\kappa|$. If we consider only those $P_{i,\kappa}$ for which $|\kappa| = T$, then they form a set of $\binom{T+s}{s}N$ homogeneous polynomials of degree $D + dT$ whose images in $\mathbf{C}[\underline{X}]/\mathfrak{q}$ are linearly independent over $\mathbf{C}$. This proves (95). $\qquad\square$

4. Statement and proof of the zero estimate

We now have all the necessary tools for proving the following homogeneous version of the zero estimate of P. Philippon (Theorem 2.1 of [**PPh3**]). See [**PPh9**] for a refined version where essentially the term $\binom{T+s}{s}$ is replaced by T^s.

THEOREM 4.1. *Let $G \subseteq \mathbf{P}^n(\mathbf{C})$ be a commutative connected algebraic group of dimension d, Σ a finite subset of G with $e \in \Sigma$, W a subspace of the tangent* space T_G of G at e, T an integer ≥ 0 and $P \in \mathbf{C}[\underline{X}]$ a homogeneous polynomial not belonging to $\mathfrak{G} = \mathfrak{I}(G)$, of degree D, which vanishes to order $> dT$ at each point of

$$\Sigma(d) = \{\sigma_1 + \cdots + \sigma_d \, ; \, \sigma_1, \ldots, \sigma_d \in \Sigma\}.$$

Then, there exists a connected algebraic subgroup H_0 of G, distinct from G, such that

$$\binom{T+s}{s} \operatorname{Card}\left((\Sigma + H_0)/H_0\right)\mathcal{H}(H_0; D) \leq \mathcal{H}(G; cD),$$

where $s = \dim_{\mathbf{C}}\left((W + T_{H_0})/T_{H_0}\right)$ and $c = c(G)$ is defined in §2.1. Moreover, one may choose H_0 so that it is an irreducible component of an algebraic subset E of G, of the same dimension as H_0, where E is the set of zeros in G of a family of homogeneous polynomials of degree $\leq cD$.

PROOF. Define recursively a sequence of ideals $(I_k)_{k \geq 1}$ of $\mathbf{C}[\underline{X}]$ by

$$I_1 = (\mathfrak{G}, P)^* = \partial^0(P) \quad \text{and} \quad I_{k+1} = \partial^T_\Sigma(I_k), \quad (k \geq 1).$$

By Proposition 3.6 (ii), this gives

$$I_{k+1} = \partial^{kT}_{\Sigma(k)}(I_1) = \partial^{kT}_{\Sigma(k)}(P), \quad (k \geq 1).$$

For each $k \geq 1$, define $X_k = G \cap Z(I_k)$ to be the set of zeros of I_k in G. Since $(I_i)_{i \geq 1}$ is an ascending chain of ideals, we get

$$G \supseteq X_1 \supseteq X_2 \supseteq \cdots \supseteq X_{d+1} \supseteq \cdots$$

We also have $X_1 \neq G$ since $P \notin \mathfrak{G}$. Moreover, Proposition 3.6 (iii) gives $e \in X_{d+1}$ since P vanishes to order $> dT$ at each point of $\Sigma(d)$. This implies

$$d > \dim(X_1) \geq \cdots \geq \dim(X_{d+1}) \geq 0.$$

Hence, there exists an integer r with $1 \leq r \leq d$ such that $\dim X_r = \dim X_{r+1}$. Let m be the common dimension of X_r and X_{r+1}, and let V be an irreducible component of X_{r+1} of dimension m. Then, V is also an irreducible component of X_r.

Since $I_{r+1} = \partial^T_\Sigma(I_r) \supseteq \partial^0_\Sigma(I_r)$, the elements of V are zeros of $\partial^0_\Sigma(I_r)$. By Proposition 3.6 (iii), this means that, for each $\sigma \in \Sigma$, the elements of I_r vanish on $\sigma + V$, and therefore $\sigma + V \subseteq X_r$.

This motivates to consider the following three sets:

$$E = \{a \in G \, ; \, a + V \subseteq X_r\}$$
$$F = \{a \in G \, ; \, a + V \subseteq X_{r+1}\}$$
$$H = \{a \in G \, ; \, a + V = V\}.$$

We showed above that E contains Σ. We now study how E, F and H are related:

a) It is clear from its definition that H is a subgroup of G. Observe also that, for any $a \in G$, the condition $a + V = V$ is equivalent to $a + V \subseteq V$. This is because τ_a is an isomorphism of G as algebraic set. So, it maps V to a closed irreducible subset of G of dimension m. Such a set, if contained in V must coincide with V. Hence, we find

$$H = \{a \in G \, ; \, a + V \subseteq V\} = \bigcap_{v \in V}(-v + V).$$

Since $-v + V$ is a closed subset of G for any $v \in V$, this shows that H is a closed subset of G. Therefore, H is an algebraic subgroup of G. Put $d = \dim(H)$.

b) It is also clear that $a + H \subseteq E$ for each $a \in E$. So, E is a union of translates of H. Moreover, for each $a \in E$, the set $a + V$ is a closed irreducible subset of X_r of dimension m, so it is a component of X_r. Since X_r admits only finitely many components, we deduce that E is a finite union of translates of H. Similarly, F is a finite union of translates of H. Hence, E and F are closed subsets of G of dimension d.

c) Using Proposition 3.6 (iii), we find

$$a \in E \iff a + V \subseteq X_r = Z(I_r) \cap G$$
$$\iff a \in Z\big(\partial_V^0(I_r)\big) \cap G.$$

Thus,

$$E = Z(\mathfrak{A}) \cap G$$

where $\mathfrak{A} = \partial_V^0(I_r)$. Similarly, we find

$$a \in F \iff a + V \subseteq X_{r+1} = Z(I_{r+1}) \cap G$$
$$\iff a \in Z\big(\partial_V^0(I_{r+1})\big) \cap G.$$

Since $\partial_V^0(I_{r+1}) = \partial_{V+\Sigma}^T(I_r) = \partial_\Sigma^T(\mathfrak{A})$, this gives

$$F = Z\big(\partial_\Sigma^T(\mathfrak{A})\big) \cap G.$$

d) Since E is a finite union of translates of H and since E contains Σ, the set $\Sigma + H$ is a union of irreducible components of $E = Z(\mathfrak{A}) \cap G$. On the other hand, since F contains H, we have $H \subseteq Z\big(\partial_\Sigma^T(\mathfrak{A})\big) \cap G$. Moreover, Proposition 3.6 (ii) gives $\partial_\Sigma^T(\mathfrak{A}) = \partial_\Sigma^0\big(\partial^T(\mathfrak{A})\big)$. By virtue of Proposition 3.6 (iii), this implies that $\Sigma + H \subseteq Z\big(\partial^T(\mathfrak{A})\big) \cap G$. Since $\mathfrak{A} \subseteq \partial^T(\mathfrak{A})$, we deduce that $\Sigma + H$ is also a union of irreducible components of $Z\big(\partial^T(\mathfrak{A})\big) \cap G$. Then, applying Proposition 3.8, we conclude that

$$(98) \qquad \mathcal{H}(\mathfrak{A}; k) \geq \binom{T + s}{s} \mathcal{H}(\Sigma + H; k)$$

for any integer $k \geq 0$.

e) By definition of I_r and part (ii) of Proposition 3.6, we have

$$\mathfrak{A} = \partial_V^0(I_r) = \partial_{V+\Sigma(r-1)}^{(r-1)T}(P).$$

So, by part (iv) of the same proposition, we get $\mathfrak{A} = (\mathfrak{G}, S)^*$ where S is a set of homogeneous polynomials of degree $\leq cD$. By Proposition 2.2, this implies

$$\mathcal{H}(\mathfrak{A}; cD) \leq \mathcal{H}(G; cD).$$

Finally, by Proposition 2.3, we have

$$\mathcal{H}(\Sigma + H; cD) \geq \mathrm{Card}\,\big((\Sigma + H_0)/H_0\big)\mathcal{H}(H_0; D),$$

where H_0 denotes the connected component of the identity in H. The main assertion of the theorem follows by combining the last two inequalities with (98) for the choice of $k = cD$. The second assertion follows simply from the fact that H_0 is an irreducible component of E and that E is the set of zeros of S in G. $\qquad \square$

REMARK. A quick analysis shows that, for each $\sigma \in \Sigma$, the set $\sigma + V$ is an irreducible component of both the set of zeros of I_r in G and the set of zeros of $\partial^T(I_r)$ in G. As in part (e) of the above proof, this gives a lower bound for $\mathcal{H}(I_r; cD)$. On the other hand, Proposition 2.2 gives the upper bound $\mathcal{H}(I_r; cD) \leq \mathcal{H}(G; cD)$. This is essentially the approach of [MW1], [MW3] and [Wus2]. In a sense, it focuses on the two sets X_r and X_{r+1}, but does not lead to the expected estimates because one knows little about V. The remarkable observation of P. Philippon is that the more abstract sets E and F are related in a similar way as X_r and X_{r+1}. The role of V is played by the group H and the role of I_r is played by $\mathfrak{A}$. Reasoning as in § 7 of [MW1], it can be shown that the above theorem of P. Philippon is best possible, up to the value of the constants (see also [PPh3], errata et addenda).

The above zero estimate can be viewed as a lower bound for the degree of a homogeneous polynomial vanishing with multiplicities on a subset of the group G, but not on whole group. By way of conclusion, we show how to derive from this result the main zero estimate of D. W. Masser and G. Wüstholz in [MW1] :

COROLLARY 4.2. *Let G be as in the theorem, and let Γ be a finitely generated subgroup of G. Define*

$$\mu = \mu(\Gamma; G) = \min_{H \neq G} \frac{\mathrm{rank}\big((\Gamma + H)/H\big)}{\dim(G/H)}$$

where the minimum is taken over all algebraic subgroups H of G which are distinct from G. Assume that Γ is generated by elements $\gamma_1, \ldots, \gamma_\ell$ and that there exists a homogeneous polynomial $P \in \mathbf{C}[X_0, \ldots, X_n]$ of degree D which, for some real number $S \geq 0$, vanishes at each point of the set

$$\Gamma(S) = \big\{ s_1 \gamma_1 + \cdots + s_\ell \gamma_\ell \, ; \, 0 \leq s_1, \ldots, s_\ell \leq S \big\}.$$

Assume further that P does not vanish identically on G. Then, we have

$$D \geq c'(S/d)^\mu$$

where c' is a positive constant depending only on G.

PROOF. In the notations of the theorem, the set $\Gamma(S)$ contains $\Sigma(d)$, where

$$\Sigma = \big\{ s_1 \gamma_1 + \cdots + s_\ell \gamma_\ell \, ; \, 0 \leq s_1, \ldots, s_\ell \leq S/d \big\}.$$

Thus, there exists a connected algebraic subgroup H_0 of G distinct from G such that

(99) $$\mathrm{Card}\,\big((\Sigma + H_0)/H_0\big)\mathcal{H}(H_0; D) \leq \mathcal{H}(G; cD).$$

Let λ be the rank of $(\Gamma + H_0)/H_0$ and let δ be the codimension of H_0 in G. By permuting the γ_i's if necessary, we may assume that $\gamma_1, \ldots, \gamma_\lambda$ have linearly independent images in G/H_0. The points of Σ of the form $s_1 \gamma_1 + \cdots + s_\lambda \gamma_\lambda$ with $s_1, \ldots, s_\lambda \in \mathbf{Z}$ then have distinct images in G/H_0 and so we get

$$\mathrm{Card}\,\big((\Sigma + H_0)/H_0\big) \geq \big([S/d] + 1\big)^\lambda.$$

Since $\mathcal{H}(G; cD) = \deg(G)(cD)^n$ and $\mathcal{H}(H_0; D) \geq D^{n-\delta}$, the inequality (99) implies

$$\big([S/d] + 1\big)^\lambda \leq c^n \deg(G) D^\delta.$$

This proves the corollary with $c' = (c^n \deg(G))^{-1}$.   □

Note that, in the case where G is the group $\mathbf{G}_m^n$ of the example of §2.1, we have $c = 1$ and $\deg(G) = 1$, so that the above corollary holds with $c' = 1$. Moreover, the description of the algebraic subgroups of $\mathbf{G}_m^n$ given in this example allows one to interpret the quantity $\mu(\Gamma; G)$ in concrete terms. For example, if we write $\gamma_j = (1 : e^{y_{1,j}} : \cdots : e^{y_{n,j}})$ for $j = 1, \ldots, \ell$, and if we denote by M the $n \times \ell$ matrix $(y_{i,j})_{1 \le i \le n,\ 1 \le j \le \ell}$, then $\mu(\Gamma; G)$ is the smallest ratio λ/δ for which there exist matrices $P \in \mathrm{GL}_n(\mathbf{Q})$ and $Q \in \mathrm{GL}_\ell(\mathbf{Q})$ such that the product PMQ decomposes as a 2×2 block matrix of the form

$$PMQ = \begin{pmatrix} A & D \\ B & C \end{pmatrix}$$

where A is a $\delta \times \lambda$ matrix and where each entry of D is a rational multiple of $2\pi i$ (see also [**Mas2**]).

CHAPTER 12

Measures of algebraic independence for Mahler functions

1. Theorems

Let $\omega_1, \ldots, \omega_m$ be algebraically independent complex numbers. If a function $\varphi(H, s)$ satisfies

$$|P(\omega_1, \ldots, \omega_m)| \geq \varphi(H, s),$$

for any nonzero polynomial $P \in \mathbf{Z}[x_1, \ldots, x_m]$ with $|P| \leq H, \deg_{\overline{x}} P \leq s$, then $\varphi(H, s)$ is called an *algebraic independence measure* of $\omega_1, \ldots, \omega_m$. Let

$$t(P) = \log|P| + \deg_{\overline{x}} P.$$

If a positive number τ satisfies

$$\log|P(\omega_1, \ldots, \omega_m)| \geq -\gamma t(P)^\tau,$$

for any nonzero polynomial P, where γ is a positive constant depending on $\omega_1, \ldots,$ ω_m, then we say that the *transcendence type* of the field $\mathbf{Q}(\omega_1, \ldots, \omega_m)$ is not greater than τ. It is proved that one always has $\tau \geq m + 1$ (see Lemma 3.2 in Chapter 15).

THEOREM 1.1. *Let K be an algebraic number field, and $f_1(z), \ldots, f_m(z) \in K[[z]]$ be power series algebraically independent over $K(z)$ converging in a disc $U \subset \{|z| < 1\}$. Suppose that for an integer $d \geq 2$, the functional equation*

$$\begin{pmatrix} f_1(z^d) \\ \vdots \\ f_m(z^d) \end{pmatrix} = A(z) \begin{pmatrix} f_1(z) \\ \vdots \\ f_m(z) \end{pmatrix} + B(z)$$

is satisfied, where $A(z)$ is an $m \times m$ matrix with entries in $K(z)$ and $B(z)$ is an m-dimensional vector with entries in $K(z)$. Suppose that α is a nonzero algebraic number, $\alpha \in U$, and α^{d^k} is not a pole of $A(z), B(z)$ for any $k \geq 0$. Then for any H and $s \geq 1$ and for any nonzero polynomial $R \in \mathbf{Z}[x_1, \ldots, x_m]$ whose degree does not exceed s and whose coefficients are not greater than H in absolute value, the following inequality holds:

$$|R(f_1(\alpha), \ldots, f_m(\alpha))| > \exp\left(-\gamma s^m (\log H + s^{m+2})\right),$$

where γ is a positive constant depending only on α and the functions $f_1, \ldots, f_m$.

If we restrict the functional equation, we have a better measure of algebraic independence.

(∗) Chapter's author : Kumiko NISHIOKA .

THEOREM 1.2. *Let K be an algebraic number field and $f_1(z), \ldots, f_m(z) \in K[[z]]$ be functions algebraically independent over $K(z)$. Suppose that for an integer $d \geq 2$, the functional equation*

$$\begin{pmatrix} f_1(z) \\ \vdots \\ f_m(z) \end{pmatrix} = C(z) \begin{pmatrix} f_1(z^d) \\ \vdots \\ f_m(z^d) \end{pmatrix} + D(z)$$

is satisfied, where $C(z)$ is an $m \times m$ matrix with entries in $K[z]$ and $D(z)$ is an m-dimensional vector with entries in $K[z]$. Suppose that α is an algebraic number, $0 < |\alpha| < 1$, and α^{d^k} is not a zero of $\det C(z)$ for any $k \geq 0$. Then for any H and $s \geq 1$ and for any nonzero polynomial $R \in \mathbf{Z}[x_1, \ldots, x_m]$ whose degree does not exceed s and whose coefficients are not greater than H in absolute value, the following inequality holds:

$$|R(f_1(\alpha), \ldots, f_m(\alpha))| > \exp\left(-\gamma s^m (\log H + s^2 \log(s+1))\right),$$

where γ is a positive constant depending only on α and the functions $f_1, \ldots, f_m$.

By this theorem we know that the transcendence type of the field $\mathbf{Q}(f_1(\alpha), \ldots, f_m(\alpha))$ is not greater than $m + 2 + \varepsilon$ for any positive ε.

For the proof, we use the following multiplicity estimate.

THEOREM 1.3. *Let $f_1(z), \ldots, f_m(z)$ be formal power series with coefficients in the field $\mathbf{C}$ satisfying the functional equation*

$$\begin{pmatrix} f_1(z^d) \\ \vdots \\ f_m(z^d) \end{pmatrix} = A(z) \begin{pmatrix} f_1(z) \\ \vdots \\ f_m(z) \end{pmatrix} + B(z),$$

where $d \geq 2$ is an integer, $A(z)$ is an $m \times m$ matrix and $B(z)$ is an m-dimensional vector with entries in $\mathbf{C}(z)$. Suppose that $Q(z, x_1, \ldots, x_m) \in \mathbf{C}[z, x_1, \ldots, x_m]$ is a polynomial with $\deg_z Q \leq M$, $\deg_{\overline{x}} Q \leq N$ where $M \geq N \geq 1$. If $Q(z, f_1(z), \ldots, f_m(z)) \neq 0$, then

$$\mathrm{ord}_{z=0} Q(z, f_1(z), \ldots, f_m(z)) \leq cMN^m,$$

where c is a positive constant independent of M, N.

These theorems are proved in [**Nis2**, Chapter 4]. In § 2, of this Chapter we present a simplified proof of Theorem 1.2, we will assume $f_1(z), \ldots, f_m(z)$ have rational integer coefficients and the entries of matrices $C(z)$, $D(z)$ are polynomials with rational integer coefficients. In § 3, we prove Theorem 1.3.

2. Proof of main theorem

Let

$$\overline{\omega} = (\omega_0, \ldots, \omega_m) = (1, f_1(\alpha), \ldots, f_m(\alpha)).$$

Let $K = \mathbf{Q}(\alpha), \nu = [K : \mathbf{Q}]$. Denote by $c_1, c_2, \ldots$ positive constants depending only on α, $f_1, \ldots, f_m$. In what follows, we use the notions and propositions from § 4 of Chapter 3.

LEMMA 2.1. *Let N and k be positive integers satisfying $N \geq c_1$ and $d^k \geq c_2 N \log N$. Then there exists a polynomial $R_k \in \mathbf{Z}[z, \overline{x}] = \mathbf{Z}[z, x_0, \dots, x_m]$ which is homogeneous in $\overline{x}$ and satisfies the inequalities*

$$\deg_z R_k \leq c_3 d^k N, \quad \deg_{\overline{x}} R_k = N,$$

$$\log |R_k| \leq c_3 N \log N + c_3 k N,$$

$$c_4 d^k N^{m+1} \leq -\log |R_k(\alpha, 1, f_1(\alpha), \dots, f_m(\alpha))| \leq c_5 d^k N^{m+1}.$$

PROOF. Let $f_i(z) = \sum_{h=0}^{\infty} a_{ih} z^h$. By using the functional equation, we obtain inductively that

$$|a_{ih}| \leq c_1 h^{c_2} \quad (h \geq 1), \quad |a_{i0}| \leq c_1.$$

(In the case of Theorem 1.1 we have the estimate : $|a_{ih}| \leq c_1^{h+1}$). Let

$$f_1^{j_1} \cdots f_m^{j_m} = \sum_{h=0}^{\infty} a_{\mathbf{j}h} z^h,$$

where $\mathbf{j} = (j_1, \cdots, j_m)$ and $|\mathbf{j}| = j_1 + \cdots + j_m$. It is easily verified that

$$|a_{\mathbf{j}h}| \leq (c_3 h^{c_4})^{|\mathbf{j}|} \quad (h \geq 1), \quad |a_{\mathbf{j}0}| \leq c_3^{|\mathbf{j}|}.$$

$\square$

By using Siegel's lemma (see Chapter 2, Lemma 2.6), we see that for each integer $N > 1$ there exists a nonzero polynomial $R \in \mathbf{Z}[z, x_1, \dots, x_m]$ which satisfies

$$\deg_z R \leq N, \quad \deg_{\overline{x}} R \leq N,$$

$$\mathrm{ord}_{z=0} R(z, f_1(z), \dots, f_m(z)) \geq N^{m+1}/2m!,$$

$$|R| \leq (N+1)^{m+1} \left(c_3 (N^{m+1}/2m!)^{c_4} \right)^N \leq N^{c_5 N}.$$

For this, put

$$E_N(z) = R(z, f_1(z), \dots, f_m(z)) = \sum_{h=0}^{\infty} b_h z^h.$$

We consider the coefficients of R as unknowns and solve linear equations $b_h = 0$ ($0 \leq h < N^{m+1}/2m!$). The coefficients of the equations are $a_{\mathbf{j}h}$ ($|\mathbf{j}| \leq N$, $h < N^{m+1}/2m!$). There are $(N+1)\binom{N+m}{m}$ ($\geq N^{m+1}/m!$) unknowns.

By the construction of $E_N(z)$, we have

$$\log |b_h| \leq c_6 N \log h.$$

Let $n = \mathrm{ord}_{z=0} E_N(z)$. We have

$$\left| (b_{n+i}/b_n) \alpha^{d^k i} \right| \leq |b_{n+i}| |\alpha|^{d^k i} \leq (n+i)^{c_6 N} |\alpha|^{d^k i}$$

$$\leq n^{c_6 N} (i+1)^{c_6 N} |\alpha|^{d^k i} \leq n^{c_6 N} \left(e^{c_6 N} |\alpha|^{d^k} \right)^i.$$

Therefore, if $d^k \geq N \log N$ and N is sufficiently large, then $e^{c_6 N} |\alpha^{d^k}| < 1/2$ and

$$\left| \sum_{i=1}^{\infty} (b_{n+i}/b_n) \alpha^{d^k i} \right| \leq 2 n^{c_6 N} e^{c_6 N} |\alpha|^{d^k}.$$

Theorem 1.3 implies

$$n = \mathrm{ord}_{z=0} E_N(z) \leq c_7 N^{m+1}.$$

Hence

$$\left| \sum_{i=1}^{\infty} (b_{n+i}/b_n) \alpha^{d^k i} \right| < \frac{1}{2},$$

if $N \geq c_8$ and $d^k \geq c_9 N \log N$. Since

$$E_N(\alpha^{d^k}) = b_n \alpha^{d^k n} \left(1 + \sum_{i=1}^{\infty} (b_{n+i}/b_n) \alpha^{d^k i} \right),$$

we obtain

$$\frac{1}{2} |b_n| |\alpha|^{d^k n} \leq |E_N(\alpha^{d^k})| \leq \frac{3}{2} |b_n| |\alpha|^{d^k n},$$

and so

$$c_{10} d^k N^{m+1} \leq -\log |E_N(\alpha^{d^k})| \leq c_{11} d^k N^{m+1}.$$

Since $f_1(z), \ldots, f_m(z)$ are algebraically independent over $\mathbf{Q}(z)$, the determinant of the matrix $C(z)$ is not zero. Therefore there exist a matrix $A(z) \in M_m(\mathbf{Q}(z))$ and a vector $B(z) \in (\mathbf{Q}(z))^m$ such that

$$\begin{pmatrix} f_1(z^d) \\ \vdots \\ f_m(z^d) \end{pmatrix} = A(z) \begin{pmatrix} f_1(z) \\ \vdots \\ f_m(z) \end{pmatrix} + B(z).$$

Let $a(z) = \det C(z)$. Then every entry of $a(z)A(z), a(z)B(z)$ is in $\mathbf{Z}[z]$ and α^{d^k} ($k \geq 0$) are not zeros of $a(z)$. Put

$$\begin{pmatrix} a(z) f_1(z^d) \\ a(z) f_2(z^d) \\ \vdots \\ a(z) f_m(z^d) \end{pmatrix} = \begin{pmatrix} a_{11}(z) & a_{12}(z) & \cdots & a_{1m}(z) \\ a_{21}(z) & a_{22}(z) & \cdots & a_{2m}(z) \\ & & \cdots & \\ a_{m1}(z) & a_{m2}(z) & \cdots & a_{mm}(z) \end{pmatrix} \begin{pmatrix} f_1(z) \\ f_2(z) \\ \vdots \\ f_m(z) \end{pmatrix} + \begin{pmatrix} b_1(z) \\ b_2(z) \\ \vdots \\ b_m(z) \end{pmatrix},$$

where $a_{ij}(z), b_i(z) \in \mathbf{Z}[z]$. Let $R_0(z, x_0, \ldots, x_m) \in \mathbf{Z}[z, x_0, \ldots, x_m]$ be a homogeneous polynomial of degree N in $x_0, \ldots, x_m$ which satisfies

$$R_0(z, 1, x_1, \ldots, x_m) = R(z, x_1, \ldots, x_m).$$

Letting

$$A_i(z, x) = A_i(z, x_0, \ldots, x_m) = b_i(z) x_0 + a_{i1}(z) x_1 + \cdots + a_{im}(z) x_m,$$

we define a sequence of polynomials $R_k(z, x_0, \ldots, x_m)$ for $k \geq 1$ by

$$R_k(z, x_0, \ldots, x_m) = R_{k-1}(z^d, a(z) x_0, A_1(z, x), \ldots, A_m(z, x)).$$

Then R_k is in $\mathbf{Z}[z, x_0, \ldots, x_m]$, homogeneous of degree N in $x_0, \ldots, x_m$, and

$$R_k(z, 1, f_1(z), \ldots, f_m(z)) = E_N(z^{d^k}) \left(\prod_{j=0}^{k-1} a(z^{d^j}) \right)^N.$$

Now on we use the usual notation $f \ll \phi$ for two series f and ϕ in the same variables, which indicates that ϕ is a majorant of f coefficients wise. Then

$$a(z) \ll c_{12}(1+z)^{c_{13}},$$

$$A_i(z, x_0, \ldots, x_m) \ll c_{12}(1+z)^{c_{13}}(x_0 + \cdots + x_m),$$

$$R_0(z, x_0, \ldots, x_m) \ll N^{c_5 N}(1+z)^N(x_0 + \cdots + x_m)^N.$$

Then it is easily proved by induction on k that

$$R_k \ll N^{c_5 N} ((m+1)c_{12})^{kN} (1 + z^{d^k})^N$$

$$\times \left((1 + z^{d^{k-1}}) \cdots (1 + z) \right)^{c_{13} N} (x_0 + \cdots + x_m)^N .$$

The lemma follows from this.

Letting $Q_k(\overline{x}) = R_k(\alpha, \overline{x})$, we have the following Lemmas.

LEMMA 2.2. *Let N and k be positive integers satisfying $N \geq c_1$ and $d^k \geq c_2 N \log N$, there exists a homogeneous polynomial $Q_k \in K[x_0, \ldots, x_m]$ which satisfies the inequalities*

$$\deg Q_k = N , \quad h(Q_k) \leq c_3 N d^k ,$$

$$c_4 N^{m+1} d^k \leq - \log \|Q_k\|_{\overline{w}} \leq c_5 N^{m+1} d^k .$$

LEMMA 2.3. *Suppose that λ is sufficiently large real number and*

(100) $$D \geq \lambda, \quad \log H \geq \lambda D^2 \log D.$$

Suppose that $I \subset K[x_0, \ldots, x_m]$ is a nonzero unmixed homogeneous ideal such that $\dim I \geq 0$ and

$$\deg I \leq \lambda^{m - \dim I - 1} D^{m - \dim I}, \quad h(I) \leq \lambda^{m - \dim I - 1} D^{m - \dim I - 1} \log H.$$

Then

$$\log |I(\overline{w})| \geq -\lambda^{\dim I + 1} D^{\dim I} (Dh(I) + \log H \deg I).$$

We now derive Theorem 1.2 from Lemma 2.3. Let $P \in \mathbf{Z}[x_0, \ldots, x_m]$ be a homogeneous polynomial such that

$$P(1, x_1, \ldots, x_m) = R(x_1, \ldots, x_m), \quad \deg P = \deg R.$$

Let $I = (P)$ be the homogeneous principal ideal in $K[x_0, \ldots, x_m]$. Then $\dim I = m - 1$ and by Proposition 4.8 from Chapter 3, we have

$$\deg I = \deg P \leq s,$$

$$h(I) \leq h(P) + \nu m^2 \deg P \leq \log H + \nu m^2 s.$$

Let λ be a sufficiently large constant and $s \geq \lambda$. Then $h(I) \leq \log H + \lambda s^2 \log s$. By Proposition 4.8 from Chapter 3 and Lemma 2.3 we have

$$\log |P(\overline{w})| + 2m^2 s \geq \log |I(\overline{w})| \geq -2\lambda^m s^m (\log H + \lambda s^2 \log s).$$

This implies Theorem 1.2.

PROOF OF LEMMA 2.3. Suppose that there exist ideals for which Lemma 2.3 is not true. Let I be an ideal with the least $\dim I$ among such ideals. Put $r = \dim I + 1$. Then $1 \leq r \leq m$. The lemma is true for all ideals J with $\dim J < \dim I$.

Let $I = I_1 \cap \ldots \cap I_s$ be a reduced primary decomposition and $\mathfrak{p}_j = \sqrt{I_j}$. By the Proposition 4.7 from Chapter 3 we get for $i = 1, \ldots, s$,

$$\deg \mathfrak{p}_i \leq \deg I \leq \lambda^{m-r} D^{m-r+1},$$

$$h(\mathfrak{p}_i) \leq h(I) + \nu m^2 \deg I \leq \lambda^{m-r} D^{m-r} (\log H + \nu m^2 D).$$

For some i $(1 \leq i \leq s)$,

$$\log |\mathfrak{p}_i(\overline{w})| \leq -\frac{1}{3} \lambda^r D^{r-1} (Dh(\mathfrak{p}_i) + \log H \deg \mathfrak{p}_i) = -S.$$

Since otherwise,

$$-\lambda^r D^{r-1}(Dh(I) + \log H \deg I) > \log|I(\overline{\omega})|$$

$$\geq \sum_{j=1}^{s} k_j \log|\mathfrak{p}_j(\overline{\omega})| - m^3 \deg I$$

$$\geq -\frac{1}{3}\lambda^r D^{r-1}\left(D\sum_{j=1}^{s} k_j h(\mathfrak{p}_j) + \log H \sum_{j=1}^{s} k_j \deg \mathfrak{p}_j\right) - m^3 \deg I$$

$$\geq -\frac{1}{3}\lambda^r D^{r-1}(Dh(I) + D\nu m^2 \deg I + \log H \deg I) - m^3 \deg I$$

$$\geq -\frac{2}{3}\lambda^r D^{r-1}(Dh(I) + \log H \deg I),$$

which is a contradiction if λ is sufficiently large. Noting $\deg \mathfrak{p}_i \geq 1$ and $\log H \geq \lambda D^2 \log D$, we have

$$\lambda^m D^m \log H \geq S \geq \frac{1}{3}\lambda^r \log H.$$

Let $\mathfrak{p}_i = \mathfrak{p}$, $\lambda = \mu^{2m+2}$ and $\rho = \min_{\overline{\beta} \in V(\mathfrak{p})}\|\overline{\omega} - \overline{\beta}\|$. By Proposition 4.13 in Chapter 3 we see

$$\log \rho \leq -\frac{1}{3r}\lambda^r D^{r-1}\left(D\frac{h(\mathfrak{p})}{\deg \mathfrak{p}} + \log H\right) + \frac{1}{r}\frac{h(\mathfrak{p})}{\deg \mathfrak{p}} + (\nu + 3)m^3$$

$$\leq -\frac{1}{4r}\lambda^r D^{r-1}\left(D\frac{h(\mathfrak{p})}{\deg \mathfrak{p}} + \log H\right)$$

$$\leq -\frac{1}{4r}\lambda^r \log H.$$

Let N be the integer satisfying

(101) $$N^m \log H \geq \mu^{-2m} \min\left(S, \log\frac{1}{\rho}\right) > (N-1)^m \log H.$$

Then we have

$$N^m \log H \geq \mu^{-2m}\frac{1}{4r}\lambda^r \log H \geq \frac{\mu^2}{4r}\log H,$$

and so $N^m \geq \mu^2/4r$. Hence $N \geq c_1$ and $(N-1)^m \geq N^m/2$ if λ is sufficiently large. By (101) we have

$$\frac{1}{2}N^m \log H < \mu^{-2m}S \leq \mu^{-2m}\lambda^m D^m \log H.$$

Therefore

(102) $$N \leq 2\mu^{2m}D.$$

Let $\eta = [4dc_5/c_4] + 1$ and k be the integer satisfying

(103) $$c_4 N^{m+1}d^{k-1} \leq \frac{2}{\eta}\min\left(S, \log\frac{1}{\rho}\right) < c_4 N^{m+1}d^k.$$

By (101) and (103),

$$N^{m+1}d^k \leq \frac{2d}{\eta c_4}\min\left(S, \log\frac{1}{\rho}\right) \leq \frac{2d}{\eta c_4}\mu^{2m}N^m \log H,$$

and

(104)
$$Nd^k \leq \frac{2d}{\eta c_4} \mu^{2m} \log H.$$

Also by (101) and (103) we have

$$c_4 N^{m+1} d^k > \frac{2}{\eta} \min \left(S, \log \frac{1}{\rho} \right) > \frac{2}{\eta} \frac{\mu^{2m}}{2} N^m \log H.$$

By (102), $\log D \geq \frac{1}{2} \log N$ since $D \geq \lambda \geq 2\mu^{2m}$. Therefore by (100) and (102),

$$d^k > \frac{\mu^{2m}}{\eta c_4} \frac{\log H}{N} \geq \frac{\mu^{2m} \lambda}{\eta c_4} \frac{D^2 \log D}{N} \geq \frac{\mu^{2m} \lambda}{4\mu^{4m} \eta c_4} \frac{1}{2} N \log N$$
$$\geq \mu N \log N \geq c_2 N \log N .$$

Since $N \geq c_1$, $d^k \geq c_2 N \log N$ there exists a polynomial Q_k such as constructed in Lemma 2.2. We have by (103),

$$\frac{2}{\eta} \min \left(S, \log \frac{1}{\rho} \right) < c_4 N^{m+1} d^k \leq -\log \|Q_k\|_{\overline{\omega}} \leq c_5 N^{m+1} d^k$$

$$\leq \frac{c_5 d}{c_4} \frac{2}{\eta} \min \left(S, \log \frac{1}{\rho} \right) < \frac{1}{2} \log \frac{1}{\rho} .$$

If $Q_k \in \mathfrak{p}$, by Corollary 4.9 in Chapter 3 we have

$$\|Q_k\|_{\overline{\omega}} \leq \rho\, e^{(2m+1)N},$$

and so

$$\frac{1}{2} \log \frac{1}{\rho} > -\log \|Q_k\|_{\overline{\omega}} \geq \log \frac{1}{\rho} - (2m+1)N,$$

and

$$\log \frac{1}{\rho} \leq (4m+2)N.$$

By (103) we have

$$c_4 N^{m+1} d^{k-1} \leq \frac{2}{\eta} \log \frac{1}{\rho} \leq \frac{8m+4}{\eta} N,$$

and this contradicts $N^m \geq \mu^2/4r$ if λ is sufficiently large. Hence $Q_k \notin \mathfrak{p}$. We apply Corollary 4.12 in Chapter 3 to the polynomial Q_k and the ideal $\mathfrak{p}$. If $r \geq 2$ and λ is sufficiently large, there exists an unmixed homogeneous ideal $J \subset K[x_0, \dots, x_m]$, $\dim J = \dim \mathfrak{p} - 1$, such that

$$\deg J \leq \eta \deg \mathfrak{p} \deg Q_k \leq \mu^{2m+1} D \deg \mathfrak{p} \leq \lambda^{m-r+1} D^{m-r+2},$$

$$h(J) \leq \eta(\deg Q_k h(\mathfrak{p}) + h(Q_k) \deg \mathfrak{p} + \nu m(r+2) \deg Q_k \deg \mathfrak{p})$$
$$\leq \mu^{2m+1}(Dh(\mathfrak{p}) + \log H \deg \mathfrak{p})$$
$$\leq \lambda^{m-r+1} D^{m-r+1} \log H ,$$

which are deduced from (102) and (104), and

$$\log |J(\overline{\omega})| \leq -S + \mu^{2m+1}(Dh(\mathfrak{p}) + \log H \deg \mathfrak{p})$$

(105)
$$\leq -\frac{1}{4} \lambda^r D^{r-1}(Dh(\mathfrak{p}) + \log H \deg \mathfrak{p}) .$$

Since the lemma is true for the ideal J,

$$\log |J(\overline{\omega})| \geq -\lambda^{r-1}D^{r-2}(Dh(J) + \log H \deg J)$$
$$\geq -2\mu^{2m+1}\lambda^{r-1}D^{r-1}(Dh(\mathfrak{p}) + \log H \deg \mathfrak{p}) .$$

This contradicts (105). If $r = 1$, the left hand side of the inequality (105) is nonnegative, a contradiction. $\square$

3. Proof of multipicity estimate

Assuming

$$\mathrm{ord}_{z=0}Q(z, f_1(z), \dots, f_m(z)) = \lambda^{m+1}MN^m,$$

where λ is larger than some constant which will be determined below, we will derive a contradiction.

Choose a polynomial $a(z) \in \mathbf{C}[z]$ such that every entry of $a(z)A(z)$, $a(z)B(z)$ is in $\mathbf{C}[z]$ and put

$$\begin{pmatrix} a(z)f_1(z^d) \\ a(z)f_2(z^d) \\ \vdots \\ a(z)f_m(z^d) \end{pmatrix}$$

$$= \begin{pmatrix} a_{11}(z) & a_{12}(z) & \dots & a_{1m}(z) \\ a_{21}(z) & a_{22}(z) & \dots & a_{2m}(z) \\ & & \dots & \\ a_{m1}(z) & a_{m2}(z) & \dots & a_{mm}(z) \end{pmatrix} \begin{pmatrix} f_1(z) \\ f_2(z) \\ \vdots \\ f_m(z) \end{pmatrix} + \begin{pmatrix} b_1(z) \\ b_2(z) \\ \vdots \\ b_m(z) \end{pmatrix} ,$$

where $a_{ij}(z)$, $b_i(z) \in \mathbf{C}[z]$. Let $R_0(z, x_0, \dots, x_m) \in \mathbf{C}[z, x_0, \dots, x_m]$ be a homogeneous polynomial of degree N in $x_0, \dots, x_m$ which satisfies

$$R_0(z, 1, x_1, \dots, x_m) = Q(z, x_1, \dots, x_m), \qquad \deg_z R_0 \leq M.$$

In the same way as in § 2, we have for $k \geq 1$ a sequence of polynomials $R_k(z, x_0, \dots, x_m)$ which are homogeneous of degree N in $x_0, \dots, x_m$ and

$$R_k(z, 1, f_1(z), \dots, f_m(z)) = Q\big(z^{d^k}, f_1(z^{d^k}), \dots, f_m(z^{d^k})\big)\Big(\prod_{j=0}^{k-1} a(z^{d^j})\Big)^N .$$

Then we have

$$\mathrm{ord}_{z=0}R_k(z, 1, f_1(z), \dots, f_m(z)) = \lambda^{m+1}d^k MN^m + c_1 N \frac{d^k - 1}{d - 1},$$

and so

$$\lambda^{m+1}d^k MN^m \leq \mathrm{ord}_{z=0}R_k(z, 1, f_1(z), \dots, f_m(z)) \leq 2\lambda^{m+1}d^k MN^m.$$

Let $K = \mathbf{C}(z)$. For every $v \in \mathbf{C}$ we define the absolute value $|\ |_v$ by $|\alpha|_v = e^{-\mathrm{ord}_{z=v}\alpha}$. When $v = \infty$, $\left|\frac{P}{Q}\right|_\infty = e^{\deg P - \deg Q}$, for $P, Q \in \mathbf{C}[z]$. Let $|\ | = |\ |_0$ and $\overline{\omega} = (1, f_1(z), \dots, f_m(z))$. We use the notations introduced in Chapter 3. Then $|\overline{\omega}| = 1$. Let $Q_k \in K[\overline{x}]$ be the polynomial such that

$$Q_k(\overline{x}) = R_k(z, x_0, \dots, x_m).$$

We see

$$\deg Q_k = N, \quad h(Q_k) \leq \deg_z R_k \leq c_2 d^k M,$$

and (see notations in Chapter 3)

(106) $$\frac{1}{2}\lambda^{m+1}d^k MN^m \leq -\log \|Q_k\|_{\varpi} \leq 2\lambda^{m+1}d^k MN^m.$$

Choose a positive integer l such that

(107) $$d^l > \lambda^{m+1}N^m.$$

Let $I = (Q_l)$ be the ideal of $K[\overline{x}]$ generated by Q_l. Then $\dim I = m - 1$ and by Proposition 4.8 in Chapter 3

$$\deg I = \deg Q_l = N, \quad h(I) = h(Q_l) \leq c_2 d^l M,$$

$$\log |I(\overline{\omega})| \leq \log \|Q_l\|_{\varpi} \leq -\frac{1}{2}\lambda^{m+1}d^l MN^m.$$

Let $I = I_1 \cap \cdots \cap I_s$ be a reduced primary decomposition. By Proposition 4.7 in Chapter 3 we get for $i = 1, \ldots, s$,

$$\deg \mathfrak{p}_i \leq \deg I, \quad h(\mathfrak{p}_i) \leq h(I) \leq c_2 d^l M.$$

For some i $(1 \leq i \leq s)$,

$$\log |\mathfrak{p}_i(\overline{\omega})| \leq -\lambda^m N^{m-1}(Nh(\mathfrak{p}_i) + c_2 d^l M \deg \mathfrak{p}_i).$$

Since otherwise,

$$-\frac{1}{2}\lambda^{m+1}d^l MN^m$$

$$\geq \log |I(\overline{\omega})| \geq \sum_{j=1}^{s} k_j \log |\mathfrak{p}_j(\overline{\omega})|$$

$$\geq -\lambda^m N^{m-1}\left(N\sum_{j=1}^{s} k_j h(\mathfrak{p}_j) + c_2 d^l M \sum_{j=1}^{s} k_j \deg \mathfrak{p}_j\right)$$

$$\geq -\lambda^m N^{m-1}(Nh(I) + c_2 d^l M \deg I)$$

$$\geq -\lambda^m N^{m-1} \cdot 2c_2 d^l MN,$$

which is a contradiction. Let $\mathfrak{p}_i = \mathfrak{p}^{(1)}$. Then $\dim \mathfrak{p}^{(1)} = m - 1$. By induction we show that for each n $(1 \leq n \leq m)$ there is a prime ideal $\mathfrak{p}^{(n)}$, $\dim \mathfrak{p}^{(n)} = m - n$, such that

$$\deg \mathfrak{p}^{(n)} \leq \lambda^{n-1}N^n, \quad h(\mathfrak{p}^{(n)}) \leq \lambda^{n-1}c_2 d^l MN^{n-1},$$

$$\log |\mathfrak{p}^{(n)}(\overline{\omega})| \leq -\lambda^{m-n+1}N^{m-n}(Nh(\mathfrak{p}^{(n)}) + c_2 d^l M \deg \mathfrak{p}^{(n)}) = -X_n.$$

When $n = 1$, the ideal $\mathfrak{p}^{(1)}$ is already given. Assuming $\mathfrak{p}^{(n)}$ exists for n $(1 \leq n \leq m - 1)$, we construct $\mathfrak{p}^{(n+1)}$. By (107) we have

(108) $$8\lambda^{m+1}MN^m \leq \lambda^{m-n+1}N^{m-n}c_2 d^l M \leq X_n \leq \lambda^{m+1}d^l MN^m.$$

Let $\rho = \min_{\overline{\beta} \in V(\mathfrak{p}^{(n)})} \|\overline{\omega} - \overline{\beta}\|$. By Proposition 4.13 in Chapter 3 and (107),

$$\log \rho \leq -\frac{1}{m-n+1}\lambda^{m-n+1}N^{m-n}\left(N\frac{h(\mathfrak{p}^{(n)})}{\deg \mathfrak{p}^{(n)}} + c_2 d^l M\right) + \frac{h(\mathfrak{p}^{(n)})}{\deg \mathfrak{p}^{(n)}}$$

$$\leq -8\lambda^{m-n}N^{m-n}d^l M$$

(109) $$\leq -8\lambda^{m+1}MN^m.$$

Let $\eta = 16d$ and k be the integer satisfying

(110) $$\frac{1}{2}\lambda^{m+1}d^{k-1}MN^m < \frac{2}{\eta}\min\left(X_n, \log\frac{1}{\rho}\right) \leq \frac{1}{2}\lambda^{m+1}d^k MN^m.$$

By (108), (109), (110) we have

$$\frac{2}{\eta}8\lambda^{m+1}MN^m \leq \frac{1}{2}\lambda^{m+1}d^k MN^m,$$

and so $k \geq 0$. By (108), (110),

$$\frac{1}{2}\lambda^{m+1}d^{k-1}MN^m < \frac{2}{\eta}X_n \leq \frac{2}{\eta}\lambda^{m+1}d^l MN^m,$$

and $k < l$. If $Q_k \in \mathfrak{p}^{(n)}$, by Corollary 4.9 in Chapter 3 we have

$$\|Q_k\|_{\overline{\omega}} \leq \rho,$$

and by (106), (110),

$$\frac{1}{2}\log\frac{1}{\rho} = 4d\frac{2}{\eta}\log\frac{1}{\rho} > 2\lambda^{m+1}d^k MN^m$$

$$\geq -\log\|Q_k\|_{\overline{\omega}} \geq \log\frac{1}{\rho}.$$

This is a contradiction. Hence $Q_k \notin \mathfrak{p}^{(n)}$. We apply Corollary 4.12 in Chapter 3 to the polynomial Q_k and the ideal $\mathfrak{p}^{(n)}$. There exists an unmixed homogeneous ideal J of $K[\overline{x}]$, $\dim J = m - n - 1$, such that

$$\deg J \leq \eta \deg \mathfrak{p}^{(n)} \deg Q_k \leq \eta N \deg \mathfrak{p}^{(n)} \leq \lambda^n N^{n+1},$$

$$h(J) \leq \eta(h(\mathfrak{p}^{(n)}) \deg Q_k + h(Q_k) \deg \mathfrak{p}^{(n)})$$
$$\leq \eta(Nh(\mathfrak{p}^{(n)}) + c_2 d^l M \deg \mathfrak{p}^{(n)})$$
$$\leq \lambda^n c_2 d^l MN^n,$$

$$\log|J(\overline{\omega})| \leq -X_n + \eta(h(\mathfrak{p}^{(n)}) \deg Q_k + h(Q_k) \deg \mathfrak{p}^{(n)})$$
$$\leq -X_n + \eta(Nh(\mathfrak{p}^{(n)}) + c_2 d^l M \deg \mathfrak{p}^{(n)})$$

(111) $$\leq -\frac{X_n}{2}.$$

Let $J = I_1 \cap \cdots \cap I_s$ be a reduced primary decomposition and $\mathfrak{p}_i = \sqrt{I_i}$. Using Proposition 4.7 in Chapter 3 we assert for some i $(1 \leq i \leq s)$,

$$\log|\mathfrak{p}_i(\overline{\omega})| \leq -\lambda^{m-n}N^{m-n-1}(Nh(\mathfrak{p}_i) + c_2 d^l M \deg \mathfrak{p}_i)$$

and so $\mathfrak{p}^{(n+1)} = \mathfrak{p}_i$ satisfies the required inequalities. Since otherwise,

$$-\frac{1}{2}\lambda^{m-n+1}N^{m-n}(Nh(\mathfrak{p}^{(n)}) + c_2 d^l M \deg \mathfrak{p}^{(n)})$$

$$\geq \log|J(\overline{\omega})| \geq \sum_{j=1}^{s}k_j\log|\mathfrak{p}_j(\overline{\omega})|$$

$$\geq -\lambda^{m-n}N^{m-n-1}(Nh(J) + c_2 d^l M \deg J)$$

$$\geq -\lambda^{m-n}N^{m-n-1}(N\eta(Nh(\mathfrak{p}^{(n)}) + c_2 d^l M \deg \mathfrak{p}^{(n)}) + c_2 d^l M\eta N \deg \mathfrak{p}^{(n)})$$

$$\geq -2\eta\lambda^{m-n}N^{m-n}(Nh(\mathfrak{p}^{(n)}) + c_2 d^l M \deg \mathfrak{p}^{(n)}),$$

a contradiction. For the ideal $\mathfrak{p}^{(m)}$, we can choose $Q_k \notin \mathfrak{p}^{(m)}$ in the same way as above. In this case, we have $0 \leq -X_n/2$ instead of (111). This is the final contradiction and the theorem is proved.

CHAPTER 13

Algebraic Independence in Algebraic Groups.
Part I: Small Transcendence Degrees

1. Introduction

Our aim in this chapter is to provide an introduction, as simple as possible, to general statements of algebraic independence in the framework of algebraic groups. We consider only small transcendence degrees, by which we mean results asserting the algebraic independence of at least two numbers belonging to some set of numbers typically defined as values of the exponential map of some commutative algebraic group. Even in this special case, a single statement covering all the expected applications, remains to be settled. Here we have restricted our insight to a rather simple situation involving only a one parameter subgroup and for simplicity, we have also neglected the periods eventually contained in this subgroup. We give two statements which contain the most classical results, such as Gel'fond's theorem.

In section 2, we set up our frame and state two general results. Next we deduce from these two theorems some standard corollaries concerning mainly algebraic independence properties of values of the usual exponential function and the Weierstrass $\wp$ function. For simplicity, we prove in §§5-8 one of these corollaries instead of the general statements. We use the method of interpolation determinants together with a criterion of algebraic independence with multiplicities (§4).

2. General statements.

Let K be a subfield in the field $\mathbf{C}$ of complex numbers and let G be a commutative connected algebraic group defined over K. It is convenient to view G as a product :

$$G \simeq G_0 \times G_1 \times G_2 \,, \quad \text{with} \quad G_0 = \mathbf{G}_{\mathrm{a}}^{d_0}, \quad G_1 = \mathbf{G}_{\mathrm{m}}^{d_1},$$

where $\mathbf{G}_{\mathrm{a}}$ and $\mathbf{G}_{\mathrm{m}}$ stand respectively for the additive and multiplicative groups over K. Note that the subscripts $0, 1$ and 2 here above, measure the order of growth (at most polynomial, exponential and of order ≤ 2) of the corresponding exponential maps. We denote by d_2 the dimension of G_2 so that $d := d_0 + d_1 + d_2$ is the dimension of G. Let $T_G(\mathbf{C})$ be the vector space of the complex points of the tangent space at the origin of the algebraic group G, viewed as an additive group. Then we have a canonical map, called the exponential map,

$$\exp_G : T_G(\mathbf{C}) \longrightarrow G(\mathbf{C})$$

which is a morphism of complex analytic group. See the appendix of [**Wal3**] and Bourbaki (LIE III §6), for a precise definition in terms of Lie derivatives. It is

---

(∗) Chapter's author : Michel LAURENT .

important to know that the exponential map is universal in the sense that each map of complex analytic groups from a complex vector space to $G(\mathbf{C})$ factors through $\exp_G$. An *analytic subgroup* of $G(\mathbf{C})$ is the image of such a map. We identify

$$T_G = T_{G_0} \times T_{G_1} \times T_{G_2}, \qquad T_{G_0}(\mathbf{C}) = \mathbf{C}^{d_0}, \qquad T_{G_1}(\mathbf{C}) = \mathbf{C}^{d_1},$$

in such a way that the map $\exp_G$ can be written

$$\mathbf{C}^{d_0} \times \mathbf{C}^{d_1} \times T_{G_2}(\mathbf{C}) \longrightarrow \mathbf{C}^{d_0} \times (\mathbf{C}^*)^{d_1} \times G_2(\mathbf{C})$$

$$(z_1, \ldots, z_{d_0}, z_{d_0+1}, \ldots, z_{d_0+d_1}, \underline{z}) \longmapsto (z_1, \ldots, z_{d_0}, e^{z_{d_0+1}}, \ldots, e^{z_{d_0+d_1}}, \exp_{G_2}(\underline{z})).$$

More generally, note that the exponential map can be concretely described in terms of polynomial, exponential and theta functions (see the appendix of [**Wal3**]).

Now, let V be a complex line contained in $T_G(\mathbf{C})$ and let Y be a subgroup of finite rank ℓ over $\mathbf{Z}$ contained in V :

$$Y \subset V \subseteq T_G(\mathbf{C}) \overset{\exp_G}{\longrightarrow} G(\mathbf{C}).$$

We suppose that

(i) $$\exp_G(Y) \subseteq G(K),$$

Since G could be a priori arbitrarily enlarged, we suppose that G is the smallest algebraic group containing the analytic subgroup $\exp_G(V)$ in the sense that

(ii) $$\exp_G(V) \quad \text{is Zariski-dense in } G(\mathbf{C}).$$

Note that this second assumption obviously implies that $d_0 \leq 1$ because the dimension of V is equal to one.

THEOREM 2.1. *Suppose that the assumptions* (i) *and* (ii) *hold and that*

$$\ell d \geq 2(\ell + d_1 + 2d_2).$$

Then the transcendence degree over $\mathbf{Q}$ *of the field* K *is* ≥ 2.

Now, we can relax the above inequality if we suppose that V is defined over K. The tangent space T_G is a K-vector space. A complex vector space V contained in $T_G(\mathbf{C})$ is said to be defined over K if we can find some basis of V whose elements belong to T_G. When V is a complex line, this simply means that if $(x_1, \ldots, x_d)$ are the coordinates of a generator of V in a K-rational basis of $T_G(\mathbf{C})$, all the ratios x_i/x_j belong to K whenever $x_j \neq 0$.

THEOREM 2.2. *Suppose that the assumptions* (i) *and* (ii) *hold and that*

$$\ell(d - 1) \geq \ell + d_1 + 2d_2 + d_0,$$

Suppose moreover that V *is defined over* K. *Then the transcendence degree over* $\mathbf{Q}$ *of the field* K *is* ≥ 2.

It is expected that d_0 can be replaced by 0 in the right hand side of the above inequality. According to the remark that d_0 is equal to 0 or 1, the assumption in Theorem 2.2 reads $\ell(d - 1) > \ell + d_1 + 2d_2$ whenever $d_0 = 1$, and we hope that this strict inequality is unnecessary.

3. Concrete applications.

The most classical corollaries of the Theorems 2.1 and 2.2 concern algebraic independence properties of the exponential function. For shortness, we shall denote by $\operatorname{trdeg} K$ the transcendence degree over $\mathbf{Q}$ of any field K contained in $\mathbf{C}$.

THEOREM 3.1. *Let m and n be integers ≥ 1 and let $\{x_i, 1 \leq i \leq m\}$ and $\{y_j, 1 \leq j \leq n\}$ be two sets of complex numbers linearly independent over $\mathbf{Q}$. Denote*

$$K_1 = \mathbf{Q}(e^{x_i y_j})_{\substack{1 \leq i \leq m \\ 1 \leq j \leq n}} \quad , \qquad K_2 = \mathbf{Q}(y_j, e^{x_i y_j})_{\substack{1 \leq i \leq m \\ 1 \leq j \leq n}}$$

$$K_3 = \mathbf{Q}(x_i, e^{x_i y_j})_{\substack{1 \leq i \leq m \\ 1 \leq j \leq n}} \quad , \qquad K_4 = \mathbf{Q}(x_i, y_j, e^{x_i y_j})_{\substack{1 \leq i \leq m \\ 1 \leq j \leq n}}.$$

i) *Suppose $mn \geq 2m + 2n$, then $\operatorname{trdeg} K_1 \geq 2$.*
ii) *Suppose $mn \geq 2m + n$, then $\operatorname{trdeg} K_2 \geq 2$.*
iii) *Suppose $mn \geq m + 2n$, then $\operatorname{trdeg} K_3 \geq 2$.*
iv) *Suppose $mn \geq m + n + 1$, then $\operatorname{trdeg} K_4 \geq 2$.*

The assertions (ii) and (iii) above are obviously equivalent. However we have kept in Theorem 3.1 the two statements for which we shall give different dual proofs.

PROOF. We shall prove simultaneously (i) and (iii) and next (ii) and (iv). Let us first consider the algebraic group $G = \mathbf{G}_m^m$ with the exponential map

$$\mathbf{C}^m \xrightarrow{\exp_G} (\mathbf{C}^*)^m, \qquad (z_1, \ldots, z_m) \longmapsto (e^{z_1}, \ldots, e^{z_m}).$$

Let $\underline{x} = (x_1, \ldots, x_m)$ and let us consider the complex line $V = \mathbf{C}\underline{x} \subseteq \mathbf{C}^m$ together with its subgroup $Y = \oplus_{j=1}^n \mathbf{Z}y_j\underline{x}$ of rank $\ell = n$. Remark that the algebraic subgroups of the linear torus $(\mathbf{C}^*)^m$ are defined by a finite set of monomial equations $\prod_{\mu=1}^m X_\mu^{a_\mu} = 1$ with $a_\mu \in \mathbf{Z}$. Since the Zariski closure of any analytic subgroup is a connected algebraic subgroup and since the components of $\underline{x}$ are linearly independent over $\mathbf{Z}$, it is clear that $\exp_G(V)$ cannot be contained in any proper algebraic subgroup of G. It follows that $\exp_G(V)$ is Zariski-dense in $G(\mathbf{C})$. Now take $K = K_1$ in the case (i) and $K = K_3$ in the case (iii). In both cases, the image $\exp_G(Y)$ is contained in $G(K)$ and in the case (iii), the line V is obviously rational over K since the components of $\underline{x}$ belong to K_3. Here we have $d_0 = d_2 = 0, d_1 = m, \ell = n$, so that the assumptions of the Theorems 2.1 and 2.2 read respectively $mn \geq 2m + 2n$ and $(m - 1)n \geq m + n$. All the assumptions of these two theorems have been checked. They imply that the transcendence degree over $\mathbf{Q}$ of the fields K_1 and K_3 is ≥ 2.

The proofs of the assertions (ii) and (iv) are quite similar. Now, we use the algebraic group $G = \mathbf{G}_a \times \mathbf{G}_m^m$ with the exponential map

$$\mathbf{C}^{m+1} \xrightarrow{\exp_G} \mathbf{C} \times (\mathbf{C}^*)^m, \qquad (z_0, z_1, \ldots, z_m) \longmapsto (z_0, e^{z_1}, \ldots, e^{z_m}).$$

We define $\underline{x} = (1, x_1, \ldots, x_m) \in \mathbf{C}^{m+1}$ and consider the line $V = \mathbf{C}\underline{x}$ containing the subgroup $Y = \oplus_{j=1}^n \mathbf{Z}y_j\underline{x}$. The algebraic subgroups of G are product of an algebraic subgroup of $\mathbf{G}_a$ (these are $\{0\}$ and $\mathbf{G}_a$) with an algebraic subgroup of $\mathbf{G}_m^m$. Since the first component of $\underline{x}$ is nonzero, the Zariski closure of $\exp_G(V)$ must contain $\mathbf{G}_a$, and the same argument as before shows that it also contains $\mathbf{G}_m^m$. Then it is necessarily equal to G. We choose $K = K_2$ in the case (ii) and $K = K_4$ in the case (iv). Now $d_0 = 1, d_1 = m, d_2 = 0, \ell = n$ so that the conditions of Theorems 2.1 and 2.2 are respectively $(m + 1)n \geq 2m + 2n$ and $mn \geq m + n + 1$. $\qquad\square$

As an example, we recover from the cases (ii) or (iii) of Theorem 3.1 the following result due to Gel'fond. Let α be a nonzero algebraic number for which we have selected a nonzero determination $\log \alpha$ of its logarithm, and let β be a cubic number. Then the two number $\alpha^\beta := e^{\beta \log \alpha}$ and $\alpha^{\beta^2} := e^{\beta^2 \log \alpha}$ are algebraically independent over $\mathbf{Q}$. We can use for instance (ii) with $m = n = 3$ and

$$x_1 = \log \alpha, \quad x_2 = \beta \log \alpha, \quad x_3 = \beta^2 \log \alpha, \quad y_1 = 1, \quad y_2 = \beta, \quad y_3 = \beta^2.$$

Note also that the conjectural improvement of Theorem 2.2 mentioned above, would imply in case (iv) that the transcendence degree of the field $\mathbf{Q}(x_i, y_j, e^{x_i y_j})$ is ≥ 2 if $mn \geq m+n$. When $m = n = 2$, we would obtain the algebraic independence of the two numbers $\log \alpha$ and α^β for any quadratic number β and any algebraic number α with $\log \alpha \neq 0$. By Nesterenko's result on Eisenstein's series, see Chapter 3, this is only known for a root of unity α and an imaginary quadratic number β.

As an example of application to non linear algebraic groups, we translate now the results of Theorem 3.1 in the context of elliptic functions. Let Λ be a lattice in $\mathbf{C}$ with algebraic invariants g_2 and g_3. We denote by $\wp$ the usual Weierstrass elliptic function associated to Λ and by $\mathbf{k} = \mathrm{End}\,\Lambda \otimes \mathbf{Q}$ the field generated by the ring of linear endomorphisms of the lattice Λ. Recall that $\mathbf{k}$ is isomorphic either to $\mathbf{Q}$ or to an imaginary quadratic field.

THEOREM 3.2. *Let* $x_1, \ldots, x_m$ *be* m *complex numbers linearly independent over* $\mathbf{k}$ *and let* $y_1, \ldots, y_n$ *be* n *complex numbers linearly independent over* $\mathbf{Q}$. *We suppose that the products* $x_i y_j$ *do not belong to* Λ. *Denote*

$$K_1 = \mathbf{Q}(\,\wp(x_i y_j)\,)_{\substack{1 \leq i \leq m \\ 1 \leq j \leq n}} \quad , \qquad K_2 = \mathbf{Q}(\,y_j,\, \wp(x_i y_j)\,)_{\substack{1 \leq i \leq m \\ 1 \leq j \leq n}}$$

$$K_3 = \mathbf{Q}(\,x_i,\, \wp(x_i y_j)\,)_{\substack{1 \leq i \leq m \\ 1 \leq j \leq n}} \quad , \qquad K_4 = \mathbf{Q}(\,x_i,\, y_j,\, \wp(x_i y_j)\,)_{\substack{1 \leq i \leq m \\ 1 \leq j \leq n}}.$$

i) *Suppose* $mn \geq 4m + 2n$, *then* $\mathrm{trdeg}\,K_1 \geq 2$.
ii) *Suppose* $mn \geq 4m + n$, *then* $\mathrm{trdeg}\,K_2 \geq 2$.
iii) *Suppose* $mn \geq 2m + 2n$, *then* $\mathrm{trdeg}\,K_3 \geq 2$.
iv) *Suppose* $mn \geq 2m + n + 1$, *then* $\mathrm{trdeg}\,K_4 \geq 2$.

Note that in the elliptic case, the x_i are supposed to be linearly independent over $\mathbf{k}$ and the y_j over $\mathbf{Q}$, so that the assertions (ii) and (iii) are no more equivalent.

PROOF. We only sketch the proof, its ingredients being the same as in the exponential case. Let E be the Weierstrass elliptic curve $Y^2 Z = 4X^3 - g_2 X Z^2 - g_3 Z^3$ whose exponential map is

$$\mathbf{C} \xrightarrow{\exp_E} E(\mathbf{C}), \qquad z \longmapsto (\wp(z) : \wp'(z) : 1).$$

We consider the algebraic groups E^m and $\mathbf{G}_a \times E^m$, and we identify the exponential maps of these two groups with the product of the exponential maps of their factors. The complex line V and its subgroup Y of rank $\ell = n$ over $\mathbf{Z}$ are defined as in the proof of Theorem 3.1. Now $d_1 = 0$ and $d_2 = m$, which has the effect to replace m by $2m$ in the right hand side of the inequalities (i) ... (iv) from Theorem 3.1. Note that the tangent space at the origin of the algebraic subgroups of E^m, viewed as subspaces of $T_{E^m} = (T_E)^m$, are defined by linear equations $\sum_{\mu=1}^m a_\mu z_\mu = 0$ with coefficients $a_\mu \in \mathbf{k}$. This is the reason why we need to assume now that the x_i are linearly independent over **k**. Note also that the algebraic subgroups of $\mathbf{G}_a \times E^m$ are products of an algebraic subgroup of $\mathbf{G}_a$ by an algebraic subgroup of E^m, so that we can repeat the arguments from the proof of Theorem 3.1. □

Suppose that the lattice Λ has complex multiplication by a quadratic integer τ and suppose that the v_j are linearly independent over the quadratic field $\mathbf{k} = \mathbf{Q}(\tau)$. Then we can double the number n of y_j by multiplying them by τ and apply Theorem 3.2 to the new set of data $(x_1, \ldots, x_m)$ and $(y_1, \ldots, y_n, \tau y_1, \ldots, \tau y_n)$. In the statements (i), (ii) and (iii), we recover exactly the same numerical constraints as in the corresponding cases from Theorem 3.1. In the remaining case (iv), the two inequalities are different, due to the undesirable term d_0 in Theorem 2.2. For instance, we get the exact elliptic analogue of Gel'fond theorem in the case of complex multiplication :

COROLLARY 3.3. *Suppose that the lattice Λ has complex multiplication in a quadratic field* **k**. *Let β be an algebraic number such that the degree $[\mathbf{k}(\beta) : \mathbf{k}] = 3$, and let u be a nonzero complex number such that $\wp(u)$ is algebraic ($u \in \Lambda\setminus\{0\}$ is also allowed). Then the two numbers $\wp(\beta u)$ and $\wp(\beta^2 u)$ are algebraically independent over* **Q**.

To conclude this section, let us mention some complements. First, the nice result of Brownawell-Waldschmidt on the transcendence of at least one of the two numbers e^e and e^{e^2} (the proof belongs actually to algebraic independence !) has been extended to algebraic groups. See [**Tub2**] for a one parameter subgroup and [**RW**] for analytic subgroups of linear algebraic groups. The main ingredient is to introduce some quotient G' of G with the property that the projection on G' of the group $\exp_G(Y)$ is contained in $G'(\overline{\mathbf{Q}})$. On the other hand, some results entering obviously in the framework of algebraic groups remain isolated. For example, let us indicate Choodnovsky's proof of the algebraic independence of the two numbers $\zeta(u) - \frac{\eta}{\omega}u$ and $\frac{\eta}{\omega}$, where $u \in \mathbf{C} \setminus \mathbf{Q}\omega$ is such that $\wp(u)$ is algebraic, $\omega \in \Lambda \setminus \{0\}$ and $\eta = \zeta(z + \omega) - \zeta(z)$ is the corresponding quasi-period of the Weierstrass zeta function *cf.* [**Chu2**]. Some results due to Reyssat *cf.* [**Rey2**] involving values of the Weierstrass sigma function remain also specific.

In the next two chapters, we shall give a complete proof of the assertion (ii) from Theorem 3.1, working directly on the specific data G, Y, V introduced there. The proof of Theorem 2.1 is formally identical, and for Theorem 2.2 we need moreover to introduce derivatives of the auxiliary functions. See respectively Theorem 1 of [**Wal6**] and Theorem 1 of [**Tub1**] with $n = 1$.

4. A criterion of algebraic independence with multiplicities.

We state in this section a criterion for algebraic independence in dimension one, which extends the classical Gel'fond criterion in two directions. First, we work directly with polynomial functions on an algebraic curve, and secondly we consider derivatives along this curve. The last ingredient provides a key for proving algebraic independence using interpolation determinants.

Let $\mathcal{C}$ be an affine curve in $\mathbf{C}^m$ defined over **Q** and let $\overline{\mathcal{C}}$ be its closure in $\mathbf{P}^m$ for the natural embedding

$$\mathbf{C}^m \subset \mathbf{P}^m(\mathbf{C}), \qquad (x_1, \ldots, x_m) \longmapsto (1 : x_1 : \cdots : x_m).$$

The Chow form of $\overline{\mathcal{C}}$, which is defined up to a rational scalar factor, is a polynomial $f(\underline{u}, \underline{v}) \in \mathbf{Q}[\underline{u}, \underline{v}]$, homogeneous with degree $\deg(\mathcal{C})$ in each set of variables $\underline{u} =$

$(u_0, \ldots, u_m)$ and $\underline{v} = (v_0, \ldots, v_m)$, such that $f(\underline{a}, \underline{b}) = 0$ whenever the linear variety defined by the two projective equations

$$a_0 X_0 + \cdots + a_m X_m = 0 \quad \text{and} \quad b_0 X_0 + \cdots + b_m X_m = 0$$

meets $\overline{C}$ in $\mathbf{P}^m(\mathbf{C})$. If we normalize f up to a sign by the properties $f \in \mathbf{Z}[\underline{u}, \underline{v}]$ and f primitive, we define the *size of the curve* C by the formula

$$t(C) := \max \left\{ \log H(f), (\log(m+1)) \deg(C) \right\}$$

where $H(f)$ stands for the usual *height* of the polynomial f, by which we mean the maximum of the absolute value of the coefficients of the involved polynomial.

Let $\partial : \mathbf{Q}(C) \to \mathbf{Q}(C)$ be any nonzero derivation of the field of rational functions on the curve C. For each integer $s \geq 1$, we define $\partial^{[s]} = \partial^s / s!$ to be the s-th iterate of ∂ divided by $s!$. If Q is a polynomial in $\mathbf{Q}[X_1, \ldots, X_m]$, we shall also denote by the same symbol Q its restriction to C.

THEOREM 4.1. *Let C be an algebraic curve in $\mathbf{C}^m$ defined over $\mathbf{Q}$ and let $\underline{\theta}$ be a point in C which does not belong to $\overline{\mathbf{Q}}^m$. Let $(s_n)_{n \in \mathbf{N}}$ be a sequence of positive integers, and $(t_n)_{n \in \mathbf{N}}$ and $(d_n)_{n \in \mathbf{N}}$ be two sequences of positive real numbers such that*

$$7m \deg(C) \leq \frac{d_n}{s_n} \leq \frac{d_{n+1}}{s_{n+1}} \leq 2\frac{d_n}{s_n}, \qquad \frac{t_n}{s_n} \leq \frac{t_{n+1}}{s_{n+1}} \leq 2\frac{t_n}{s_n},$$

and

$$t_n \geq 8\frac{t(C)}{\deg(C)} d_n,$$

for every $n \geq 0$. We suppose moreover that the sequence t_n/s_n tends to infinity with n. Then there does not exist any sequence $(Q_n)_{n \in \mathbf{N}}$ of polynomials in $\mathbf{Z}[X_1, \ldots, X_m]$ non identically zero on C such that

$$\deg(Q_n) \leq d_n , \log H(Q_n) \leq t_n ,$$

and

$$\log \max_{0 \leq \sigma < s_n} \left| \partial^{[\sigma]} Q_n(\underline{\theta}) \right| \leq -cd_n t_n / s_n ,$$

with $c = 40000(\deg(C))^2$, for each integer n sufficiently large.

PROOF. We only explain its principle and refer to Corollary 3 of [**LR1**] for a complete proof. Let us introduce the two sequences

$$D_n = c'\frac{d_n}{s_n} , \qquad T_n = c'\frac{t_n}{s_n} , \qquad (n \geq 0),$$

for a suitable positive constant c'. For infinitely many values of n, there exists an algebraic approximation $\underline{\alpha}_n \in C(\overline{\mathbf{Q}})$ of the point $\underline{\theta}$ such that

$$c''D_n \leq d(\underline{\alpha}_n) \leq D_n , \quad t(\underline{\alpha}_n) \leq T_n \quad \text{and} \quad \log \|\underline{\theta} - \underline{\alpha}_n\| \leq -c''D_n T_n ,$$

where $d(\underline{\alpha}_n)$ and $t(\underline{\alpha}_n)$ stand respectively for the *degree* and the *size* of the algebraic m-tuple $\underline{\alpha}_n$, and where $\| \ \|$ denotes the maximum of the absolute values of the coordinates. Now, suppose that such a sequence Q_n does exist. The above lower bound for $d(\underline{\alpha}_n)$ together with Bezout's theorem, implies easily that for some integer σ with $0 \leq \sigma \leq s_n/3$, the value $\partial^{[\sigma]} Q_n(\underline{\alpha}_n)$ is nonzero. Since the point $\underline{\alpha}_n$ is close to the point $\underline{\theta}$, an argument of continuity shows that the quantity $|\partial^{[\sigma]} Q_n(\underline{\alpha}_n)|$ is also very small. On the other hand, Liouville's inequality provides an upper bound for this quantity, which reveals to be contradictory with the lower bound for a proper choice of the constant c'. □

5. Introducing a matrix $\mathcal{M}$.

From now on, we shall be concerned with a complete proof of the assertion (ii) of Theorem 3.1. Let us first recall its statement

PROPOSITION 5.1. *Let m and n be integers ≥ 2 such that $mn \geq 2m + n$. Let $x_1, \ldots, x_m$ and $y_1, \ldots, y_n$ be families of $\mathbf{Q}$-linearly independent complex numbers. Then, the transcendence degree over $\mathbf{Q}$ of the field*

$$\mathbf{Q}\left(y_\nu, e^{x_\mu y_\nu} ; 1 \leq \mu \leq m, 1 \leq \nu \leq n\right)$$

is ≥ 2.

We introduce the point

$$\underline{\theta} = (y_1, \ldots, y_n, e^{x_1 y_1}, \ldots, e^{x_m y_n}) \in \mathbf{C}^{n+mn},$$

after having chosen an arbitrary ordering of the coordinates. By the usual Gel'fond's theorem on the transcendence of α^β for any algebraic number α such that $\log \alpha \neq 0$ and any algebraic irrational number β, we know that $\underline{\theta} \notin \overline{\mathbf{Q}}^{n+mn}$. Fix an arbitrary algebraic curve $\mathcal{C} \subset \mathbf{C}^{n+mn}$ defined over $\mathbf{Q}$ and any nonzero derivation ∂ of the field $\mathbf{Q}(\mathcal{C})$. Suppose that the point $\underline{\theta}$ belongs to the curve $\mathcal{C}$. Then we shall construct a sequence of polynomials $(Q_N)_{N \geq 0}$ satisfying all the conditions prescribed in Theorem 4.1. We are led to a contradiction. The conclusion is that $\underline{\theta}$ cannot belong to any curve $\mathcal{C}$, which obviously implies Proposition 5.1. The remaining of this chapter is devoted to the construction of such a sequence $(Q_N)_{N \geq 0}$, together with some comments on related topics.

Let K, L, R be integers ≥ 1. We consider the $(K+1)\binom{L+m}{m}$ functions

$$\varphi(z) = z^k \exp\left(\left(\sum_{\mu=1}^m \ell_\mu x_\mu\right) z\right),$$

where

$$0 \leq k \leq K, \quad \sum_{\mu=1}^m \ell_\mu \leq L, \quad \ell_\mu \geq 0, \quad (1 \leq \mu \leq m),$$

and the R^n complex numbers

$$\zeta = \sum_{\nu=1}^n r_\nu y_\nu, \quad (0 \leq r_\nu < R, 1 \leq \nu \leq n).$$

Let us fix an ordering of the two sets $\{\varphi\}$ and $\{\zeta\}$ and define the matrix of evaluation

$$\mathcal{M} = \left(\varphi(\zeta)\right),$$

where the rows of $\mathcal{M}$ are indexed by the ordered set of functions $\{\varphi\}$ while the columns are indexed by the set of points $\{\zeta\}$. Explicitely, we have

$$\mathcal{M} = \left(\left(\sum_{\nu=1}^n r_\nu y_\nu\right)^k \prod_{\mu=1}^m \prod_{\nu=1}^n (e^{x_\mu y_\nu})^{\ell_\mu r_\nu}\right)_{\substack{(k, \ell_1, \ldots, \ell_m) \\ (r_1, \ldots, r_n)}}.$$

The entries of the matrix $\mathcal{M}$ belong clearly to the ring $\mathbf{Z}[\underline{\theta}]$. For suitable values K, L and R of the parameters, we shall verify in the next sections that a minor of maximal order from the associated matrix $\mathcal{M}$ is a convenient candidate for our polynomial Q.

6. The rank of the matrix $\mathcal{M}$.

In this section, we prove the following

PROPOSITION 6.1. *Assume $R^n > (m+1)^{n+1}KL^m$. Then, the matrix $\mathcal{M}$ has maximal rank equal to the number $M := (K+1)\binom{L+m}{m}$ of its rows.*

Up to the value of the constant factor $(m+1)^{n+1}$, this statement is best possible since the matrix $\mathcal{M}$ has R^n columns and $M \sim KL^m/m!$ rows. Let us introduce again the algebraic group $G = \mathbf{G}_a \times \mathbf{G}_m^m$ with exponential map

$$\mathbf{C}^{m+1} \overset{\exp_G}{\longrightarrow} \mathbf{C} \times (\mathbf{C}^*)^m, \qquad (z_0, z_1, \ldots, z_m) \longmapsto (z_0, e^{z_1}, \ldots, e^{z_m}),$$

and the analytic subgroup $\exp_G(V)$, where we have set

$$V = \mathbf{C}\underline{x}, \qquad \underline{x} = (1, x_1, \ldots, x_m).$$

Recall also that a connected algebraic subgroup of G is isomorphic to a product $\mathbf{G}_a^{\delta_0} \times \mathbf{G}_m^{\delta_1}$. We have shown in Section 3 that $\exp_G(V)$ is Zariski dense in G. The following lemma refines this result.

LEMMA 6.2. *For any proper connected algebraic subgroup G' of G, the intersection*

$$\exp_G(V) \cap G'(\mathbf{C})$$

is a free abelian group whose rank over $\mathbf{Z}$ is ≤ 1. This rank is one exactly when $G' \simeq \mathbf{G}_a \times \mathbf{G}_m^{m-1}$.

PROOF. We check the assertion for all possible exponents (δ_0, δ_1). If $\delta_0 = 0$, the subgroup $G'(\mathbf{C})$ is contained in $\{0\} \times (\mathbf{C}^*)^m$ and it is clear that

$$G'(\mathbf{C}) \cap \exp_G(V) = \{0\} \qquad \text{since} \qquad \exp_G(\underline{x}z) = (z, e^{x_1 z}, \ldots, e^{x_m z}).$$

Suppose now that $\delta_0 = 1$. Then $G'(\mathbf{C}) = \mathbf{C} \times \mathbf{T}$ for some algebraic torus $\mathbf{T} \subset (\mathbf{C}^*)^m$. The group $\mathbf{T}$ is defined by monomial equations

$$\mathbf{T} = \left\{ (Z_1, \ldots, Z_m) \in (\mathbf{C}^*)^m \, ; \, \prod_{\mu=1}^m Z_\mu^{t_\mu} = 1 \quad \text{for any } (t_1, \ldots, t_m) \in \mathcal{T} \right\},$$

where $\mathcal{T}$ is a subgroup of $\mathbf{Z}^m$ with rank $m - \delta_1 \geq 1$. Now, the point $\exp_G(z\underline{x})$ belongs to $G'(\mathbf{C})$ exactly when

$$z \left(\sum_{\mu=1}^m t_\mu x_\mu \right) \in 2\pi\sqrt{-1}\, \mathbf{Z} \quad \text{for any } (t_1, \ldots, t_m) \in \mathcal{T}.$$

If the rank of $\mathcal{T}$ is ≥ 2, we deduce from these relations that $z = 0$, since the complex numbers $x_1, \ldots, x_m$ are supposed to be linearly independent over $\mathbf{Z}$. In the remaining case $\delta_0 = 1$, $\delta_1 = m - 1$, we have clearly

$$\exp_G(V) \cap G'(\mathbf{C}) = \exp_G(\mathbf{Z}\underline{x}z') \quad \text{with} \quad z' = \frac{2\pi\sqrt{-1}}{\sum_{\mu=1}^m t_\mu x_\mu},$$

if $(t_1, \ldots, t_m)$ is a generator of $\mathcal{T}$. To conclude the proof, we have only to remark that the map $z \mapsto \exp_G(\underline{x}z)$ is injective. $\qquad\square$

We are now able to prove the proposition. Let us introduce the subgroup

$$\Gamma = \exp_G(\underline{x}Y) \subset G(\mathbf{C}) \qquad \text{where} \qquad Y = \oplus_{\nu=1}^n \mathbf{Z}y_\nu \,,$$

and for any real positive real number S, denote as usual

$$\Gamma(S) = \left\{ \exp_G\left(\underline{x}\left(\sum_{\nu=1}^n s_\nu y_\nu \right) \right) ; \quad 0 \le s_\nu < S \,, \nu = 1, \dots, n \right\}.$$

We will argue by contradiction, assuming that there is a non-trivial relation of linear dependence between the rows of $\mathcal{M}$. This translates into the existence of a nonzero polynomial of $\mathbf{C}[Z_0, Z_1, \dots, Z_m]$ with degree $\le K$ in Z_0 and degree $\le L$ in the variables $Z_1, \dots, Z_m$, which vanishes at all points of $\Gamma(R)$. Put $R' = R/(m+1)$. The sum $\Gamma(R') + \cdots + \Gamma(R')$ of $m+1$ copies of $\Gamma(R')$ in the group G is clearly contained in $\Gamma(R)$. Here is now the key point of the proof where we use a zero lemma. Under these assumptions, we can assert the existence of a connected algebraic subgroup $G' \simeq (\mathbf{G}_a^{\delta_0} \times \mathbf{G}_m^{\delta_1})$ of G, distinct from G, such that

(112) $$\frac{(\delta_0 + \delta_1)!}{\delta_0! \delta_1!} \operatorname{Card}\left(\frac{\Gamma(R') + G'}{G'} \right) \le (m+1) K^{1-\delta_0} L^{m-\delta_1}.$$

We refer to Proposition 8.1 of [**Wal8**] for a detailed proof of this assertion, and we postpone to the remark below some explanations about the meaning of the fundamental inequality (112). When $(\delta_0, \delta_1) \ne (1, m-1)$, we know by the lemma that $\exp_G(V) \cap G'(\mathbf{C}) = \{0\}$ so that

$$\operatorname{Card}\left(\frac{\Gamma(R') + G'}{G'} \right) = \operatorname{Card}(\Gamma(R')) \ge (R')^n = \frac{R^n}{(m+1)^n}.$$

Substituting this lower bound into (112), we get

$$\frac{R^n}{(m+1)^n} \le (m+1) K L^m$$

in contradiction with the hypothesis. In the remaining case $\delta_0 = 1, \delta_1 = m-1$, we know from the lemma that

$$\Gamma \cap G'(\mathbf{C}) \subseteq \exp_G(V) \cap G'(\mathbf{C}) = \exp_G(\mathbf{Z}\underline{x}z')$$

for some nonzero complex number z'. Now, we select $n-1$ elements from the set $\{y_1, \dots, y_n\}$ such that these $n-1$ complex numbers together with z' are linearly independent over $\mathbf{Z}$. We denote by Y' the subgroup in $\mathbf{C}$ generated by these $n-1$ elements and we set $\Gamma' = \exp_G(\underline{x}Y')$. Then, it is clear that $\Gamma' \cap G'(\mathbf{C}) = \{0\}$, so that

$$\operatorname{Card}\left(\frac{\Gamma(R') + G'}{G'} \right) \ge \operatorname{Card}(\Gamma'(R')) \ge (R')^{n-1} = \frac{R^{n-1}}{(m+1)^{n-1}}.$$

From the inequality (112) we deduce now that $R^{n-1} \le \frac{(m+1)^n}{m} L$. On the other hand, the hypothesis provides us the lower bound

$$R^{n-1} > (m+1)^{(n^2-1)/n} L^{(mn-m)/n} \ge (m+1)^{(n^2-1)/n} L,$$

which is contradictory with the upper bound since m and n are ≥ 2. $\square$

REMARK. The inequality (112) is a comparison of degrees, analogous to Bézout's theorem. We indicate here some recipe in order to recover easily (112). Let n_1 and n_2 be two integers ≥ 1 and let $A = \mathbf{Z}[e_1, e_2]$ be the commutative ring generated by e_1 and e_2 with relations $e_1^{n_1+1} = e_2^{n_2+1} = 0$. We graduate A by deciding that e_1 and e_2 have degree one. Then it is clear that $A_{n_1+n_2} = \mathbf{Z}e_1^{n_1}e_2^{n_2}$. By abuse of notation, we identify the element $ae_1^{n_1}e_2^{n_2}$ in $A_{n_1+n_2}$ with the integer a. To each irreducible subvariety $Z \subseteq \mathbf{P}_{\mathbf{C}}^{n_1} \times \mathbf{P}_{\mathbf{C}}^{n_2}$ of dimension δ, we associate

$$[Z] = \sum_{i+j=\delta} a_{i,j} e_1^{n_1-i} e_2^{n_2-j} \in A_{n_1+n_2-\delta},$$

where the coefficient $a_{i,j} = [Z]e_1^i e_2^j$ is the intersection number of Z with i generic linear varieties of the type $H_1 \times \mathbf{P}^{n_2}$ and j varieties of the shape $\mathbf{P}^{n_1} \times H_2$ and where H_1 and H_2 denote hyperplanes in $\mathbf{P}^{n_1}$ and $\mathbf{P}^{n_2}$ respectively. We extend by linearity the definition to cycles whose components have the same dimension δ. In this way $a_1 e_1 + a_2 e_2$ is associated to any hypersurface whose defining equation is a polynomial homogeneous of degree a_1 (resp. a_2) in the projective variables corresponding to the factor $\mathbf{P}^{n_1}$ (resp. $\mathbf{P}^{n_2}$). Moreover, when two varieties Z and Z' intersect properly and are of complementary dimensions $\delta + \delta' = n_1 + n_2$, the product $[Z][Z']$ is their intersection number.

Taking now $n_1 = 1$ and $n_2 = m$, we can reformulate more precisely (112) as an inequality in A_{m+1} :

(113)
$$\operatorname{Card}\left(\frac{\Gamma(R') + G'}{G'}\right)(Ke_1 + Le_2)^{\dim G'}[G']$$
$$\leq (Ke_1 + Le_2)^{m+1} = (m+1)KL^m.$$

Under the hypotheses of the zero lemma, we can construct a connected algebraic subgroup G' such that the union of all the translates of G' by elements of $\Gamma(R')$ are contained in an algebraic subset defined by polynomial equations of bidegree bounded respectively by K and L. Now (113) can be read as a Bézout's inequality in such a situation. For a proof in the homogeneous case, we refer to Chapter 11 in this volume, and to the original paper [**PPh3**] in the multihomogeneous case.

7. Analytic upper bound.

Now we go back to the notations from section 4. We assume that the point $\underline{\theta}$ belongs to an algebraic curve $\mathcal{C}$ and we select a nonzero derivation ∂ in the function field $\mathbf{Q}(\mathcal{C})$. Let K, L, R be parameters with $R^n \geq M = (K+1)\binom{L+m}{m}$. Let Δ be an $M \times M$ minor from $\mathcal{M}$. Since the entries of $\mathcal{M}$ belong to $\mathbf{Z}[\underline{\theta}]$, the number Δ also belongs to this ring. Let us write

$$\Delta = Q(\underline{\theta}) \quad \text{with} \quad Q \in \mathbf{Z}[X_1, \ldots, X_n, Y_1, \ldots, Y_{mn}].$$

As in section 3, we do not distinguish a polynomial Q with its restriction to $\mathcal{C}$, viewed as an element of $\mathbf{Q}(\mathcal{C})$. The aim of this section is to bound non trivially the values of derivatives $\partial^{[s]}Q$ at the point $\underline{\theta}$. We denote as usual by the symbol $\mathcal{O}(\ )$ some positive factor of the quantity $(\ )$ which is independent of the parameters K, L, R.

PROPOSITION 7.1. *For any integer σ with $0 \leq \sigma < s := [M/2]$, and any real number $\rho \geq 1$, we have the estimate*

$$\log |\partial^{[\sigma]} Q(\underline{\theta})| \leq -\frac{M^2 \log \rho}{8} + \mathcal{O}\Big(M\big(K \log(\rho R) + \rho L R\big)\Big).$$

PROOF. Let $(k_i, \ell_{1i}, \ldots, \ell_{mi})$, $(1 \leq i \leq M)$, be the row indices of the matrix $\mathcal{M}$, and let $(r_{1j}, \ldots, r_{nj})$, $(1 \leq j \leq M)$, be the indices of the M columns of $\mathcal{M}$ which enter into the minor Δ. Then, we have the formula

$$\Delta = \det\Big(\varphi_i(\zeta_j)\Big)_{1 \leq i,j \leq M} = \det\Big(Q_{i,j}(\underline{\theta})\Big)_{1 \leq i,j \leq M},$$

where

$$\varphi_i(z) = z^{k_i} \exp\left(\Big(\sum_{\mu=1}^{m} \ell_{\mu i} x_\mu\Big) z\right), \qquad \zeta_j = \sum_{\nu=1}^{n} r_{\nu j} y_\nu, \qquad (1 \leq i,j \leq M),$$

$$Q_{i,j}(X_1, \ldots, X_n, Y_1, \ldots, Y_{mn}) = \left(\sum_{\nu=1}^{n} r_{\nu j} X_\nu\right)^{k_i} \prod_{\mu=1}^{m} \prod_{\nu=1}^{n} (Y_{\mu,\nu})^{\ell_{\mu i} r_{\nu j}},$$

for $1 \leq i,j \leq M$. Note that obviously the partial degrees of the polynomials $Q_{i,j}$ are bounded above by K in the variables X_μ, by LR in the variables $Y_{\mu,\nu}$, $(1 \leq \mu \leq m, 1 \leq \nu \leq n)$, and that we have the following bound of height:

$$\log H(Q_{i,j}) \leq \mathcal{O}(K \log R).$$

Let us denote $Q = \det(Q_{i,j})$. We use now the well known rule of derivation of a determinant according to its lines. Leibniz's formula yields the expression

$$\partial^{[\sigma]} Q = \sum_{\sigma_1 + \cdots + \sigma_M = \sigma} \det\Big(\partial^{[\sigma_i]} Q_{i,j}\Big)_{\substack{1 \leq i \leq M \\ 1 \leq j \leq M}}.$$

For each summand $\det\big(\partial^{[\sigma_i]} Q_{i,j}\big)$, consider the set $I \subset \{1, \ldots, M\}$ of row indices i for which $\sigma_i = 0$. Since $\sigma_1 + \cdots + \sigma_M \leq s \leq M/2$, it is clear that the cardinality of I is $\geq M/2$. Laplace's expansion formula enables us to isolate the lines with index $i \in I$, so that we can write $\partial^{[\sigma]} Q$ as a linear combination

$$\partial^{[\sigma]} Q = \sum_{\substack{(I,J) \\ M/2 \leq |I| = |J| \leq M}} C_{I,J} \det\Big(Q_{i,j}\Big)_{\substack{i \in I \\ j \in J}},$$

whose coefficients $C_{I,J}$ are polynomial functions in the entries $\partial^{[\sigma_i]} Q_{i,j}$. It can be easily shown by induction on σ, that there exists a nonzero polynomial function A on $\mathcal{C}$ such that for any integer $\sigma \geq 0$ and any polynomial P, the product $A^\sigma \partial^{[\sigma]} P$ is the restriction on the curve $\mathcal{C}$ of some polynomial P_σ satisfying

$$\deg P_\sigma \leq \deg P + \mathcal{O}(\sigma), \qquad \log H(P_\sigma) \leq \log H(P) + \mathcal{O}(\sigma + \deg P).$$

For a detailed proof, we refer to Lemma 8 of [**LR1**] where the constants are computed explicitly for a specific derivation ∂. Here the involved polynomials A, P and P_σ have rational coefficients. For all subsets I and J as above, it follows that $A^s C_{I,J}$ is the restriction on $\mathcal{C}$ of a polynomial $P_{I,J}$ such that

$$\deg P_{I,J} \leq \mathcal{O}\Big(M(K + LR)\Big), \qquad \log H(P_{I,J}) \leq \mathcal{O}\Big(M(K \log R + LR)\Big).$$

Since $A(\underline{\theta})$ is nonzero, by evaluating the function $\partial^{[\sigma]}Q$ at the point $\underline{\theta}$, we obtain the formula

$$\partial^{[\sigma]}Q(\underline{\theta}) = \sum_{\substack{(I,J) \\ M/2 \le |I|=|J| \le M}} c_{I,J} \det\left(\varphi_i(\zeta_j)\right)_{\substack{i \in I \\ j \in J}},$$

with coefficients $c_{I,J}$ such that

$$\log|c_{I,J}| \le \mathcal{O}\left(M(K\log R + LR)\right).$$

On the other hand, if we put $\nu = \operatorname{Card} I = \operatorname{Card} J$, the function

$$z \mapsto \det\left(\varphi_i(z\zeta_j)\right)_{\substack{i \in I \\ j \in J}}$$

of the complex variable z admits a zero at the origin of multiplicity $\ge (\nu^2 - \nu)/2$ as easily seen by expanding the functions φ_i in Taylor's series at the origin and using the property of multilinearity of the determinant relative to its lines. The usual Schwarz' lemma then yields the upper bound

$$\log\left|\det\left(\varphi_i(\zeta_j)\right)_{\substack{i \in I \\ j \in J}}\right| \le -\frac{\nu^2 - \nu}{2}\log\rho + \log(\nu!)$$

$$+ \sum_{i \in I} \log \max_{|z|=\rho \max_j |\zeta_j|} |\varphi_i(z)|,$$

for any real number $\rho \ge 1$. See for example Lemma 7 of [**Lau2**] for more details on this argument which is fundamental to the method of interpolation determinants. The required estimate follows from the trivial bounds

$$\max_{|z|=\rho \max|\zeta_j|} |\varphi_i(z)| \le \max_{|z|=\rho R(\sum_\nu |y_\nu|)} |\varphi_i(z)|$$

$$\le \left(\rho R. \sum_\nu |y_\nu|\right)^K \times \exp\left(\rho LR. \sum_\mu |x_\mu|. \sum_\nu |y_\nu|\right)$$

$$\le \exp\left(\mathcal{O}\left(K\log(\rho R) + \rho LR\right)\right),$$

and the lower bound $\nu \ge M/2$. $\square$

8. Proof of Proposition 5.1.

We have only to verify that the polynomial Q introduced in the preceding section, satisfies all the conditions required in the criterion of algebraic independence, whenever the parameters K, L and R are chosen in a suitable sequence indexed by an integer denoted by N here. Actually the constraints are essentially the same as in the classical proof with auxiliary functions. For any integer $N \ge 3$, let us define

$$K = \left[N^{m+n}(\log N)^{-1/2}\right],$$

$$L = \left[N^{n-1}(\log N)^{1/2m}\right],$$

$$R = (m+1)^2 N^{m+1}.$$

The condition $R^n > (m+1)^{n+1}KL^m$ is then satisfied since $n \ge 2$. It follows from Proposition 6.1 that there exists some nonzero $M \times M$ minor $\Delta = Q(\underline{\theta})$ extracted from the matrix $\mathcal{M}$. Consequently, the restriction of the polynomial Q to the curve $\mathcal{C}$ is non identically zero. With this choice of parameters, remark that the dimension $M = (K+1)\binom{L+m}{m}$ is equivalent to $N^{mn+n}/m!$ when N tends to infinity. Proposition 7.1 with $\rho = N$ yields now the upper bound

$$\log\left|\partial^{[\sigma]}Q(\underline{\theta})\right| \le -\left(\frac{1}{8(m!)^2} - o(1)\right)N^{2mn+2n}\log N,$$

for all $\sigma \le [M/2]$. Define

$$d = MN^{m+n}(\log N)^{2/7}, \qquad t = MN^{m+n}(\log N)^{4/7}, \qquad s = [M/2].$$

From the crude estimates

$$\deg Q \le \mathcal{O}\Big(M(K + LR)\Big), \qquad \log H(Q) \le \mathcal{O}\Big(M(K\log R + \log M)\Big),$$

immediately deduced from section 7, we easily check that $\deg Q \le d$ and $\log H(Q) \le t$, for large values of N. On the other hand, the ratio

$$\frac{N^{2mn+2n}\log N}{8(m!)^2 dt/s}$$

tends to infinity with N provided $mn \ge 2m + n$, and becomes larger than the constant c from Theorem 4.1 when N is large enough. Finally, the growth's conditions appearing in Theorem 4.1 are obviously satisfied for N large enough.

CHAPTER 14

Algebraic Independence in Algebraic Groups. Part II: Large Transcendence Degrees

1. Introduction

This chapter is a continuation of chapter 13 on small transcendence degrees.

Our first goal is to introduce conjectures. We are not looking for the most general ones (see [**And2**]): we only propose some open problems which might be easier to prove, given the currently available methods.

Our next purpose is to present a proof of a result of algebraic independence which shows that some fields have a "large transcendence degree": these fields are generated by numbers of the form $\alpha^{\beta}, \alpha^{\beta^2}, \ldots$, when α and β are algebraic. We do not give the proof of the best known result on this topic (due to G. Diaz [**Dia2**]), but only of a weaker statement for which the arguments may look more transparent. We shall explain how to use the transcendence criterion (chapter 8) and the zero estimate (chapter 11) together with an auxiliary function.

2. Conjectures

2.1. Commutative Algebraic Groups. We keep the following notation already introduced in chapter 13. Let K be a subfield of $\mathbf{C}$ of transcendence degree t over $\mathbf{Q}$ and G a commutative connected algebraic group of dimension d defined over K. Assume $G = G_0 \times G_1 \times G_2$ where $G_0 = \mathbf{G}_{\mathrm{a}}^{d_0}$, $G_1 = \mathbf{G}_{\mathrm{m}}^{d_1}$ and the dimension d_2 of G_2 satisfies $d = d_0 + d_1 + d_2$. Denote by T_G the tangent space at the origin of the algebraic group G. The set $T_G(K)$ of K-rational points of T_G is a K-vector space of dimension d, and the set $T_G(\mathbf{C})$ of complex points of T_G is the Lie algebra of the Lie group $G(\mathbf{C})$. Let $\exp_G : T_G(\mathbf{C}) \longrightarrow G(\mathbf{C})$ denote the exponential map of $G(\mathbf{C})$. Let V be a subspace of $T_G(\mathbf{C})$ of dimension $n < d$ such that $\exp_G V$ is Zariski dense in $G(\mathbf{C})$. Let $Y = \mathbf{Z}\eta_1 + \cdots + \mathbf{Z}\eta_\ell$ be a finitely generated subgroup of V of rank $\ell \geq 1$ such that $\exp_G Y \subset G(K)$.

A $\mathbf{C}$-vector subspace W of $T_G(\mathbf{C})$ is *defined over K* if it is spanned by a K-vector subspace of $T_G(K)$.

2.1.1. *One Parameter Subgroups.* We first assume $n = 1$ — this is the situation which is considered in chapter 13.

CONJECTURE 2.1. *Assume $d\ell > \ell + d_1 + 2d_2$. Then*

$$ t > \frac{d\ell}{\ell + d_1 + 2d_2} - 1. $$

(∗) Chapter's author : Michel WALDSCHMIDT.

EXAMPLE. Theorem 2.1 in chapter 13 proves the following special case of Conjecture 2.1 concerning small transcendence degrees:

$$d\ell \geq 2(\ell + d_1 + 2d_2) \Longrightarrow t \geq 2.$$

Conjecture 2.1 is related with Schneider's solution of Hilbert's seventh problem. Here is a conjecture related with Gel'fond's solution of the same problem.

CONJECTURE 2.2. *Assume further that V is defined over K. Then*

$$t > \frac{(d-1)\ell}{\ell + d_1 + 2d_2}.$$

EXAMPLE. A consequence of Conjecture 2.2 for small transcendence degree is

$$(d-1)\ell \geq \ell + d_1 + 2d_2 \Longrightarrow t \geq 2.$$

The state of the art on this question is described in chapter 13, § 2: the result is proved in the case $d_0 = 0$, while for $d_0 = 1$ the conclusion $t \geq 2$ is reached only under the stronger assumption $(d-1)\ell > \ell + d_1 + 2d_2$.

Conjectures 2.1 and 2.2 in the special case of a linear algebraic group reduce to the following one, which includes also the Four Exponentials Conjecture:

CONJECTURE 2.3. *Let $x_1, \ldots, x_d$ be complex numbers which are $\mathbf{Q}$-linearly independent, $y_1, \ldots, y_\ell$ be also $\mathbf{Q}$-linearly independent complex numbers and K a subfield of $\mathbf{C}$ which contains the $d\ell$ numbers $e^{x_i y_j}$ $(1 \leq i \leq d, 1 \leq j \leq \ell)$. Assume $d \geq 2$ and $\ell \geq 2$. Denote by t the transcendence degree over $\mathbf{Q}$ of K. Then*

$$t > \frac{d\ell}{\ell + d} - 1.$$

Moreover the transcendence degree t_1 of the field $K_1 = K(x_1, \ldots, x_d)$ is bounded from below by

$$t_1 > \frac{(d-1)\ell}{\ell + d}$$

and the transcendence degree t_2 of the field $K_2 = K_1(y_1, \ldots, y_\ell)$ by

$$t_2 > \frac{d\ell}{\ell + d}.$$

2.1.2. *Several Parameters Subgroups.* We consider now the general case $n \geq 1$. We define

$$\mu = \mu(Y, V) = \min_{V' \neq V} \left\{ \frac{\operatorname{rank}_{\mathbf{Z}}(Y/Y \cap V')}{\dim_{\mathbf{C}}(V/V')} \right\}$$

where V' runs over the set of $\mathbf{C}$-vector subspaces of V of dimension $< n$. Hence $\mu \leq \ell/n$ with $\ell = \operatorname{rank}_{\mathbf{Z}} Y$. On the other hand the condition $\mu > 0$ means that V is the $\mathbf{C}$-vector space spanned by Y; in this case $\mu \geq 1$. For a subgroup Y of rank $\ell \geq n$ of V, the condition $\mu(Y, V) = \ell/n$ is not a very strong assumption; roughly speaking, it only means that no set of $\mathbf{Q}$-linearly independent points of Y is contained in a subspace of V of too small a dimension.

Here is the extension of Conjecture 2.1 to n variables:

CONJECTURE 2.4. *Assume $d\mu > n\mu + d_1 + 2d_2$. Then*

$$t > \frac{d\mu}{n\mu + d_1 + 2d_2} - 1.$$

Conjecture 2.4 corresponds to Schneider's method in several variables. We can extend Conjecture 2.2 as follows: *if V is defined over K and $\mu > 0$, then*

$$t > \frac{(d-n)\mu}{n\mu + d_1 + 2d_2}.$$

This corresponds to Gel'fond's method in several variables. However there are intermediate situations which may be compared with Baker's method:

CONJECTURE 2.5. *Assume that there exists a* **C**-*vector subspace W of $T_G(\mathbf{C})$, of dimension $n' < d$, which is defined over K and contains Y. Assume also $\mu > 0$. Then*

$$t > \frac{(d-n')\mu}{n\mu + d_1 + 2d_2}.$$

REMARK. Several variations are interesting. In particular one may expect refinements of the conjectural estimates for the transcendence degree when Y contains periods of $\exp_G$. For this purpose it is useful to introduce a new parameter $\kappa = \mathrm{rank}_{\mathbf{Z}}(Y \cap \ker \exp_G)$ (see [**Wal7**], §16).

2.2. Results: Large Transcendence Degree for the Values of the Exponential Function in One Variable. From now on we restrict the discussion to one parameter subgroups of linear algebraic groups.

Partial results are known concerning the above conjectures, under a so called "Technical Hypothesis". For simplicity, we shall use only the following assumption, which is a measure of linear independence. One should stress that most known results actually involve a much weaker hypothesis, while much sharper estimates are valid for concrete applications. Therefore this condition is in fact not too strong.

DEFINITION 2.6. We shall say that a set $\{u_1, \ldots, u_n\}$ of **Q**-linearly independent complex numbers *satisfies the Technical Hypothesis* (T.H.) if, for any $\epsilon > 0$, there exists a positive number H_0 such that, for any $H \geq H_0$ and n-tuple $(h_1, \ldots, h_n)$ of rational integers satisfying $0 < \max\{|h_1|, \ldots, |h_n|\} \leq H$, the inequality

$$|h_1 u_1 + \cdots + h_n u_n| \geq \exp\{-H^\epsilon\}$$

holds.

Here is the main result of G. Diaz in [**Dia2**] :

THEOREM 2.7. *Let $x_1, \ldots, x_d$ be complex numbers which are* **Q**-*linearly independent and satisfy* (T.H.) *and $y_1, \ldots, y_\ell$ be also complex numbers which are* **Q**-*linearly independent and satisfy* (T.H.). *Let K a subfield of* **C** *which contains the $d\ell$ numbers $e^{x_i y_j}$ $(1 \leq i \leq d, 1 \leq j \leq \ell)$. Assume $d\ell > \ell + d$ and denote by t the transcendence degree over* **Q** *of K. Then*

$$t > \frac{d\ell}{\ell + d} - 1.$$

Moreover the transcendence degree t_1 of the field $K_1 = K(x_1, \ldots, x_d)$ is bounded from below by

$$t_1 > \frac{(d-1)\ell}{\ell + d},$$

and the transcendence degree t_2 of the field $K_2 = K_1(y_1, \ldots, y_\ell)$ by

$$t_2 \geq \frac{d\ell}{\ell + d}.$$

REMARK. For $t \in \mathbf{Z}$ and $x \in \mathbf{R}$, we have $t > x - 1$ if and only if $t \geq [x]$. For instance the conclusions for t and t_1 can be written

$$t \geq \left[\frac{d\ell}{\ell + d}\right] \quad \text{and} \quad t_1 \geq \left[\frac{d(\ell + 1)}{\ell + d}\right].$$

In some cases G. Diaz [**Dia1**] succeeds to prove also

$$t_2 > \frac{d\ell}{\ell + d}, \quad \text{that is} \quad t_2 \geq \left[\frac{d\ell}{\ell + d}\right] + 1.$$

The lower bound for t_1 yields the following partial answer to the Gel'fond-Schneider problem on the algebraic independence of α^{β^i}:

COROLLARY 2.8. *Let α be a non zero complex algebraic number, $\log \alpha$ a non zero logarithm of α and β an algebraic number of degree $d \geq 2$. For $z \in \mathbf{C}$, define $\alpha^z := \exp(z \log \alpha)$. Then*

$$\mathrm{trdeg}_{\mathbf{Q}}\mathbf{Q}\left(\alpha^{\beta}, \dots, \alpha^{\beta^{d-1}}\right) \geq \left[\frac{d+1}{2}\right].$$

PROOF. Take $\ell = d$ and define

$$x_i = \beta^{i-1} \quad (1 \leq i \leq d) \quad \text{and} \quad y_j = \beta^{j-1} \log \alpha \quad (1 \leq j \leq d). \quad \square$$

As pointed out in [**Nes7**], A.O. Gel'fond already anounced Corollary 2.8 in 1948 [**Gel1**], but his papers [**Gel2**] and [**Gel3**] contain a proof only of the small transcendence degree result: *for $d \geq 3$, the transcendence degree is at least 2* (see chapter 13). Hence Corollary 2.8, claimed by Gel'fond, has been proved by Diaz only 40 years later.

2.3. Historical Sketch. We refer to [**FN**], Chapter 6 for a survey on *Gel'-fond's method for algebraic independence and its recent developments* together with plenty of references. See also [**Wal5**] and [**Wal11**].

The first step is due to A.O. Gel'fond at the end of the 40's [**Gel1**], [**Gel2**], who proved results of small transcendence degree under a technical hypothesis (T.H.) for $x_1, \dots, x_d$ as well as $y_1, \dots, y_\ell$. This technical hypothesis was removed by R. Tijdeman 1970 [**Tij2**], by means of a sharper analytic zero estimate for exponential polynomials [**Tij1**].

Here is the state of our knowledge concerning "small transcendence degree" for the values of the exponential function in one variable:

THEOREM 2.9. *Let $x_1, \dots, x_d$ be complex numbers which are $\mathbf{Q}$-linearly independent and $y_1, \dots, y_\ell$ be also complex numbers which are $\mathbf{Q}$-linearly independent. Denote by t the transcendence degree of the field K generated over $\mathbf{Q}$ by the $d\ell$ numbers $e^{x_i y_j}$ $(1 \leq i \leq d, 1 \leq j \leq \ell)$, by t_1 the transcendence degree of the field $K_1 = K(x_1, \dots, x_d)$ and by t_2 the transcendence degree of the field $K_2 = K_1(y_1, \dots, y_\ell)$. Then*

$$d\ell \geq 2(\ell + d) \Rightarrow t \geq 2$$
$$d\ell \geq d + 2\ell \Rightarrow t_1 \geq 2$$
$$d\ell > \ell + d \Rightarrow t_2 \geq 2$$

Moreover, if $d = \ell = 2$, and if the two numbers $e^{x_1 y_1}$ and $e^{x_1 y_2}$ are algebraic, then $t_2 \geq 2$.

Further developments are described in W.D. Brownawell's survey [**Bro2**]. These are the only known results which do not require a Technical Hypothesis.

The first results on large transcendence degree are due to G.V. Chudnovsky [**Chu2**] who proved, under the assumptions of Theorem 2.7, the lower bound $2^t \geq d\ell/(\ell + d)$. Chudnovsky's method has been worked out by P. Warkentin, P. Philippon, E. Reyssat, R. Endell, W.D. Brownawell and Yu.V. Nesterenko (see [**Wal5**]). P. Philippon introduced the trick of *redundant variables* (which is a variant of Landau's method, that is an homogeneity argument which we are going to describe in §3.2). By means of a sharp criterion of his own [**PPh2**](cf. chapter 8), he then succeeded to replace 2^t by $t + 1$ in Chudnovsky's result and to prove, under (T.H.), the inequalities

$$t \geq \frac{d\ell}{\ell + d} - 1, \quad t_1 \geq \frac{(d-1)\ell}{\ell + d} \quad \text{and} \quad t_2 \geq \frac{d\ell}{\ell + d}.$$

Finally G. Diaz [**Dia2**] obtained Theorem 2.7, which is the sharpest known result to date, apart from a weakening of the technical hypothesis which arises from the work of W.D. Brownawell [**Bro5**], and also apart from the quantitative aspect of the subject (see [**Ably**], [**Jab**], [**Del**], [**Cav**] and [**FN**]). Notice also that in [**Nes7**] Yu.V. Nesterenko removed the use of Philippon's criterion from the proof of Theorem 2.7.

3. Proofs

3.1. Statement. We shall prove the following result:

THEOREM 3.1. *Let α be a positive real number, $\alpha \neq 1$. Let β be a real algebraic number of degree $d \geq 2$. Then the transcendence degree t of the field $\mathbf{Q}\left(\alpha, \alpha^\beta, \dots, \alpha^{\beta^{d-1}}\right)$ over $\mathbf{Q}$ satisfies*

$$t \geq \frac{d}{2} - 1$$

REMARK. The proof of this result will involve several variables. As we shall see, the same method restricted to functions of a single variable yields only the weaker estimate

$$t \geq \frac{d}{4} - \frac{1}{2}.$$

The refinement to $(d/2) - 1$ will come from Philippon's "redundant variables".

We are going to use Schneider's method for the torus $\mathbf{G}_m^d$. If we were using either Gel'fond's method (i.e. including derivatives) for $\mathbf{G}_m^d$, or else Schneider's method for the product $\mathbf{G}_a \times \mathbf{G}_m^d$, we would get $t \geq (d-1)/2$. In order to reach $t > (d-1)/2$, Diaz [**Dia2**] replaces the auxiliary polynomial $P \in \mathbf{Z}[X_1, \dots, X_d]$ by a polynomial with coefficients in the ring generated over $\mathbf{Z}$ by the numbers $e^{x_i y_j}$. Finally the restrictions $\beta \in \mathbf{R}$ and $\alpha > 0$ provide a slight simplification, but is it an easy exercise to remove them.

3.2. Tools. Let $\underline{x}_1, \dots, \underline{x}_d$ be elements in $\mathbf{C}^n$ and also $\underline{y}_1, \dots, \underline{y}_\ell$ elements in $\mathbf{C}^n$. Denote by $\underline{x}_i \underline{y}_j$ the standard scalar product in $\mathbf{C}^n$. Let K be a subfield of $\mathbf{C}$ of transcendence degree t over $\mathbf{Q}$ containing all numbers $e^{\underline{x}_i \underline{y}_j}$.

Goal: Under "suitable assumptions", we want to prove

$$t \geq \frac{d\ell + \ell + d}{(n+1)(\ell + d)} - 1.$$

Here is how Philippon's redundant variables occur. For a positive integer k, we take the k-th Cartesian powers and we replace n by kn, d by kd, ℓ by $k\ell$: for large k, we deduce

$$t \geq \frac{d\ell}{n(\ell + d)} - 1.$$

Of course one shall need to check that the above mentioned "suitable assumptions" are satisfied for Cartesian products.

3.2.1. *Criterion of Algebraic Independence.* Our first tool is the following special case of Philippon's criterion for algebraic independence [**PPh2**] and chapter 8, § 1. We denote by $\mathrm{H}(P)$ the *usual height of a polynomial P* with, say, complex coefficients, that is the maximum absolute value of its coefficients.

PROPOSITION 3.2. *Let $a \geq 1$ be a real number and $\theta = (\theta_1, \ldots, \theta_q)$ an element in $\mathbf{C}^q$. There exists a positive number C having the following property. Assume that for all sufficiently large integer N, there exist a positive integer $m = m(N) \geq 1$ and polynomials $Q_{N1}, \ldots, Q_{Nm}$ in $\mathbf{Z}[X_1, \ldots, X_q]$ with*

$$\max_{1 \leq j \leq m} \deg Q_{Nj} \leq N, \qquad \max_{1 \leq j \leq m} \mathrm{H}(Q_{Nj}) \leq e^N$$

and

(114)
$$\max_{1 \leq j \leq m} |Q_{Nj}(\theta_1, \ldots, \theta_q)| \leq e^{-CN^a},$$

such that the polynomials $Q_{N1}, \ldots, Q_{Nm}$ have no common zero in the domain

$$\left\{ z \in \mathbf{C}^q ; \max_{1 \leq i \leq q} |z_i - \theta_i| \leq e^{-3CN^a} \right\}.$$

Then the transcendence degree t of the field $\mathbf{Q}(\theta_1, \ldots, \theta_q)$ over $\mathbf{Q}$ satisfies

$$t > a - 1.$$

3.2.2. *Auxiliary Function.* For $r > 0$, and for an entire function φ of n variables, we denote by $|\varphi|_r$ the number

$$|\varphi|_r = \sup \left\{ |\varphi(\underline{z})| ; \underline{z} = (z_1, \ldots, z_n) \in \mathbf{C}^n, \max_{1 \leq i \leq n} |z_i| \leq r \right\}.$$

PROPOSITION 3.3. *Let M be a positive integer, r, Δ, U be positive real numbers and $\varphi_1, \ldots, \varphi_M$ entire functions in $\mathbf{C}^n$. Assume*

(115)
$$(8U)^{n+1} \leq M\Delta, \qquad \Delta \leq U$$

and

$$\sum_{\mu=1}^{M} |\varphi_\mu|_{er} \leq e^U.$$

Then there exists rational integers $p_1, \ldots, p_M$ in $\mathbf{Z}$ with

$$0 < \max\{|p_1|, \ldots, |p_M|\} \leq e^\Delta$$

such that the function $F = p_1\varphi_1 + \cdots + p_M\varphi_M$ satisfies

$$|F|_r \leq e^{-U}.$$

SKETCH OF PROOF. Let $T \in \mathbf{Z}$ satisfy $4U \leq T < 4U + 1$. Use Dirichlet's pigeonhole principle to solve the following system of inequalities involving M unknowns $p_1, \ldots, p_M$ in $\mathbf{Z}$:

$$\sum_{\|\tau\| < T} \frac{r^{\|\tau\|}}{\tau!} \cdot \left| \frac{d^\tau}{dz^\tau} F(0) \right| \leq \frac{1}{2} e^{-U}$$

for $F = p_1 \varphi_1 + \cdots + p_M \varphi_M$. Next use an interpolation formula

$$|F|_r \leq \left(1 + \sqrt{T}\right) e^{-T} |F|_{er} + \sum_{\|\tau\| < T} \frac{r^{\|\tau\|}}{\tau!} \cdot \left| \frac{d^\tau}{dz^\tau} F(0) \right|.$$

$\square$

(For more details, see [**Wal4**]).

3.2.3. Zeros Estimate.

a) Degree of hypersurfaces and condition (Z.E.)

NOTATION. For Σ a finite subset of $\mathbf{C}^m$, define $\omega(\Sigma)$ as the smallest total degree of a non zero polynomial in $\mathbf{C}[z_1, \ldots, z_m]$ which vanishes on Σ:

$$\omega(\Sigma) = \min\{\deg P \, ; \, P \in \mathbf{C}[z_1, \ldots, z_m], \, P \neq 0, \, P(\sigma) = 0 \text{ for any } \sigma \in \Sigma\}.$$

DEFINITION 3.4. Let $X = \mathbf{Z}\underline{x}_1 + \cdots + \mathbf{Z}\underline{x}_d$ and $Y = \mathbf{Z}\underline{y}_1 + \cdots + \mathbf{Z}\underline{y}_\ell$ be two finitely generated subgroups of $\mathbf{C}^n$ of rank d and ℓ respectively. Denote by $\theta_1, \ldots, \theta_q$ the distinct elements of the set

$$\left\{ e^{\underline{x}_i \underline{y}_j} \, ; \, 1 \leq i \leq d, \, 1 \leq j \leq \ell \right\}.$$

We shall say that (X, Y) satisfies the condition (Z.E.) if for each $\eta > 0$ there exists a positive real number $R_0 > 0$ with the following property: for any positive integer $R \geq R_0$ and any q-tuple $(\mu_1, \ldots, \mu_q)$ of complex numbers satisfying

$$\max_{1 \leq h \leq q} \left| \theta_h - \mu_h \right| < e^{-R^\eta},$$

if we set $\mu_{ij} = \mu_h$ for $e^{\underline{x}_i \underline{y}_j} = \theta_h$ and

$$\Sigma = \left\{ \left(\prod_{j=1}^{\ell} \mu_{ij}^{r_j} \right)_{1 \leq i \leq d} \, ; \, 0 \leq r_j < R \, (1 \leq j \leq \ell) \right\} \subset (\mathbf{C}^\times)^d,$$

then

$$\omega(\Sigma) \geq (R/d)^{\ell/d}.$$

One should remark that this condition (Z.E.) is not only about (X, Y) but rather concerns $(\underline{x}_1, \ldots, \underline{x}_d)$ and $(\underline{y}_1, \ldots, \underline{y}_\ell)$. However in our situation the choice of the bases of these $\mathbf{Z}$-modules will be clear from the context.

We shall check that the condition (Z.E.) holds for $n = 1$ when the two tuples of real numbers $(x_1, \ldots, x_d)$ and $(y_1, \ldots, y_\ell)$ satisfy (T.H.). Next we show that this condition (Z.E.) is stable under Cartesian products.

b) *Consequence of the Technical Hypothesis*

LEMMA 3.5. *Let* $x_1, \ldots, x_d$ *in* $\mathbf{R}$ *satisfy* (T.H.), *and let* $y_1, \ldots, y_\ell$ *in* $\mathbf{R}$ *also satisfy* (T.H.). *For each* $\eta_0 > 0$ *there exists* $L_0 > 0$ *and* $R_0 > 0$ *with the following property. Let* $L \geq L_0$ *and* $R \geq R_0$ *be positive integers and let* μ_{ij} *be complex numbers* $(1 \leq i \leq d, 1 \leq j \leq \ell)$ *satisfying*

$$\max_{\substack{1 \leq i \leq d \\ 1 \leq j \leq \ell}} \left| e^{x_i y_j} - \mu_{ij} \right| < e^{-(LR)^{\eta_0}}.$$

Then for any $\underline{\lambda} = (\lambda_1, \ldots, \lambda_d) \in \mathbf{Z}^d$ *and* $\underline{r} = (r_1, \ldots, r_\ell) \in \mathbf{Z}^\ell$ *with*

$$0 < \max_{1 \leq i \leq d} |\lambda_i| \leq L \quad and \quad 0 < \max_{1 \leq j \leq \ell} |r_j| \leq R$$

we have

$$\prod_{i=1}^{d} \prod_{j=1}^{\ell} \mu_{ij}^{\lambda_i r_j} \neq 1.$$

PROOF. We start the proof with the following remark. Let v and w be two complex numbers which satisfy $|we^{-v} - 1| \leq 1/2$. Set $z = v + \log(we^{-v})$, where log denotes the principal branch of the logarithm. Then $e^z = w$ and $|z - v| \leq 2|we^{-v} - 1|$.

Therefore for $1 \leq i \leq d$ and $1 \leq j \leq \ell$ we can find a complex logarithm z_{ij} of μ_{ij} such that

$$\max_{\substack{1 \leq i \leq d \\ 1 \leq j \leq \ell}} |x_i y_j - z_{ij}| < e^{-(1/2)(LR)^{\eta_0}}.$$

Suppose

$$\prod_{i=1}^{d} \prod_{j=1}^{\ell} \mu_{ij}^{\lambda_i r_j} = 1.$$

Since the imaginary part of z_{ij} has absolute value $< e^{-(1/2)(LR)^{\eta_0}}$, and since, for sufficiently large L and R,

$$d\ell LR e^{-(1/2)(LR)^{\eta_0}} < 2\pi,$$

the absolute value of the imaginary part of the number

$$\sum_{i=1}^{d} \sum_{j=1}^{\ell} \lambda_i r_j z_{ij}$$

is $< 2\pi$. Therefore

$$\sum_{i=1}^{d} \sum_{j=1}^{\ell} \lambda_i r_j z_{ij} = 0.$$

We deduce

$$\left| \sum_{i=1}^{d} \lambda_i x_i \right| \left| \sum_{j=1}^{\ell} r_j y_j \right| = \left| \sum_{i=1}^{d} \sum_{j=1}^{\ell} \lambda_i r_j x_i y_j \right|$$

$$\leq \sum_{i=1}^{d} \sum_{j=1}^{\ell} \lambda_i r_j |x_i y_j - z_{ij}|$$

$$\leq d\ell LR e^{-(1/2)(LR)^{\eta_0}} < e^{-(1/3)(LR)^{\eta_0}}$$

for L and R large enough. Now, for L and R sufficiently large, the left hand side is bounded from below by $e^{-L^{\eta_0} - R^{\eta_0}}$, and we get the desired contradiction as soon as $3(L^{\eta_0} + R^{\eta_0}) < (LR)^{\eta_0}$.   □

c)*Single Variable: $n = 1$*
Applying Philippon's zero estimate (theorem 5.1 from chapter 11), we now prove:

PROPOSITION 3.6. *Let $x_1, \ldots, x_d$ in $\mathbf{R}$ satisfy (T.H.), and let $y_1, \ldots, y_\ell$ in $\mathbf{R}$ also satisfy (T.H.). Let $X = \mathbf{Z}\underline{x}_1 + \cdots + \mathbf{Z}\underline{x}_d$ and $Y = \mathbf{Z}\underline{y}_1 + \cdots + \mathbf{Z}\underline{y}_\ell$. Then (X, Y) satisfies the condition (Z.E.).*

PROOF. Let η be a positive real number, R a sufficiently large positive real number and $\mu_1, \ldots, \mu_q$ complex numbers satisfying

$$\max_{1 \leq h \leq q} |\theta_h - \mu_h| < e^{-R^\eta}.$$

For $1 \leq j \leq \ell$ and $1 \leq i \leq d$, define μ_{ij} by $\mu_{ij} = \mu_h$ where $h \in \{1, \ldots, q\}$ is the index for which $e^{x_i y_j} = \theta_h$. Moreover, for $\underline{r} = (r_1, \ldots, r_\ell) \in \mathbf{Z}^\ell$ and $1 \leq i \leq d$, define

$$\mu_{i\underline{r}} = \prod_{j=1}^{\ell} \mu_{ij}^{r_j}.$$

Let

$$\Sigma = \left\{ (\mu_{1\underline{r}}, \ldots, \mu_{d\underline{r}}) \,;\, 0 \leq r_j < R \, (1 \leq j \leq \ell) \right\} \subset (\mathbf{C}^\times)^d.$$

Assume $\omega(\Sigma) < (R/d)^{\ell/d}$, there exists a non zero polynomial $P \in \mathbf{C}[X_1, \ldots, X_d]$ of total degree $D < (R/d)^{\ell/d}$ such that the R^ℓ numbers

$$P(\mu_{1\underline{r}}, \ldots, \mu_{d\underline{r}}) \qquad (0 \leq r_j < R, \ 1 \leq j \leq \ell)$$

vanish. We use theorem 5.1 from chapter 11 with $G = \mathbf{G}_m^d$, $W = 0$, $c = 1$,

$$\Sigma_0 = \left\{ (\mu_{1\underline{r}}, \ldots, \mu_{d\underline{r}}) \,;\, 0 \leq r_j < R/d \, (1 \leq j \leq \ell) \right\} \subset \Sigma \subset (\mathbf{C}^\times)^d.$$

Since P vanishes on the set

$$\Sigma \supset \Sigma_0(d) = \left\{ \sigma_1 \cdots \sigma_d \,;\, (\sigma_1, \ldots, \sigma_d) \in \Sigma^d \right\},$$

we deduce that there exists a connected algebraic subgroup H of G, of dimension $< d$, such that

$$\text{Card}\big((\Sigma_0 + H)/H\big) \mathcal{H}(H, D) \leq \mathcal{H}(G, D).$$

Here $\mathcal{H}(G, D) = D^d$. Moreover H is contained in a hypersurface of equation

$$X_1^{\lambda_1} \cdots X_d^{\lambda_d} = 1,$$

where $(\lambda_1, \ldots, \lambda_d) \in \mathbf{Z}^d$ satisfies $0 < \max_{1 \leq i \leq d} |\lambda_i| \leq D$.
    Define $\eta_0 = \eta d/(\ell + d)$ and $L = D$, so that

$$(LR)^{\eta_0} \leq R^\eta.$$

Using the condition (T.H.), we may apply Lemma 3.5: we get an injective mapping from the set

$$\{ (r_1, \ldots, r_\ell) \,;\, 0 \leq r_j < R/d \, (1 \leq j \leq \ell) \}$$

into $(\Sigma_0 + H)/H$ by mapping $(r_1, \ldots, r_\ell)$ onto the image of $(\mu_{1\underline{r}}, \ldots, \mu_{d\underline{r}})$ in $(\Sigma_0 + H)/H$. Therefore

$$\text{Card}\big((\Sigma_0 + H)/H\big) \geq (R/d)^\ell > D^d,$$

which yields the desired contradiction. □

d) *Cartesian Products*

The next lemma is well known (but we include the easy proof).

LEMMA 3.7. *Let* $\Sigma_1, \ldots, \Sigma_k$ *be finite subsets of* $\mathbf{C}^n$. *Then*

$$\omega(\Sigma_1 \times \cdots \times \Sigma_k) = \min_{1 \le h \le k} \omega(\Sigma_h).$$

PROOF. The lower bound

$$\omega(\Sigma_1 \times \cdots \times \Sigma_k) \le \min_{1 \le h \le k} \omega(\Sigma_h).$$

is easy: if $h_0 \in \{1, \ldots, k\}$ satisfies $\omega(\Sigma_{h_0}) = \min_{1 \le h \le k} \omega(\Sigma_h)$ and if $P \in \mathbf{C}[\underline{z}]$ is a non zero polynomial in n variables of total degree $\omega(\Sigma_{h_0})$ which vanishes on Σ_{h_0}, then the image of P in $\mathbf{C}[\underline{z}_1, \ldots, \underline{z}_k]$ under the morphism $\mathbf{C}[\underline{z}] \longrightarrow \mathbf{C}[\underline{z}_1, \ldots, \underline{z}_k]$ which maps $\underline{z}$ onto $\underline{z}_{h_0}$ is a non zero polynomial in nk variables of total degree $\omega(\Sigma_{h_0})$ which vanishes on $\Sigma_1 \times \cdots \times \Sigma_k$.

We prove the upper bound

$$\omega(\Sigma_1 \times \cdots \times \Sigma_k) \ge \min_{1 \le h \le k} \omega(\Sigma_h).$$

by induction on k. For $k = 1$ there is nothing to prove. Let $k \ge 2$. Assume that the result is true for $k - 1$. Let $P \in \mathbf{C}[\underline{z}_1, \ldots, \underline{z}_k]$ be a polynomial of total degree $< \min_{1 \le h \le k} \omega(\Sigma_h)$ which vanishes on $\Sigma_1 \times \cdots \times \Sigma_k$. For each $\underline{z} \in \mathbf{C}^n$, the polynomial $P_{\underline{z}} = P(\underline{z}_1, \ldots, \underline{z}_{k-1}, \underline{z}) \in \mathbf{C}[\underline{z}_1, \ldots, \underline{z}_{k-1}]$ has total degree $< \min_{1 \le h \le k-1} \omega(\Sigma_h)$ and vanishes on $\Sigma_1 \times \cdots \times \Sigma_{k-1}$. By the induction hypothesis $P_{\underline{z}} = 0$. Hence for each $(\underline{z}_1, \ldots, \underline{z}_{k-1}) \in \mathbf{C}^{n(k-1)}$ the polynomial $P(\underline{z}_1, \ldots, \underline{z}_{k-1}, \underline{z}) \in \mathbf{C}[\underline{z}]$ has total degree $< \omega(\Sigma_k)$ and vanishes on Σ_k. From the definition of $\omega(\Sigma_k)$ it follows that this polynomial is 0, hence $P = 0$. □

PROPOSITION 3.8. *Let* $X = \mathbf{Z}\underline{x}_1 + \cdots + \mathbf{Z}\underline{x}_d$ *and* $Y = \mathbf{Z}\underline{y}_1 + \cdots + \mathbf{Z}\underline{y}_\ell$ *be two finitely generated subgroups of* $\mathbf{C}^n$ *of rank* d *and* ℓ *respectively such that* (X, Y) *satisfies the condition* (Z.E.). *Let* k *be a positive integer. Define* X^k *and* Y^k *in* $\mathbf{C}^{nk}$ *by*

$$X^k = \sum_{h=1}^{k} \sum_{i=1}^{d} \mathbf{Z}\underline{x}_{hi} \quad and \quad Y^k = \sum_{h=1}^{k} \sum_{j=1}^{\ell} \mathbf{Z}\underline{y}_{hj}$$

where, for $1 \le h \le k$,

$$\underline{x}_{hi} = (\delta_{h1}\underline{x}_i, \ldots, \delta_{hk}\underline{x}_i) \in \mathbf{C}^{nk} \qquad (1 \le i \le d)$$

and

$$\underline{y}_{hj} = (\delta_{h1}\underline{y}_j, \ldots, \delta_{hk}\underline{y}_j) \in \mathbf{C}^{nk} \qquad (1 \le j \le \ell),$$

($\delta_{h,m}$ *is Kronecker's symbol). Then* (X^k, Y^k) *satisfies* (Z.E.).

PROOF. This is a consequence of Lemma 3.7. □

3.3. Proof of Theorem 3.1. We proceed in four steps.

• <u>Step</u> 0: *Data* We start with two finitely generated subgroups $X = \mathbf{Z}\underline{x}_1 + \cdots + \mathbf{Z}\underline{x}_d$ and $Y = \mathbf{Z}\underline{y}_1 + \cdots + \mathbf{Z}\underline{y}_\ell$ of $\mathbf{C}^n$, of rank d and ℓ respectively, such that (X, Y) satisfies (Z.E.). We denote by $\{\theta_1, \ldots, \theta_q\}$ the set $\{e^{\underline{x}_i \underline{y}_j} ; 1 \le i \le d, 1 \le j \le \ell\}$, by K the field $\mathbf{Q}(\theta_1, \ldots, \theta_q)$ and by t be the transcendence degree of K over $\mathbf{Q}$.

Define
$$\eta = \frac{d\ell + \ell + d}{(n+1)(\ell + d)}.$$

Our first goal is to check
$$t \ge \eta - 1.$$
We shall prove that the inequality $t > a - 1$ holds for any $a < \eta$, which will yield the desired conclusion. There is obviously no loss of generality to assume $\eta > 1$, that is $d\ell > n(\ell + d)$.

• <u>Step</u> 1: *Choice of parameters*

We fix $1 \le a < \eta$ and we denote by c_1, c_2, c_3 suitable positive real numbers. An admissible choice is to select first a "large constant" c_0 (sufficiently large with respect to the previous data), and then to define $c_1 = 1$, $c_2 = 1/c_0$, $c_3 = 1/c_0^d$.

Next let N be a sufficiently large positive integer (that is, large with respect to c_0). Define
$$R = \left[c_1 N^{d/(\ell+d)}\right], \quad L = \left[c_2 N^{\ell/(\ell+d)}\right], \quad U = \left[c_3 N^\eta\right].$$

• <u>Step</u> 2: *Construction of an auxiliary function*

We show that there exists a non zero polynomial $P \in \mathbf{Z}[X_1, \ldots, X_d]$ of degree $\le L$ in each X_i $(1 \le i \le d)$ and usual height $\le e^{N/2}$, such that the exponential sum in n variables
$$F(\underline{z}) = P(e^{\underline{x}_1 \underline{z}}, \ldots, e^{\underline{x}_d \underline{z}}) \qquad (\underline{z} \in \mathbf{C}^n)$$
satisfies
$$|F(r_1 \underline{y}_1 + \cdots + r_\ell \underline{y}_\ell)| \le e^{-U}$$
for all $(r_1, \ldots, r_\ell) \in \mathbf{Z}^\ell$ with $0 \le r_j < R$ $(1 \le j \le \ell)$.

The construction of this auxiliary function rests on Proposition 3.3: set
$$r = R(|\underline{y}_1| + \cdots + |\underline{y}_\ell|), \qquad \Delta = N$$
and
$$\{\varphi_1, \ldots, \varphi_M\} = \{e^{(\lambda_1 \underline{x}_1 + \cdots + \lambda_d \underline{x}_d)\underline{z}} ; 0 \le \lambda_i < L \, (1 \le i \le d)\},$$
so that $M = L^d$. The main condition (3.2) follows from the bound
$$(8U)^{n+1} \le L^d N/2,$$
which holds as soon as
$$2(8c_3)^{n+1} < c_2^d.$$

Notice also that the inequality
$$\sum_{\mu=1}^{L} |\varphi_\mu|_{er} < e^U$$
is satisfied, since the condition $\eta > 1$ implies
$$\log L + c_4 LR < U,$$
where
$$c_4 = e(|\underline{x}_1| + \cdots + |\underline{x}_d|)(|\underline{y}_1| + \cdots + |\underline{y}_\ell|).$$

- <u>Step</u> 3: *Using the criterion for algebraic independence and the zero estimate*
  For each $\underline{r} \in \mathbf{Z}^\ell$ with $0 \leq r_j < R$ $(1 \leq j \leq \ell)$ we define

$$Q_{N\underline{r}}(X_{11}, \dots, X_{d\ell}) = P\left(\prod_{j=1}^{\ell} X_{1j}^{r_j}, \dots, \prod_{j=1}^{\ell} X_{dj}^{r_j}\right).$$

This is a polynomial in $d\ell$ variables, with rational integer coefficients, total degree $\leq LR$ and usual height $\leq e^{N/2}$, which satisfies

$$Q_{N\underline{r}}(e^{\underline{x}_1\underline{y}_1}, \dots, e^{\underline{x}_d\underline{y}_\ell}) = F(r_1\underline{y}_1 + \dots + r_\ell\underline{y}_\ell).$$

This number can be written as

$$Q_{N\underline{r}}(e^{\underline{x}_1\underline{y}_1}, \dots, e^{\underline{x}_d\underline{y}_\ell}) = \widetilde{Q}_{N\underline{r}}(\underline{\theta}),$$

for some polynomial $\widetilde{Q}_{N\underline{r}} \in \mathbf{Z}[X_1, \dots, X_r]$ of total degree $\leq N$ and usual height $\leq e^N$. We are going to use Proposition 3.2 with $n = d\ell$ for these polynomials and the point $\underline{\theta} = (\theta_1, \dots, \theta_q)$. The required upper bound (3.1), which reads

$$\max_{\underline{r}} |\widetilde{Q}_{N\underline{r}}(\underline{\theta})| \leq e^{-CN^a},$$

where C is the constant in Proposition 3.2 associated to $\underline{\theta}$ and a, follows from step 2.

We now check that the polynomials $Q_{N\underline{r}}$ have no common zero in the domain

$$\left\{ \underline{\mu} = (\mu_1, \dots, \mu_q) \in \mathbf{C}^q \ ; \ \max_{1 \leq h \leq q} |\theta_h - \mu_h| \leq e^{-3CN^a} \right\}.$$

Choose $\underline{\mu} = (\mu_1, \dots, \mu_q) \in \mathbf{C}^q$ with

$$\max_{1 \leq h \leq q} |\theta_h - \mu_h| \leq e^{-3CN^a}.$$

Since c_1 and c_2 satisfy

$$c_1^\ell > d^{\ell+d} c_2^d,$$

we have

$$R^\ell > d^{\ell+d} L^d,$$

and the total degree of P is $\leq dL < (R/d)^{\ell/d}$. From condition (Z.E.) we deduce that there exists $\underline{r} \in \mathbf{Z}^\ell$ with $0 \leq r_j < R$ $(1 \leq j \leq \ell)$ such that, if we set

$$\mu_{i\underline{r}} = \prod_{j=1}^{\ell} \mu_{ij}^{r_j} \qquad (1 \leq i \leq d),$$

then the number

$$\widetilde{Q}_{N\underline{r}}(\underline{\theta}) = P(\mu_{1\underline{r}}, \dots, \mu_{d\underline{r}})$$

does not vanish.

- <u>Step</u> 4: *Conclusion of the proof*
  Since we have checked all hypotheses of Proposition 3.2, we deduce

$$t \geq \frac{d\ell + \ell + d}{(n+1)(\ell+d)} - 1.$$

As explained before, using redundant variables together with Proposition 3.8, we obtain

$$t \geq \frac{d\ell}{n(\ell+d)} - 1.$$

We apply this result to the case $n = 1$, writing x for $\underline{x}$ and y for $\underline{y}$. In order to complete the proof of Theorem 3.1, we take $\ell = d$,

$$x_i = \beta^{i-1} \quad (1 \le i \le d), \qquad y_j = \beta^{j-1} \log \alpha \quad (1 \le j \le \ell),$$

so that

$$e^{x_i y_j} = \alpha^{\beta^{i+j-2}} \qquad \text{for } 1 \le i \le d \text{ and } 1 \le j \le \ell.$$

In this case (T.H.) for $x_1, \ldots, x_d$ as well as for $y_1, \ldots, y_\ell$ is satisfied by Liouville's inequality: there exists $L_0 > 0$ so that, for any $(\lambda_1, \ldots, \lambda_d) \in \mathbf{Z}^d \setminus \{0\}$, we have

$$|\lambda_1 + \lambda_2 \beta + \cdots + \lambda_d \beta^{d-1}| \ge L^{-c} \ge e^{-L^c},$$

where L stands for $\max\{|\lambda_1|, \ldots, |\lambda_d|, L_0\}$. $\square$

CHAPTER 15

Some metric results in Transcendental Numbers Theory

1. Introduction

In this Chapter we describe some results in the metric theory of transcendental numbers. Let begin with some notation. If $P \in \mathbb{Z}[x_1, \ldots, x_m]$ is a non – zero polynomial, we define its *size* $t(P)$ as $h(P) + \deg(P)$. Here, $h(P)$ is the Weil's logarithmic height of P (so, if the gcd of the coefficients of P is 1, then $h(P)$ is the logarithm of the maximum module of the coefficients of P) and $\deg(P)$ is the total degree of P. Let $\alpha = (\alpha_1, \ldots, \alpha_m) \in \mathbb{C}^m$ with $\alpha_1, \ldots, \alpha_m$ algebraically dependent: we define $t(\alpha)$ as the minimum size of a non – zero polynomial $P \in \mathbb{Z}[x_1, \ldots, x_m]$ such that $P(\alpha) = 0$.

We shall prove the following conjecture of Chudnovsky (see [**Chu2**, Problem 1.3, page 178]):

CONJECTURE 1.1. *Let m be an integer, $m \geq 1$. Then for almost all $\omega \in \mathbf{C}^m$ there exists a positive constant C such that*

$$\log |P(\omega)| \geq -Ct(P)^{m+1}$$

for any non – zero $P \in \mathbb{Z}[x_1, \ldots, x_m]$.

Let $\alpha \in \mathbf{C}^m$ and let $r > 0$; we denote by $\mathcal{B}_m(\alpha, r)$ the set of points $\omega \in \mathbb{C}^m$ such that $|\omega - \alpha| = \max_{1 \leq j \leq n} |\alpha_j - \omega_j| \leq r$. Given a positive real number τ we also denote by $\mathbf{T}_\tau^m$ the set of $\omega \in \mathcal{B}_m(0, 1)$ such that the inequality

$$|P(\omega)| < \exp\{-Ct(P)^\tau\}$$

has nontrivial solutions $P \in \mathbb{Z}[x_1, \ldots, x_m]$ for any $C > 0$. Therefore, Chudnovsky's conjecture is equivalent to the statement: $\mathrm{meas}(\mathbf{T}_{m+1}^m) = 0$. We also remark that $\mathbf{T}_{\tau'}^m \subset \mathbf{T}_\tau^m$ if $\tau \leq \tau'$. Moreover, by the box principle (see Lemma 3.2), $\mathbf{T}_\tau^m = \mathcal{B}_m(0, 1)$ for any $\tau \in (0, m+1)$.

Define an other subset of the unit ball as follow. Let $\eta > 0$ and let $\mathbf{A}_\eta^m$ be the set of $\omega \in \mathcal{B}_m(0, 1)$ such that for any $C' > 0$ the inequality

$$0 < |\omega - \alpha| < \exp\{-C't(\alpha)^\eta\}$$

has a solutions $\alpha = (\alpha_1, \ldots, \alpha_m) \in \mathbf{C}^m$ with $\alpha_1, \ldots, \alpha_m$ algebraically dependent. As before, we have $\mathbf{A}_{\eta'}^m \subset \mathbf{A}_\eta^m$ if $\eta \leq \eta'$. Moreover, it is easy to see that

$$\mathbf{A}_\eta^m \subset \mathbf{T}_\eta^m$$

(∗) Chapter's author : Francesco AMOROSO.

for $\eta \geq n+1$. Indeed, let ω be in the unit ball of $\mathbb{C}^m$ and assume that there exists $\alpha = (\alpha_1, \ldots, \alpha_m) \in \mathbb{C}^m$ with $\alpha_1, \ldots, \alpha_m$ algebraically dependent such that

$$0 < |\alpha - \omega| < \exp\{-Ct(\alpha)^\eta\}$$

for some $\eta \geq m+1$. Let $P \in \mathbb{Z}[x_1, \ldots, x_m]$ be a non-zero polynomial with integer coefficients, vanishing at α and sucht that $t(P) = t(\alpha)$. Then, using the relation

$$|\alpha_1^{\lambda_1} \cdots \alpha_m^{\lambda_m} - \omega_1^{\lambda_1} \cdots \omega_m^{\lambda_m}| \leq D|\alpha - \omega| \max\{|\alpha|, |\omega|, 1\}^{D-1},$$

which holds for any multiindex lg of weight $\lambda_1 + \cdots + \lambda_m \leq D$, we can easily prove that

$$|P(\omega)| < \exp\{-B^{-1}Ct(P)^\eta\}$$

for some positive constant $B = B(m)$, provided that $C \geq B$.

The opposite inclusion $\mathbf{T}_{m+1}^m \subset \mathbf{A}_{m+1}^m$ is also true (Theorem 4.2), but the proof is much more difficult. The proof of the Conjecture 1.1 easily follows from this last inclusion and from the fact that $\mathbf{A}_{m+1}^m$ is a negligeable set (Theorem 4.2).

More generally we can define, for $m \in \mathbb{N}$ and $\tau \geq m+1$ a real number $\eta(\tau, m)$ as

$$\eta(\tau, m) = \sup\{\eta; \quad \mathbf{T}_\tau^m \subset \mathbf{A}_\eta^m\}$$

and ask whenever we have $\eta(\tau, m) = \tau$ ("comparison problem"). The conjectural answer is $\eta(\tau, m) = \tau$, but this is still an open question. If $m = 1$ the conjecture $\eta(\tau, 1) = \tau$ is true. In several dimension, only partial results are known. In [**Amo2**] it is proved that

$$\eta(\tau, m) \geq \max\left\{m + 1 + \frac{\tau - (m+1)}{m}, \tau - 1\right\},$$

which gives $\eta(m + 1, m) = m + 1$ (and therefore prove Chudnovsky's conjecture). Moreover, if $m \geq 2$

$$\eta(\tau, m) \geq \max\left\{m + \frac{\tau - 2}{m-1}, \tau - 1\right\}$$

which implies $\eta(\tau, 2) = \tau$.

2. One dimensional results

As mentioned in the introduction, the answer to the comparison problem in dimension 1 is easy:

PROPOSITION 2.1. *Let ω be a complex number. Assume that there exists a polynomial $P \in \mathbb{Z}[x]$ such that*

$$|P(\omega)| < \exp\{-Ct(P)^\tau\}$$

for some $\tau \geq 2$ and some $C > 5 \cdot 2^{\tau+1}$. Then we can find a root α of P such that

$$|\alpha - \omega| < \exp\{-2^{-\tau-1}Ct(\alpha)^\tau\}.$$

PROOF. If $P(\omega) = 0$ the result is obvious, so assume $P(\omega) \neq 0$. Let $P = P_1^{e_1} \cdots P_k^{e_k}$ be the factorisation of P into irreducible factors. We first prove that

(116) $$|P_j(\omega)| < \exp\{-C(t(P_j)/2)^\tau\}$$

for at least one index j. Assume the contrary: then,

$$-\log|P(\omega)| \leq C \sum_{j=1}^{k} e_j (t(P_j)/2)^\tau \leq C \left(\frac{1}{2} \sum_{j=1}^{k} e_j t(P_j) \right)^\tau.$$

Gelfond's inequality (see [**Gel2**, Ch.3,§4, Lemma 2])

$$e_1 h(P_1) + \cdots + e_k h(P_k) \leq \deg(P) \log 2 + h(P) \leq t(P)$$

gives $e_1 t(P_1) + \cdots + e_k t(P_k) \leq 2t(P)$. Hence we obtain $-\log|P(\omega)| \leq Ct(P)^\tau$ which contradict our assumption.

Let now $P_j(x) = a(x - \alpha_1) \cdots (x - \alpha_d)$ (where $d = \deg(P_j)$) and assume

$$|\omega - \alpha_1| \leq \cdots \leq |\omega - \alpha_d|.$$

From the inequality $|\omega - \alpha_i| \geq |\alpha_1 - \alpha_i|/2$ $(i = 2, \ldots, d)$ we easily obtain

(117) $$|P_j(\omega)| \geq 2^{-d+1}|\omega - \alpha_1| \cdot |P_j'(\alpha_1)|.$$

To find a lower bound for $|P_j'(\alpha_1)|$, we quote the following well known resultant inequality (see [**Wal1**, p.5.5]): *Let F, G be two co-prime polynomials with integer coefficients and let z be any complex number. Then:*

$$1 < (\deg(F) + \deg(G))\|F\|^{\deg(G)}\|G\|^{\deg(F)} \max\{|F(z)|, |G(z)|\},$$

where $\|\cdot\|$ *is the euclidean norm, i.e. the square root of the sum of the square of the coefficients.* This inequality, with $F = P_j$, $G = P_j'$ and $z = \alpha_1$, gives (using the upper bounds $\log\|P\| \leq \frac{1}{2}\log d + h(P)$ and $h(P') \leq \log d + h(P)$ which hold for any polynomial $P \in \mathbf{Z}[x]$ of degree $\leq m$):

$$-\log|P_j'(\alpha_1)| \leq 3d \log d + 2dh(P_j);$$

Hence, using (116) and (117),

$$\begin{aligned}
\log|\omega - \alpha_1| &\leq (d-1)\log 2 - \log|P_j'(\alpha_1)| + \log|P_j(\omega)| \\
&\leq d + 3d\log d + 2dh(P_j) - C(t(P_j)/2)^\tau \\
&\leq 6t(P_j)^2 - C(t(P_j)/2)^\tau \leq -2^{-\tau-1}Ct(\alpha_j)^\tau.
\end{aligned}$$

$\square$

As a corollary, we find that $\mathbf{T}_\tau^1 = \mathbf{A}_\tau^1$ for any $\tau \geq 2$. We now recall the definition of Hausdorff's dimension.

DEFINITION 2.2. Let Ω be a subset of $\mathbb{R}^n$ and let d a positive integer. We say that Ω has Hausdorff's dimension $< d$ if for any $\varepsilon > 0$ we can find a denumerable set of balls B_i of radii $r_i < \varepsilon$ such that

$$\Omega \subset \bigcup_{i \in \mathbb{N}} B_i \qquad \sum_i r_i^d < \varepsilon.$$

We also define the Hausdorff's dimension of a set Ω as the infimum of the set of $d > 0$ for which Ω has Hausdorff's dimension $< d$.

If Ω has Hausdorff's dimension $< n$, then Ω is a negligeable set (in the sense of the Lebesgue measure in $\mathbb{R}^n$).

We can now prove the main result of this section (see [**Amo1**, Theorem 2]):

THEOREM 2.3. $\mathbf{T}_2^1$ *has Hausdorff's dimension* 0.

PROOF. Since $\mathbf{T}_2^1 = \mathbf{A}_2^1$, we will show that $\mathbf{A}_2^1$ has Hausdorff's dimension 0. For $t \in \mathbb{N}$ let

$$\Lambda_t = \{P \in \mathbb{Z}[x] \text{ such that } [t(P)] = t\}.$$

Let also, for $k \in \mathbb{N}$,

$$\Omega_k = \bigcup_{t \in \mathbb{N}} \bigcup_{P \in \Lambda_t} \bigcup_{\substack{\alpha \in \mathbf{C} \\ P(\alpha)=0}} B_1(\alpha, \exp\{-kt^2\}).$$

We have

$$\mathbf{A}_2^1 = \bigcap_{k \in \mathbb{N}} \Omega_k.$$

Let d, ε be two positive real numbers. Since

$$\log \mathrm{Card}(\Lambda_t) \le ct^2$$

for some absolute constant c, we also have

$$\sum_{t \in \mathbb{N}} \sum_{P \in \Lambda_t} \sum_{\substack{\alpha \in \mathbf{C} \\ P(\alpha)=0}} \exp\{-dkt^2\} < \varepsilon.$$

This proves that $\mathbf{A}_2^1$ has Hausdorff's dimension $< d$ for any $d > 0$.

□

Using [**Fed**][1]Corollary 2.10.12, p.176, and [**Rog**][2], Theorem 4, p.48, and Theorem VII 3, p.104, we deduce:

COROLLARY 2.4. *The set* $\mathbb{C}\backslash\mathbf{T}_2^1$ *is totally disconnected. The set* $\mathbf{T}_2^1$ *is arcwise connected.*

3. Several dimensional results: "comparison Theorem"

The aim of this section is the proof of the following theorem, which is a special case of main result of [**Amo2**] :

THEOREM 3.1. *There exists a constant* $A = A(m) \ge 1$ *having the following property. Let* $k \le m$ *be a positive integer and let* $\omega \in \mathbb{P}^m(\mathbb{C})$ *Let us assume that there exists an homogeneous unmixed ideal* $I \subset \mathbb{Q}[x_0, \dots, x_m]$ *of rank* k *such that*

$$|I(\omega)| < \exp\left\{-Ct(I)^{(m+1)/k}\right\}$$

for some $C \ge (2A)^{(m^2-km+m)/k}$. *Then,* $\exists \alpha \in V(I)$ *such that*

$$\|\alpha - \omega\| < \exp\left\{-\frac{1}{2}A^{-2}C^{k/(m^2-km+m)}t(\alpha)^{m+1}\right\}.$$

We recall that, for $\alpha = [\alpha_0 : \cdots : \alpha_m]$, $\omega = [\omega_0 : \cdots : \omega_m] \in \mathbb{P}^m(\mathbb{C})$,

$$\|\alpha - \omega\| = \frac{\displaystyle\max_{0 \le i < j \le m} |\omega_i \alpha_j - \omega_j \alpha_i|}{\displaystyle\max_{0 \le i \le m} |\alpha_i| \max_{0 \le i \le m} |\omega_i|}$$

---

[1][**Fed**] M. Federer. *Geometric measure theory*, Springer, Berlin, (1969).
[2][**Rog**] C.A. Rogers, *Hausdorff measures*, Cambridge Univ. Press, Cambridge, (1970).

and $|\alpha| = \max_{0 \le j \le m} |\alpha_j|$ (see Chapter 3). We also refer to Chapter 3 for the definitions of $|I(\omega)|$. Finally, the quantity $t(I)$ in the Theorem 3.1 is the *size* of I, i.e. $\deg(I) + h(I)$ (see again Chapter 3 for the definitions of $\deg I$ and $h(I)$).

In the sequel of this section we denote by $c_1, \ldots, c_6$ positive constants depending only on m. The proof of Theorem 3.1 splits in several lemmas. We start with an easy consequence of the box – principle. For a non – zero homogeneous polynomial $P \in \mathbb{Q}[x_0, \ldots, x_m]$ of degree d and for $\omega \in \mathbb{P}^m(\mathbb{C})$, we denote, as in Chapter 3,

$$\|P\|_\omega = |P(\omega)| \cdot |P|^{-1} \cdot |\omega|^{-d},$$

where $|P|$ is the maximum absolute value of the coefficients of P.

LEMMA 3.2. *Let $m \ge 1$ be an integer and let $\omega \in \mathbb{P}^m(\mathbb{C})$. Then for any real number $T \ge c_1$ there exists a non – zero homogeneous polynomial $P \in \mathbb{Q}[x_0, \ldots, x_m]$ with size $\le T$ satisfying*

$$\|P\|_\omega \le \exp\left\{ - c_1^{-1} T^{m+1} \right\}.$$

PROOF. Let H and d be two positive integers which will be choosen later and let Λ be the set of homogeneous polynomials $P \in \mathbb{Z}[x_0, \ldots, x_m]$ of degree d with non negative coefficients bounded by H. Let $D = \binom{d+m}{m}$ be the number of monomials of total degree d, and remark for further references that

$$(d+1)^m/m! \le D \le (d+m)^m/m!.$$

Let also

$$\delta = |\omega|^{-d} \min_{\substack{P_1, P_2 \in \Lambda, \\ P_1 \ne P_2}} |P_1(\omega) - P_2(\omega)|.$$

Since for any $P \in \Lambda$ we have $|\omega|^{-d}|P(\omega)| \le DH$, the ball of $\mathbb{C}$ with centre at the origin and radius $DH + \delta/2$ contains the disjoint union of the open balls of centre $|\omega|^{-d}P(\omega)$ and radius $\delta/2$, where P runs on Λ. Comparing the areas, we obtain

$$\mathrm{Card}(\Lambda)(\delta/2)^2 \le (DH + \delta/2)^2.$$

The cardinality of Λ is $(H+1)^D$; hence:

$$\delta \le \frac{2DH}{(H+1)^{D/2} - 1} \le 2DH^{1-D/2}$$

and so there exist two polynomials $P_1, P_2 \in \Lambda$, $P_1 \ne P_2$, such that

$$\|P_1 - P_2\|_\omega \le 2DH^{1-D/2}.$$

The polynomial $P = P_1 - P_2$ has degree d, maximum absolute value of the coefficients $\le H$ and satisfies $\|P\|_\omega \le 2DH^{1-D/2}$. The lemma easily follows taking $d = [T/2]$ and $H = [\exp\{T/2\}]$. $\square$

LEMMA 3.3. *Let $C \ge c_2$ be a positive real number and $I \subset \mathbb{Q}[x_0, \ldots, x_m]$ be an homogeneous unmixed ideal of rank $k \le m$ such that*

$$|I(\omega)| < \exp\left\{ - Ct(I)^{(m+1)/k} \right\}.$$

Then there exists a homogeneous prime ideal $\mathfrak{p} \subset I$ of rank k such that

$$|\mathfrak{p}(\omega)| < \exp\left\{ - c_2^{-1} Ct(\mathfrak{p})^{(m+1)/k} \right\}.$$

PROOF. Let $I = I_1 \cap \ldots \cap I_s$ be the reduced primary decomposition of I and let $\mathfrak{p}_j = \sqrt{I_j}$ and e_j be the exponent of I_j. We can assume $\omega \notin V(\mathfrak{p}_j)$ for $j = 1, \ldots, s$. From the Proposition 4.7 of Chapter 3 we obtain:

$$\sum_{j=1}^{s} e_j t(\mathfrak{p}_j) \leq (m^2 + 1) t(I)$$

and

$$\sum_{j=1}^{s} e_j \log |\mathfrak{p}_j(\omega)| \leq \log |I(\omega)| + m^3 t(I) \leq -C t(I)^{(m+1)/k} + m^3 t(I).$$

Let now assume $\log |\mathfrak{p}_j(\omega)| > -\tilde{C} t(\mathfrak{p}_j)^{(m+1)/k}$ for $j = 1, \ldots, s$ and for some $\tilde{C} \geq 1$. Then:

$$Ct(I)^{(m+1)/k} < m^3 t(I) + \tilde{C} \sum_{j=1}^{s} e_j t(\mathfrak{p}_j)^{(m+1)/k}$$

$$\leq m^3 t(I) + \tilde{C} \left(\sum_{j=1}^{s} e_j t(\mathfrak{p}_j) \right)^{(m+1)/k}$$

$$\leq m^3 t(I) + \tilde{C}(m^2 + 1)^{m+1} t(I)^{(m+1)/k}$$

$$\leq 2(m^2 + 1)^{m+1} \tilde{C} t(I)^{(m+1)/k}.$$

Hence $\tilde{C} > \frac{1}{2}(m^2 + 1)^{-m-1} C$. The lemma follows choosing $c_2 = 2(m^2 + 1)^{m+1}$. $\square$

We also recall two results of Chapter 3. The first one follows immediately from Corollary 4.12 in Chapter 3, while the second one follows from Proposition 4.13 in Chapter 3.

LEMMA 3.4. *Let $\mathfrak{p} \subset \mathbb{Q}[x_0, \ldots, x_m]$ be a homogeneous prime ideal of rank $k \leq n$ and let $Q \in \mathbb{Q}[x_0, \ldots, x_m]$ be a homogeneous polynomials such that $Q \notin \mathfrak{p}$. Let also $\omega \in \mathbb{P}^m(\mathbb{C})$ such that $\omega \notin V(\mathfrak{p})$ and define*

$$\rho = \min_{\alpha \in V(\mathfrak{p})} \|\omega - \alpha\|.$$

Assume that there exist $S > 0$ and $\theta \in \mathbb{N}$ such that

$$|\mathfrak{p}(\omega)| \leq e^{-S}, \qquad \|P\|_\omega \leq e^{-2m \deg P}$$

and

$$-\theta \log \|P\|_\omega \geq 2 \min(S, \log \frac{1}{\rho}).$$

Then, if $k \leq m - 1$, there exists a homogeneous unmixed ideal J of rank $k + 1$ such that $V(J) = V((\mathfrak{p}, P))$ and

$$t(J) \leq m(m + 2)\theta t(Q) t(\mathfrak{p}),$$

$$\log |J(\omega)| \leq -S + 13m^2 \theta t(Q) t(\mathfrak{p}).$$

Moreover, if $k = m$ we have

$$S \leq 13m^2 \theta t(Q) t(\mathfrak{p}).$$

LEMMA 3.5. *Let $I \subset \mathbb{Q}[x_0, \ldots, x_m]$ be a homogeneous unmixed ideal of rank k. Then for any $\omega \in \mathbb{P}^m(\mathbb{C})$ such that $\omega \notin V(I)$ there exists a zero $\alpha \in V(I)$ such that*

$$(\deg I) \log \|\omega - \alpha\| \leq \frac{1}{m+1-k} \log |I(\omega)| + 4m^3 t(I).$$

We shall prove Theorem 3.1 by induction on the rank k of I. The following lemma resume the inductive step.

LEMMA 3.6. *There exists a constant $A = A(m) \geq 1$ having the following property. Let C be a positive real number and θ be a positive integer such that $C \geq \theta A$. Let us assume that there exists a homogeneous unmixed ideal $I \subset \mathbb{Q}[x_0, \ldots, x_m]$ of rank $k \leq m$ such that*

$$|I(\omega)| < \exp\left\{-Ct(I)^{(m+1)/k}\right\}.$$

Then, either $\exists \alpha \in V(\mathfrak{p})$, such that

$$\|\alpha - \omega\| < \exp\left\{-A^{-1}\theta t(\alpha)^{m+1}\right\}$$

or there exists an homogeneous unmixed ideal $J \subset \mathbb{Q}[x_0, \ldots, x_n]$ of rank $k+1$ such that $V(J) \subset V(I)$ and

$$|J(\omega)| < \exp\left\{-A^{-1}\theta^{-m/(k+1)}C^{k/(k+1)}t(J)^{(m+1)/(k+1)}\right\}.$$

Moreover, if $k = m$ only the first case can occur.

PROOF. We can assume $\omega \notin V(I)$, otherwise we choose $\alpha = \omega$. Lemma 3.3 gives a homogeneous prime ideal $\mathfrak{p} \supset I$ of rank k such that $\log |\mathfrak{p}(\omega)| < -S$ where

$$S = c_2^{-1}Ct(\mathfrak{p})^{(m+1)/k}.$$

Let $\alpha \in V(\mathfrak{p})$ such that $\rho = \|\alpha - \omega\|$ is minimal, and define a real number T by the following equality:

$$\theta c_1^{-1}T^{m+1} = 2\min\left(S, \log\frac{1}{\rho}\right).$$

Assume

$$C \geq 8m^4 c_2;$$

then, by Lemma 3.5,

$$\deg \mathfrak{p} \cdot \log\frac{1}{\rho} \geq -\frac{1}{m}\log|\mathfrak{p}(\omega)| - 4m^3 t(\mathfrak{p})$$

$$\geq \frac{1}{2}m^{-1}c_2^{-1}Ct(\mathfrak{p})^{(m+1)/k}$$

$$\geq \frac{1}{2}m^{-1}c_2^{-1}C\deg\mathfrak{p}.$$

Hence,

$$\min\left(S, \log\frac{1}{\rho}\right) \geq \frac{1}{2}m^{-1}c_2^{-1}C$$

and

(118) $$T \geq \left(m^{-1}c_1 c_2^{-1}(C/\theta)\right)^{1/(m+1)} = c_3(C/\theta)^{1/(m+1)}.$$

We also remark the inequalities:

(119)
$$\log \frac{1}{\rho} \geq \frac{1}{2} c_1^{-1} \theta T^{m+1},$$

(120)
$$T \leq \left(2c_1 \theta^{-1} S\right)^{1/(m+1)} = c_4 (C/\theta)^{1/(m+1)} t(\mathfrak{p})^{1/k}.$$

If
$$A \geq (c_1/c_3)^{m+1},$$
we have $T \geq c_1$ by (118); hence we can apply Lemma 3.2, which gives a polynomial P of size $\leq T$ such that:

(121)
$$\log \|P\|_\omega \leq -c_1^{-1} T^{m+1}.$$

We now distinguish two cases.
- First case: $P \in \mathfrak{p}$.

   Then $t(\alpha) \leq t(P) \leq T$, hence, by (119),
$$\log \frac{1}{\rho} \geq \frac{1}{2} c_1^{-1} \theta t(\alpha)^{m+1}.$$

Therefore in this case, the first assertion of Lemma 3.6 is satisfied (if we choose $A \geq \max(1, \, 8m^4 c_2, \, (c_1/c_2)^{m+1}, \, 1/(2c_1))$.
- Second case: $P \notin \mathfrak{p}$ and $k \leq m - 1$.

   From the choice of S and from (121) we see that $|\mathfrak{p}(\omega)| \leq e^{-S}$ and $-\theta \log \|P\|_\omega \geq 2 \min(S, \log(1/\rho))$. Moreover the other hypothesis of Lemma 3.4,
$$\|P\|_\omega \leq e^{-2m \deg P}$$
is satisfied if $T^m \geq 2c_1 m$ (see (121)) which certainly occur if we choose
$$A \geq c_3^{-m-1} (2c_1 m)^{(m+1)/m}$$

(see (118)). Hence we can apply Lemma 3.4.

   Assume first $k \leq m - 1$. Then Lemma 3.4 gives a homogeneous unmixed ideal J of rank $k + 1$ such that $V(J) = V((\mathfrak{p}, P))$ and

(122)
$$\log |J(\omega)| \leq -S + 13m^2 \theta T t(\mathfrak{p}),$$

(123)
$$t(J) \leq m(m+2) \theta T t(\mathfrak{p}).$$

Assume
$$A \geq (26 c_2 c_4 m^2)^{(m+1)/m}.$$

Then, by (120),
$$13m^2 \theta T t(\mathfrak{p}) \leq 13 c_4 m^2 \theta^{m/(m+1)} C^{1/(m+1)} t(\mathfrak{p})^{(k+1)/k}$$
$$\leq \frac{1}{2} c_2^{-1} C t(\mathfrak{p})^{(k+1)/k} \leq \frac{1}{2} S.$$

Henceforth, from (122),

(124)
$$\log |J(\omega)| \leq -\frac{1}{2} c_2^{-1} C t(\mathfrak{p})^{(m+1)/k}.$$

On the other hand, again by (120),
$$m(m+2) \theta T t(\mathfrak{p}) \leq m(m+2) c_4 \theta^{m/(m+1)} C^{1/(m+1)} t(\mathfrak{p})^{(k+1)/k}.$$

Therefore, from (123) we obtain the following lower bound for $t(\mathfrak{p})$:
$$t(\mathfrak{p}) \geq c_5 \theta^{-mk/((m+1)(k+1))} C^{-k/((m+1)(k+1))} t(J)^{k/(k+1)}$$

Inserting this lower bound in (124), we finally found

$$\log |J(\omega)| \leq -c_6 \theta^{-m/(k+1)} C^{k/(k+1)} t(J)^{(m+1)/(k+1)}.$$

Therefore, in this case, the second assertion of Lemma 3.6 holds, choosing

$$(125) \quad A \geq \max\left\{ 1, 8m^4 c_2, \left(\frac{c_1}{c_3}\right)^{m+1}, c_3^{-m-1}(2c_1 m)^{\frac{m+1}{m}}, (26 c_2 c_4 m^2)^{\frac{m+1}{m}}, \frac{1}{c_6} \right\}.$$

• Third case: $P \not\subseteq \mathfrak{p}$ and $k = m$.

As before, we can apply Lemma 3.4 provided that $A \geq c_3^{-m-1}(2c_1 m)^{\frac{m+1}{m}}$. Since $k = m$, this lemma gives

$$S \leq 13m^2 \theta T t(\mathfrak{p}).$$

The same computation as before shows that this relation is inconsistent if we assume

$$A > (13 c_2 c_4 m^2)^{(m+1)/m}.$$

So this case can not occurs if A satisfies (125).

Lemma 3.6 follows, choosing

$$A = \max\left\{ 1, 8m^4 c_2, \left(\frac{c_1}{c_3}\right)^{m+1}, (2c_1)^{-1}, c_3^{-m-1}(2c_1 m)^{\frac{m+1}{m}}, (26 c_2 c_4 m^2)^{\frac{m+1}{m}}, \frac{1}{c_6} \right\}.$$

□

We can now prove Theorem 3.1 by induction on k. Obviously, we can assume $\omega \notin V(I)$. Let A be the constant which appear in Lemma 3.6.

• $k = m$. Let $\theta = [A^{-1}C]$. By assumption, $C \geq 2A$, hence $\theta \geq \frac{1}{2}A^{-1}C$; moreover $C/\theta \geq A$. Lemma 3.6 gives $\alpha \in V(I)$ such that

$$\log \|\alpha - \omega\| < -A^{-1} \theta t(\alpha)^{m+1} \leq -\frac{1}{2} A^{-2} C t(\alpha)^{m+1}.$$

• $k < m$. Let

$$\theta = \left[A^{-1} C^{k/(m^2 - km + m)} \right].$$

By assumption $C \geq (2A)^{(m^2 - km + m)/k}$, hence

$$(126) \qquad \frac{1}{2} A^{-1} C^{k/(m^2 - km + m)} \leq \theta;$$

moreover we also have

$$(127) \qquad \theta \leq A^{-1} C^{k/(m^2 - km + m)}.$$

This last inequality gives $C/\theta \geq A$, thus we can apply Lemma 3.6. If there exists $\alpha \in I$ such that

$$\log \|\alpha - \omega\| < -A^{-1} \theta t(\alpha)^{m+1}$$

our assertion follows, since, by (126),

$$A^{-1} \theta \geq \frac{1}{2} A^{-2} C^{k/(m^2 - km + m)}.$$

Otherwise, there exists an homogeneous unmixed ideal J of rank $k + 1$ such that $V(J) \subset V(I)$ and

$$\log |J(\omega)| < -A^{-1} \theta^{-m/(k+1)} C^{k/(k+1)} t(J)^{(m+1)/(k+1)}.$$

Let $\tilde{C} = A^{-1} \theta^{-m/(k+1)} C^{k/(k+1)}$. By (127) and by the assumption

$$C \geq (2A)^{(m^2 - km + m)/k},$$

we have

$$\tilde{C} \geq A^{-1} \left(A^{-1} C^{k/(m^2 - km + m)} \right)^{-m/(k+1)} C^{k/(k+1)}$$

$$= A^{(m-k-1)/(k+1)} C^{k(m^2 - km)/((k+1)(m^2 - km + m))}$$

(128)
$$\geq C^{k(m^2 - km)/((k+1)(m^2 - km + m))}$$

(129)
$$\geq (2A)^{(m^2 - km)/(k+1)}.$$

Hence, by inductive hypothesis, we can find $\alpha \in V(J)$ such that

$$\log \|\alpha - \omega\| < -\frac{1}{2} A^{-2} \tilde{C}^{(k+1)/(m^2 - km)} t(\alpha)^{m+1}$$

$$\leq -\frac{1}{2} A^{-2} C^{k/(m^2 - km + 1)} t(\alpha)^{m+1},$$

where in the last inequality we have used again (128). $\qquad\square$

4. Several dimensional results: proof of Chudnovsky's conjecture

From Theorem 3.1 we easily deduce the following result concerning polynomials:

COROLLARY 4.1. *For any integer $m \geq 1$ and for any real number $\tau \geq m + 1$ there exists a positive constant $\tilde{B}$ having the following property. Let ω be in the unit ball of $\mathbb{C}^m$ and assume that there exists a non – zero polynomial $P \in \mathbb{Z}[x_1, \ldots, x_m]$ such that*

$$|P(\omega)| < \exp\left\{ -Ct(P)^{m+1} \right\}$$

for some $C \geq \tilde{B}$. Then, there exists $\alpha \in \mathbb{C}^m$ such that $P(\alpha) = 0$ and

$$|\alpha - \omega| < \exp\left\{ -\tilde{B}^{-1} C^{1/m^2} t(\alpha)^{m+1} \right\}.$$

PROOF. We can assume $P(\omega) \neq 0$, otherwise we choose $\alpha = \omega$; we also denote by c_7, c_8 two positive constants depending only on m. Let I be the principal ideal generated by the homogenization $^h P$ of P and let $\omega' = (1, \omega)$; by Proposition 4.8 of Chapter 3 we have

$$\log |I(\omega')| \leq \log \|P\|_{\omega'} + 2m^2 \deg(P) \leq -Ct(P)^{m+1} + 2m^2 t(P)$$

and $t(I) \leq (m^2 + 1)t(P)$. Hence

$$\log |I(\omega')| \leq \frac{1}{2}(m^2 + 1)^{-m-1} Ct(I)^{m+1},$$

provided that $C \geq 4m^2$. Theorem 3.1 gives α' such that $^h P(\alpha') = 0$ and

(130)
$$\log \|\alpha' - \omega'\| < -B^{-1}(C/2)^{1/m^2} t(\alpha')^{m+1}.$$

If $C \geq c_7$, then $\|\alpha' - \omega'\| < 1/2$, hence $\alpha'_0 \neq 0$ and the vector $\alpha \in \mathbb{C}^m$ defined by $\alpha_i = \alpha'_i / \alpha'_0$ $(i = 1, \ldots, m)$ satisfies $P(\alpha) = 0$ and

$$|\alpha - \omega| \leq \max\{1, |\alpha|\} \|\alpha' - \omega'\|.$$

Since $\|\alpha' - \omega'\| < 1/2$, this gives at once $|\alpha| \leq 2$. Thus $|\alpha - \omega| \leq 2\|\alpha' - \omega'\|$, and we deduce from (130) that

$$\log |\alpha - \omega| < -B^{-1}(C/2)^{1/m^2} t(\alpha)^{m+1} + \log 2 \leq -(4B)^{-1} C^{1/m^2} t(\alpha)^{m+1},$$

if $C \geq c_8$. Corollary 4.1 follows, choosing $\tilde{B} = \max\{4m^2, c_7, c_8, 2B\}$. $\qquad\square$

As a corollary, we see that $\mathbf{T}_{m+1}^m = \mathbf{A}_{m+1}^m$. We can now prove Chudnovsky's conjecture:

THEOREM 4.2. *The set $\mathbf{T}_{m+1}^m$ is negligeable.*

PROOF. By the previous result, it is enough to show that $\mathbf{A}_{m+1}^m$ is negligeable. Given $P \in \mathbb{Z}[x_1, \ldots, x_m]$ and $\varepsilon > 0$, denote

$$U_P(\varepsilon) = \{\omega \in \mathcal{B}_m(0,1) \text{ such that } \min_{\substack{\alpha \in \mathbb{C}^{m+1} \\ P(\alpha)=0}} |\omega - \alpha| \leq \varepsilon\}.$$

For $t \in \mathbb{N}$ let

$$\Lambda_t = \{P \in \mathbb{Z}[x_1, \ldots, x_m] \text{ such that } [t(P)] = t\}.$$

Let also, for $k \in \mathbb{N}$,

$$\Omega_{m,k} = \bigcup_{t \in \mathbb{N}} \bigcup_{P \in \Lambda_t} U_P(\exp\{-kt^{m+1}\}).$$

We have

$$\mathbf{A}_{m+1}^m = \bigcap_{k \in \mathbb{N}} \Omega_{m,k}.$$

Now, we quote the following lemma:

LEMMA 4.3. *For any $\varepsilon > 0$ we have*

$$\mathrm{meas}(U_P(\varepsilon)) \leq \frac{\pi^{2m}}{(m-1)!}\varepsilon^2 \deg(P).$$

PROOF. By a theorem of Lelong

$$\mathrm{meas}(\{P = 0\} \cap \mathcal{B}_m(0,1)) \leq \frac{\pi^{2(m-1)}}{(m-1)!} \deg(P)$$

(see [**Lel1**][3]Théorème 7). Hence, by a Fubini – Tonelli argument,

$$\mathrm{meas}(U_P(\varepsilon)) \lesssim \pi^2 \varepsilon^2 \cdot \frac{\pi^{2(m-1)}}{(m-1)!} \deg(P).$$

$\square$

Since

$$\log \mathrm{Card}(\Lambda_t) \leq c(m)t^{m+1}$$

for some constant $c(m)$ depending only on m, we deduce from the lemma above that

$$\mathrm{meas}(\Omega_{m,k}) \leq \sum_{t=1}^{+\infty} \exp\{c(m)t^{m+1}\}\frac{\pi^{2m}}{(m-1)!} \exp\{-kt^{m+1}\}t \to 0$$

as $k \to +\infty$. Theorem 4.2 is proved.

$\square$

---

[3][**Lel1**] P. Lelong. Propriétés métriques des variétés analytiques complexes définies par une équation, *Ann. École Norm. Sup.* 67, (1950), 393–419.

CHAPTER 16

The Hilbert Nullstellensatz, Inequalities for Polynomials, and Algebraic Independence

1. The Hilbert Nullstellensatz and Effectivity

Given a set of polynomials, $P_1(\mathbf{x}), \ldots, P_m(\mathbf{x})$, one of the most basic questions one can ask about them is whether they have common zeros or not. Hilbert gave a comprehensive answer to this sort of question in his famous theorem. To state it, we will use the following notation:

NOTATION. Let $\mathfrak{A}$ be the polynomial ideal generated in $R := k[x_1, \ldots, x_n]$ by the polynomials $P_1, \ldots, P_m$ of respective degrees $D_2(=: D) \geq D_3 \geq \cdots \geq D_m \geq D_1$.

1.1. Classical (Ineffective) Versions.
The following statement is the now usual formulation of Hilbert's theorem. See [Hil][1] for the original statement.

THEOREM 1.1. (Hilbert Nullstellensatz) *There is a $\rho > 0$ such that if $Q(\mathbf{x})$ vanishes on the zeros (say over an algebraic closure of k) of the ideal $\mathfrak{A} < R$, then $Q^\rho \in \mathfrak{A}$.*

Although it is often said that, at least until Hilbert, algebra was in principle concerned with calculation, Hilbert makes no mention at all of a bound for ρ nor how to determine the A_i in an relation:

$$A_1 P_1 + \cdots + A_m P_m = Q^\rho.$$

One special case is particularly interesting because it provides a test of whether the polynomials P_i have a common zero at all. We give it a common name to distinguish it from the general case.

THEOREM 1.2. (Bézout Version) $P_1(\mathbf{x}), \ldots, P_m(\mathbf{x})$ *have no common zeros if and only if there exist* $A_1, \ldots, A_m \in R$ *such that*

(131) $$A_1 P_1 + \cdots + A_m P_m = 1.$$

The ineffectivity of Hilbert's theorem gave no clue as to what bound $\delta(D, n)$ might suffice to guarantee the existence of the A_i in (131) satisfying $\deg A_i \leq \delta(D, n)$.

The first step in this direction was taken in 1926 by Grete Hermann, who developed linear algebra over R, based on previous work by her thesis advisor, Emmy Noether, and Noether's student Henzelt, who died in WWI. She did not bound ρ, but gave an algorithm which, given ρ, would produce the A_i.

(∗) Chapter's author : W. DALE BROWNAWELL.

[1][Hil] D. Hilbert. Über die vollen Invariantensysteme. *Math. Ann.* 42, (1883), 313–373.

The next step was taken in 1929 by J.L. Rabinowitsch:

THEOREM 1.3. (Rabinowitsch) *The Bézout version of the Nullstellensatz is equivalent to the full version.*

The proof is so simple and beautiful that every student should be exposed to it. Clearly we have only to establish that the Bézout version implies the full version.

PROOF. Since Q vanishes on all the common zeros of $P_1, \ldots, P_m$, the polynomials $P_1, \ldots, P_m, 1 - x_{n+1}Q$ have no common zeros. Consequently, there exist polynomials $B_1, \ldots, B_{m+1} \in R[x_{n+1}]$ such that

$$1 = B_1 P_1 + \cdots + B_m P_m + B_{m+1}(1 - x_{n+1}Q).$$

As this is a formal identity, it will continue to hold under any specialization of the variable x_{n+1}. So we can set $x_{n+1} = 1/Q$ and clear out the denominators to find that

$$Q^e = B_1^* P_1 + \cdots + B_m^* P_m,$$

where $e = \max_i \deg_{x_{n+1}} B_i$, and

$$B_i^* = Q^e B_i|_{x_{n+1}=1/Q}.$$

$\square$

After some investigations by M. Reufel, A. Seidenberg, and M. Lazard, this line of investigation was taken up again in 1982 by D.W. Masser and G. Wüstholz in their fundamental paper [**MW2**] on the algebraic independence of values of elliptic functions.

1.2. Relevance of Sharp Nullstellensatz for Independence. The motivation arose almost as soon as Masser and Wüstholz began to develop their sharp zero estimates for commutative algebraic groups. Let us review the salient points of their independence proof for values of $\wp(\alpha_i \beta_j)$, $1 \le i \le r$, $1 \le j \le s$ to see why: (We assume that $g_2, g_3 \in \overline{\mathbb{Q}}$ and delay making further hypotheses explicit for the moment.)

Outline of Transcendence Machine

- Siegel's Lemma $\rightsquigarrow$ Auxiliary polynomial in the $\wp(\alpha_i z)$, whose
- Values at $\sum k_j \beta_j$ $\rightsquigarrow$ polynomials P_κ in $\wp(\alpha_i \beta_j)$ (and derivatives).
- Schwarz Lemma $\rightsquigarrow$ Upper bounds on values.
- Zero Estimates $\rightsquigarrow$ The P_κ have no common (nearby) zeros.

If the P_κ had *no* common zeros, then this last step would be a veritable invitation to apply the following corollary of the Bézout form of the Nullstellensatz. (This approach is still a bit weaker than through the effective version of Theorem 1.1 actually developed in [**MW2**], but will serve to illustrate the principle.)

THEOREM 1.4. *If $P_1, \ldots, P_m \in \mathbb{Z}[\mathbf{x}]$ have no common zeros and there exist $A_1, \ldots, A_m \in \mathbb{Z}[\mathbf{x}]$ and non-zero $N \in \mathbb{Z}[\mathbf{x}]$ such that*

$$A_1(\mathbf{x})P_1(\mathbf{x}) + \ldots A_m(\mathbf{x})P_m(\mathbf{x}) = N,$$

with $\deg A_i(\mathbf{x}) \le \delta(D, n)$, then there exist such with

$$\log N, \log \operatorname{Height} A_i \le C_{\text{eff}} \delta(D, n)^n (h + \delta(D, n)),$$

where $C_{\text{eff}} > 0$ is explicit, $h \ge \log \operatorname{Height} P_i$, and $D \ge \deg P_i$.

PROOF. Cramer's Rule $\square$

COROLLARY 1.5. *For P_i as above and $\theta \in \mathbf{C}^n$,*

$$\max |P_i(\theta)| \geq e^{-C_{\mathrm{eff}}\delta(D,n)^n(h+\delta(D,n))}\|\theta\|^{-(\delta(D,n)+D)},$$

where $C_{\mathrm{eff}} > 0$ is explicit and $\|\theta\| = \max\{1, |\theta_i|\}$, for $\theta = (\theta_1, \ldots, \theta_n)$.

This result was applied by Masser and Wüstholz in the case that:

- $\wp$ has no CM
- $\alpha_1, \ldots, \alpha_r$ linearly independent, satisfying "technical hypotheses"
- $\beta_1, \ldots, \beta_s$ linearly independent, satisfying "technical hypotheses"
- $\theta_1, \ldots, \theta_n$ a transcendence basis for the field generated by $\wp(\alpha_i\beta_j)$

In this situation, the Transcendence Machine outlined above produces, for every large T, a family of polynomials, say over $\mathbf{Z}$, with

(132) $$\deg, \log \mathrm{Height}\, P_i \ll T^{r+2s}$$

(133) $$\log |P_i(\theta)| \ll -cT^{rs}\log T.$$

Let me simplify the considerations of [**MW2**] by assuming that the P_κ are without common zeros (not literally true, but "virtually" so).

1.3. Application of Nullstellensatz for Algebraic Independence. At this point, Masser and Wüstholz worked though Hermann's algorithm carefully to deduce the following:

THEOREM 1.6. (Hermann, Masser-Wüstholz)

$$\delta(D,n) \leq 2(2D)^{2^{n-1}}.$$

At the same time they obtained somewhat better bounds in a quantitative analogue of the full Hilbert Nullstellensatz, Theorem 1.1, for $\log N$ and $\log \mathrm{Height}\, A_i$ than in our quantitative Bézout Theorem 1.4.

That allowed them to deal with any common zeros which were not nearby, but such technicalities would not add much to our discussion here.

As a consequence, they could deduce the following result:

THEOREM 1.7. (Masser-Wüstholz)

$$n + \log_2 n - 1 \geq \log_2 \frac{rs}{r+2s}.$$

Even with the weaker bounds we have in Corollary 1.5 (under the purely technical simplifying assumptions on the zeros), we obtain directly that

$$n + \log_2 n \geq \log_2 \frac{rs}{r+2s}.$$

In light of today's results, which are the subject of this book, these inequalities look comparatively weak. But at the time, Chudnovsky's analogous announcements for the the exponential function had only just been established by Philippon [**PPh1**] over $\mathbb{C}_p$ and by E. Reyssat [**Rey1**] over $\mathbb{C}$. Thus the technique of [**MW2**] for establishing the transcendence bound for elliptic functions was a real breakthrough in regard to its general algebraic approach to independence criteria as well as its general zero estimates.

Consequently it became important to establish better bounds for $\delta(D,n)$. The best that one could hope for is $\delta(D,n) \leq D^n$, as shown by an example due to Masser-Philippon, Mora, and Lazard:

Example:

$$x_1^D, x_2^D - x_1 x_n^{D-1}, x_3^D - x_2 x_n^{D-1}, \ldots, x_{n-1}^D - x_{n-2} x_n^{D-1}, 1 - x_{n-1} x_n^{D-1}$$

have no common zeros. However, for a new variable t, evaluation at the point

$$\mathbf{t} = (t^{1-D^{n-1}}, t^{1-D^{n-2}}, \ldots, t^{1-D}, t)$$

shows that $1 = A_1(\mathbf{t}) t^{D-D^n}$, so $\deg_{x_n} A_1 \geq D^n - D$.

As a cautionary note, however we also knew the plausibly related example of E.W. Mayr and A.R. Meyer [MM][2], where huge coefficients were necessary.

Still, the Nullstellensatz offered great promise for independence investigations as a clean and more powerful replacement for the rather unwieldy iterative application of quantifier elimination via ordinary resultants.

2. Liouville-Łojasiewicz Inequality

I did not know how to prove the Nullstellensatz, but I was impressed with the tools for algebraic independence developed by Yu.V. Nesterenko [Nes2], [Nes3], [Nes4] and P. Philippon [PPh2] described in this book. While trying to master these tools, I grew convinced that, if one is not close to common zeros of polynomials, at least one of the polynomials must not be small. As we saw above, this is the functionality of the Nullstellensatz for algebraic independence.

2.1. The Inequality. I was able to establish the following inequality [Bro6], which I give in a slightly improved form. I use the notation

$$\|\mathfrak{A}\|_\omega := \frac{\text{Height } F(\omega^{tr} S^{(0)}, \ldots, \omega^{tr} S^{(d)})}{(\text{Height } F) \|\omega\|^{(d+1) \deg \mathfrak{A}}},$$

for any unmixed homogeneous ideal $\mathfrak{A}$ of dimension d, with Chow form F, where $S^{(0)}, \ldots, S^{(d)}$ are generic skew-symmetric matrices. In other words, I use $\|\mathfrak{A}\|_\omega$ for Nesterenko's $|\mathfrak{A}(\omega)|$. I also use Nesterenko's

$$\|P\|_\omega := \frac{|P(\omega)|}{(\text{Height } P) \|\omega\|^{\deg P}}.$$

THEOREM 2.1. *Let $\mathfrak{A} \subset \mathbf{Z}[\mathbf{x}] := \mathbf{Z}[x_0, \ldots, x_n]$ be an unmixed homogeneous ideal. Let the homogeneous polynomials $P_1, \ldots, P_m \in \mathbf{Z}[\mathbf{x}]$ have no common zero with each other and the polynomials of $\mathfrak{A}$ inside a ball $B_{\leq \rho}$ of radius ρ centered at ω. Then*

$$\max\{\|\mathfrak{A}\|_\omega, \|P_i\|_\omega\} \geq e^{-CD^\mu (D \log \text{ht } \mathfrak{A} + D \deg \mathfrak{A} + h \deg \mathfrak{A})} \left(\frac{\rho}{\|\omega\|}\right)^{2D^\nu},$$

where $\mu = \min\{n-1, \dim \mathfrak{A}\}$, $\nu = \min\{m, \dim \mathfrak{A}\}$, $\log \text{Height } P_i \leq h$, $\deg P \leq D$, $\|\omega\| = \max\{1, \max|\omega_i|\}$, and $C = C(n) > 0$ is explicit.

This is a generalization of the fundamental Liouville inequality to higher dimensions.

---

[2][MM] E.W. Mayr, A.R. Meyer. The complexity of the word problem for commutative semigroups and polynomial ideals, *Adv. in Math.* 46, (1982), 305–329.

Outline of Proof. Create, with coefficients which are not very large, a sequence $h_1, \ldots, h_\nu$ of homogeous linear combinations of the P_i which is generic enough to allow the following :

Define F_1 to be the Chow form for the principal ideal (h_1) and, inductively, to define F_{i+1} we decompose F_i into two parts,

$$F_i =: G_i \cdot H_i,$$

where

1. G_i is a product of Chow forms $F_{\mathfrak{P}}$ of prime ideals $\mathfrak{P}$ of dimension $n - d$ which do not have zeros within ρ of ω, whereas
2. H_i is a product of Chow forms $F_{\mathfrak{P}}$ of prime ideals $\mathfrak{P}$ of dimension $n - d$ which have zeros within ρ of ω.

Then we define

$$F_{i+1} := \mathrm{Res}(F_i, h_i).$$

Using standard bounds on the sizes of the coefficients of the h_i, it follows readily [Bro6] from the basic properties of Chow forms and valuations of Section 4 of Chapter 3 in this book, that

$$\max\{\|P_i\|_\omega\} \geq C_{\mathrm{eff}} \max\{\|h_i\|_\omega\} \geq C_{\mathrm{eff}} \prod \|G_i\|_\omega \geq C_{\mathrm{eff}} \left(\frac{\rho^{D^\nu}}{\|\omega\|^{2D^\nu}} \right).$$

$\square$

2.2. Application of Inequality. If we apply this inequality in the Masser-Wüstholz situation, with $\mathfrak{A}$ defined to be the (homogenization of the) ideal of relations, we find that

$$cN^{rs} \leq c'N^{(r+2s)\mu}N^{r+2s} + c'N^{(r+2s)\mu}(c' - \log \epsilon'),$$

where the last term comes from the technical hypothesis (used to establish the zero-free region as in [MW2]). We thus obtain immediately the result from [PPh2]:

Corollary 2.2. (Philippon)

$$\mu \geq \frac{rs}{r + 2s} - 1$$

We even have an automatic measure of algebraic independence:

Corollary 2.3.

$$\log \|\mathfrak{A}\|_\omega \geq -c\sigma^{\frac{\Delta - d}{\Delta - d - 1} + \epsilon},$$

for all homogeneous ideals of dimension $d < \Delta - 1$ and size at most σ.

The result also applies effortlessly to give quantitative forms of thetheorems of Philippon's thesis, of [Wal7], and of the cases of [Bro5], [BT] which were the natural siblings of the case considered above.

2.3. Comparison with Philippon's Independence Criteria. This inequality is so simple to state and easy to use in comparison with the other available techniques that the reader will suspect some hidden draw-back.

Advantage: Simplicity and ease of use.

Drawback: Need for a comparatively large zero-free region. For efficient application we need a zero-free region which is about the reciprocal of the size of the polynomials arising from the generating functions. This requirement makes it useless in certain situations, such as the Chudnovsky-Stéphanois derivation of coefficients, cf. [**Dia2**], to obtain non-zero values. There the smallness of these values comes from continuity. So in that situation, there is no way to obtain that the radius of the zero-free region should be about the $D - \nu$th root of the absolute values. Nor can it be applied to the obtain Nesterenko's results on the values of modular functions. I know of no reason why this drawback should persist in principle. That is, it may only be an artifact of our present technical clumsiness.

Marginal Advantage: It can be useful when the Technical Hypothesis does not hold sufficiently uniformly. See [**Bro5**],[**BT**],[**Del**] for more details.

With the Liouville-Lojasiewicz inequality, I imagined the link with the Nullstellensatz to have been broken. I was mistaken.

3. The Łojasiewicz Inequality Implies the Nullstellensatz

The bridge was the following result [**Bro4**], where there are no common (finite) zeros, and we forget the possible arithmetic information.

THEOREM 3.1. *If the* $P_1, \ldots, P_m$ *have no common zeros, then*

$$\max \frac{|P_i(\omega)|}{\|\omega\|^{\deg P_i}} \geq C \frac{1}{\|\omega\|^{nD_1 \ldots D_\mu}},$$

where $\mu = \min\{m, n\}$ *and* $C > 0$ *depends solely on* $P_1, \ldots, P_m$.

The significance of this inequality was explained to me by Carlos Berenstein and Alain Yger. (In the meantime I have removed the factor n on the right hand side.) See also Theorem 4.5 below. We use the following result [**Sko**][3]:

THEOREM 3.2. (Skoda) *If*

$$\int_{\mathbf{C}^n} \frac{d\lambda}{|Q|^{2(1+\epsilon)q+2}\|\mathbf{x}\|^{2K}} =: I < \infty,$$

for $|Q| := \sum |Q_i|$, *then there exist holomorphic* $A_1, \ldots, A_m$ *in* (131) *such that*

$$\int_{\mathbf{C}^n} \frac{|A|^2 d\lambda}{|Q|^{2(1+\epsilon)q}\|\mathbf{x}\|^{2K}} < \frac{1+\epsilon}{\epsilon} I < \infty.$$

The convergence of the latter integral bounds the growth of the A_i to give [**Bro4**]:

THEOREM 3.3. $\delta(D, n) \leq \mu n D^n + \mu D$.

Comparing this with the example above shows that, except for the factor μn (which, by the remark above can be reduced to μ unconditionally) and the summand μD, this bound is sharp. At that time, it was established only in characteristic 0, but the recent extension of the fundamental inequality to arbitrary fields and recent work of Berenstein and Yger carry over this approach to arbitrary fields.

---

[3][**Sko**] H. Skoda. Applications des techniques L^2 à la théorie des idéaux d'algèbres de fonctions homomorphes avec poids, *Ann. Sci. École Norm. Sup. (4ème sér.)* 5, (1972), 545–579.

4. Geometric Version of the Nullstellensatz or Irrelevance of the Nullstellen Inequality for the Nullstellensatz

After some work by Caniglia, Galligo, and Heintz [**CGH**][4] in arbitrary characteristic, a sharp general Nullstellensatz in arbitrary characteristic was established by J. Kollár [**Kol1**][5]. The statement requires the use of the notion of the radical of an ideal: The *radical* of an ideal $\mathfrak{A}$ is defined as follows:

$$\sqrt{\mathfrak{A}} := \cap_{\substack{\mathfrak{P}\,\mathrm{prime} \\ \mathfrak{P} \supset \mathfrak{A}}} \mathfrak{P} = \{F \in R : F^e \in \mathfrak{A}, \text{some } e\}$$

THEOREM 4.1. (Kollár) *If $D \geq 3$ and the P_i are homogenous (in $n+1$ variables) with $\deg P_i \leq D$, then*

$$\sqrt{\mathfrak{A}}^{D^n} \subset \mathfrak{A}.$$

COROLLARY 4.2. *If $D \geq 3$, then*

1. $\rho \leq D^n$ and
2. $\deg A_i P_i \leq D^n$ in the Bézout form of the Nullstellensatz.

P. Philippon has given an excellent exposition [**PPh4**] of Theorem 4.1, using the homology of the Koszul complex rather than local cohomology. In [**Bro7**], I have sharpened the statement to contain the following result:

THEOREM 4.3. *If k is infinite, then there exist homogeneous prime ideals $\mathfrak{P}_1$, $\ldots, \mathfrak{P}_r \supset \mathfrak{A}$ and exponents $e_0, e_1, \ldots, e_r$ such that*

$$(134) \qquad \mathfrak{M}^{e_0} \mathfrak{P}_1^{e_1} \ldots \mathfrak{P}_r^{e_r} \subset \mathfrak{A},$$

$$(135) \qquad e_0 + \sum e_i \deg \mathfrak{P}_i \leq D^n,$$

where $\mathfrak{M} = (x_0, \ldots, x_n)$.

4.1. Outline of Method of Proof. Replace $P_1, \ldots, P_m$ by sufficiently general linear homogeneous combinations $h_1, \ldots, h_l$, $l = \min\{m, n\}$.

Define $I_0 := (0)$ and inductively, using the primary decomposition,

$$(I_{i-1}, h_i) = I_i \cap K_i \cap E_i,$$

where

$$I_i = \cap_{\substack{\dim \mathfrak{Q} = n-i \\ \mathfrak{P} \not\supset \mathfrak{A}}} \mathfrak{Q}, \quad K_i = \cap_{\substack{\dim \mathfrak{Q} = n-i \\ \mathfrak{P} \supset \mathfrak{A}}} \mathfrak{Q}, \quad E_i = \cap_{\dim \mathfrak{Q} < n-i} \mathfrak{Q},$$

with the $\mathfrak{Q}$ being $\mathfrak{P}$-primary ideals.

Bézout's theorem controls $\deg I_i$ and $\deg K_i$, but not $\deg E_i$:

$$\deg I_i + \deg K_i = \deg(I_i \cap K_i) = \deg I_{i-1} \cdot \deg h_i.$$

However, in a sense, Kollár finds a replacement for E_i whose degree can be controlled:

PROPOSITION 4.4. (Key Proposition) *If we let $N_i := K_1^{3^{i-1}} K_2^{3^{i-2}} \ldots K_i$, then*

$$N_{i-1}(I_i \cap K_i) \subset (I_{i-1}, h_i).$$

---

[4][**CGH**] L. Caniglia, A. Galligo, J. Heintz. Borne simple exponentielle pour le théorème des zéros sur un corps de caractéristique quelconque, *C.R. Acad. Sci. Paris, Sér. I* 307-(6), (1988), 255–258.

[5][**Kol1**] J. Kollár. Sharp effective Nullstellensatz. *J. Amer. Math. Soc.* 1, (1988), 963–975.

Exercise: Show that

$$K_1^{\frac{3^{l-1}+1}{2}} K_2^{\frac{3^{l-2}+1}{2}} \ldots K_l \cdot I_l \subset (h_1, \ldots, h_l).$$

This containment gives Theorem 4.3. Note that Theorem 4.3 implies Kollár's theorem 4.1, since

$$\sqrt{\mathfrak{A}}^{\sum e_i} \subset \mathfrak{M}^{e_0} \prod \mathfrak{P}_i^{e_i} \subset \mathfrak{A},$$

as $\mathfrak{M}$ and each $\mathfrak{P}_i$ contains $\sqrt{\mathfrak{A}}$.

4.2. The Nullstellensatz Implies the Łojasiewicz Inequality.

Theorem 4.3 plays a key role in the following generalization [**JKS**][6] of some results of [**Bro4**], [**Bro6**]:

THEOREM 4.5. (Ji, Kollár, Shiffman) *If* $\text{dist}(\omega, V)$ *denotes the distance from the point* $\omega \in \mathbf{C}^n$ *and the common zeros* V *of* $P_1, \ldots, P_m$ *and* $D \geq 3$, *then*

$$(136) \qquad \text{dist}(\omega, V)^j \leq C \max\{\frac{|P_i(\omega)|}{(1 + \|\omega\|)^{\deg P_i}}\}(1 + \|\omega\|)^{D^\mu},$$

$\mu = \min\{m, n\}$.

In other words, the earlier implication in the proof of Theorem 3.3 has been turned around! (Still, going back to the original proofs, as in [**Bro6**], the condition $D \geq 3$ can be removed; when $m < n$, this introduces a factor of n into the rightmost exponent of (136). Thus it is not without interest that the continuation of the original approach to the Bézout form via inequalities and now algebraic versions of Grothendieck residues as mentioned in the next section affords some effective forms of the Nullstellensatz that are not contained in any others.)

5. Arithmetic Aspects of the Bézout Version

The developments touched upon in this section have been astounding. The roots go back to the approach that Berenstein and Yger suggested to me for [**Bro6**].

The underlying surprise is that Berenstein and Yger developed techniques from the analytic theory of Grothendieck residues to establish good bounds, not only on the size of the degrees of the polynomials A_i produced by these analytic methods, but also on the coefficients, say, when working over $\mathbb{Q}$. They did this first when the zero set at infinity is finite in their paper [**BY1**][7] and then more generally in [**BGVY**][8].

A major role was played by fundamental work [**PPh6**] of Philippon, who devised a method for controlling the sizes of denominators when working over $\mathbb{Z}$ and much more general rings. He did this through several important innovations:

- An axiomatic theory of a (semi-)regular, factorial ring with size.
- Precise Chow form inequalities in this very general setting.
- Utilization of the valuative criterion for an element to be integral over an ideal.

---

[6][**JKS**] S. Ji, J. Kollár, B. Shiffman. A global Łojasiewicz inequality for algebraic varieties. *Trans. Amer. Math. Soc.* 329, (1992), 813–818.

[7][**BY1**] C.A. Berenstein, A. Yger. Effective Bézout identities in $\mathbb{Q}[X_1, \ldots, X_n]$, *Acta Math.* 166, (1991), 69–120.

[8][**BGVY**] C.A. Berenstein, R. Gay, A. Vidras, A. Yger. *Residue Currents and Bézout Identities*, Progr. Math. 114, Birkhäuser, Basel-Boston, (1993).

- Application of the result of Lipman and Teissier, which is an algebraic analogue of a result by Briançon and Skoda.

Philippon uses the result of Lipman and Teissier to show the following result
(Theorem 4 of [**PPh6**]):

THEOREM 5.1. (Philippon) *If the x_0 vanishes on all the common zeros of
the homogenizations $P_i^h(\mathbf{x}) \in \mathbf{A}[\mathbf{x}]$, where $\mathbf{A}$ is a regular factorial ring equipped
with a size t, then there is a non-zero element $\alpha \in \mathbf{A}$ with size $t(\alpha) \leq C(D^\nu +
D^{\nu-1}\max\{t(P_i)\})$ such that*

$$(\alpha x_0)^{D^\mu(n+1+\kappa)} \in (P_1^h, \ldots, P_m^h),$$

where κ is the Krull dimension of $\mathbf{A}$ and $\nu = \min\{m, n+1\}$.

See also Theorem 5 of [**PPh6**].

Using formal Grothendieck residues developed by Lipman and Teissier in an
impressive manner analogous to what they had done earlier with the analytic
Grothendieck residues, Berenstein and Yger [**BY2**][9]obtain:

THEOREM 5.2. (Berenstein and Yger) *Let $P_1(\mathbf{x}), \ldots, P_m(\mathbf{x}) \in \mathbf{A}[x_1, \ldots, x_n]$
have no common zeros, where $\mathbf{A}$ is a factorial regular ring with a size t and
the $\deg P_i \leq D$. Then there are $A_1(\mathbf{x}), \ldots, A_m(\mathbf{x}) \in \mathbf{A}[x_1, \ldots, x_n]$ and non-zero
$A_0 \in \mathbf{A}$ such that*

$$\deg A_i P_i \leq CD^n, \quad \max\{t(A_i)\} \leq C_{\text{eff}} D^{4n+2} \max\{t(P_i) + D^n\},$$

and

$$A_1(\mathbf{x})P_1(\mathbf{x}) + \cdots + A_m(\mathbf{x})P_m(\mathbf{x}) = A_0.$$

6. Some Algorithmic Aspects of the Bézout Version

Many interesting aspects of the Nullstellensatz have been illuminated by considerations arising from theoretical computer science. I will mention only a few of
them in this section and encourage you to consult the literature for many more. The
basic question of complexity is to ask how many arithmetic operations are involved
in computing an algebraic expression, starting from something given. In one model,
an *arithmetic operation* involves addition, substraction, or multiplication of what
has already been calculated (or of 1), and (in positive characteristic p) extraction
of pth roots.

If no branching is involved, then the calculation is said to be a *straight-line
program*; the number of arithmetic operations is said to be its *length*. Seen in this
way, 2^{2^k} is only of length k (actually $k+1$ if we had followed the more usual
definition of counting what is given as the first thing calculated). In particular, a
straight-line program for a polynomial $P(x) \in \mathbb{Z}[x]$ is simply a sequence

$$1(= u_0), x(= u_1), u_2, u_3, \ldots, u_k = P(x),$$

where each $u_i = u_r \circ u_s$, $r, s < i$, and $\circ$ is an addition, substraction, or multiplication.
Let $\lambda(P)$ denote the minimal length of such a program.

If we allow parallel arithmetic operations, then we will have a so-called *arithmetic network*. In this situation, the total number of operations will be called the
size and the longest sequence of operations the *depth*, notions which coincide in the
straight-line case.

---

[9][**BY2**] C.A. Berenstein, A. Yger. Residue calculus and effective Nullstellensatz, *Amer. J.
Math.* 121, (1999), 723–796.

6.1. Nullstellensatz and Straight-Line Programs. As a particularly nice exposition of this point of view with further references, we mention [**FGS**][10], where the following result is proven in Section 5.2.

THEOREM 6.1. (Fitchas, Giusti, Smietanski) *Let $D \geq n$.*

- *There is an arithmetic network over k of size $m^{O(1)} D^{O(n)}$, i.e. polynomial in the size $O(mD^n)$ of the input, deciding whether the ideal generated, $\mathfrak{A}$, is trivial.*

- *If the ideal is trivial, the arithmetic network produces a straight-line program of length $m^{O(1)} D^{O(n)}$ producing coefficients A_i for the Bézout identity with $\deg A_i \leq D^{O(1)}$.*

- *Furthermore, this network and the straight-line program can be parallelized to have depth $O(n^2 \log^2 mD)$.*

6.2. Complexity Conjectures. We note that some fascinating complexity conjectures related to the Nullstellensatz appear in Chapter 6 of [**BCSS**][11]. The relevance of the first one to the Nullstellensatz is not immediately apparent.

CONJECTURE. (Blum, Cucker, Shub, Smale) *There is a universal constant c such that for all $P(x) \in \mathbb{Z}[x]$, the number of integral zeros of P is at most $\lambda(P)^c$.*

However Blum, Cucker, Shub, and Smale show that this intriguing conjecture would imply the following:

CONJECTURE. (Blum, Cucker, Shub, Smale) *The Hilbert Nullstellensatz is intractible. In other words, given polynomials P_i with N (scalar) coefficients and no common zeros, there is no program to find the coefficients A_i in the Bézout form of the Nullstellensatz in at most N^c steps.*

They show that, if this conjecture is true, then $N \neq NP$.

6.3. Diophantine Questions. It is a very interesting challenge for those of us in Diophantine Approximations to prove quantitative approximation results using the measure of complexity of polynomials introduced in this section.

For example, it is easy to establish that for $n \geq 4$,

$$1 + \log_2 \log_2 n \leq \tau(n) \leq 2 \log_2 n.$$

¿From the irrationality measures for π, we know that, if p, q are positive integers with $\tau(p), \tau(q) \leq \tau$, then

$$-c2^\tau < \log_2 |q\pi - p|.$$

It would be interesting to know whether one could do better. For example, can one approach the bound given by the Box Principle?

Perhaps this is not even the appropriate question in this setting. I mention it in hopes that someone might actually formulate the correct questions and begin the investigation of Diophantine approximations from this new and appealing point of view.

---

[10][FGS] N. Fitchas, M. Giusti, F. Smietanski. Sur la complexité du théorème des zéros, (with the collaboration of J. Heintz, L.M. Pardo, J. Sabia, and P. Salerno.), *Approx. Optim.* 8, (1995), 274–329.

[11][BCSS] L. Blum, F. Cucker, M. Shub, S. Smale. *Complexity and Real Computation*, Springer-Verlag, New York, (1998).

Bibliography

[Ably] M. Ably. Résultats quantitatifs d'indépendance algébrique pour les groupes algébriques, *J. Number Theory* 42, (1992), 194–231.

[Amo1] F. Amoroso. On the distribution of complex numbers according to their transcendence types, *Ann. Mat. Pura Appl.* 161, (1988), 359–368.

[Amo2] F. Amoroso. Values of polynomials with integer coefficients and distance to their common zeros, *Acta Arith.* 68/2, (1994), 101–112.

[And1] Y. André. *G*-fonctions et transcendance. *J. reine angew. Math.* 476, (1996), 95–125.

[And2] Y. André. Quelques conjectures de transcendance issues de la géométrie algébrique. *Manuscript,* (1997).

[Bar1] K. Barré. Mesures de transcendance pour l'invariant modulaire, *C. R. Acad. Sci. Paris, Ser.1* 323, (1996), 447–452.

[Bar2] K. Barré. Mesure d'approximation simultanée de q et $J(q)$, *J. Number Theory* 66, (1997), 102–128.

[Bar3] K. Barré. Fonctions modulaires et transcendance, *Thèse* Université de St-Étienne, (17/6/1997).

[BDGP] K. Barré, G. Diaz, F. Gramain, G. Philibert. Une preuve de la conjecture de Mahler-Manin, *Invent. Math.* 124, (1996), 1–9.

[Ber1] D. Bertrand. Séries d'Eisenstein et transcendance, *Bull. Soc. Math. France* 104, (1976), 309–321.

[Ber2] D. Bertrand. Fonctions modulaires, courbes de Tate et indépendance algébrique, *Sém. Delange-Pisot-Poitou (Théorie des Nombres),* 19ème Année (1977-78), n° 36.

[Ber3] D. Bertrand. Fonctions modulaires et indépendance algébrique, *Astérisque* 61, (1979), 29–54.

[Ber4] D. Bertrand. Valeurs de fonctions theta et hauteurs p-adiques. *Prog.Math.* 2, Birkhäuser, (1982), 1–11.

[Ber5] D. Bertrand. Lemmes de zéros et nombres transcendants, *Sém. Bourbaki 1985/86,* Astérisque 145–146, (1987), 21–44.

[Ber6] D. Bertrand. Minimal heights and polarizations on group varieties. *Duke Math. J.* 80, (1995), 223–250.

[Ber7] D. Bertrand. Theta functions and transcendence, *The Ramanujan J.* 1, (1997), 339–350.

[Bro1] W. D. Brownawell. Zero estimates for solutions of differential equations, in *Approximations diophantiennes et nombres transcendants, Luminy, 1980,* eds. D.Bertrand, M.Waldschmidt, Progr. Math. 31, (1983), Birkhäuser.

[Bro2] W.D. Brownawell. On the development of Gel'fond's method. *Proc. Southern Illinois Conf., Southern Illinois Univ., Carbondale, 1979,* Lecture Notes in Math. 751, Springer, Berlin, (1979), 18–44.

[Bro3] W. D. Brownawell. Note on a paper of P. Philippon, *Michigan Math. J.* 34, (1987), 461–464.

[Bro4] W.D. Brownawell. Bounds for the degrees in the Nullstellensatz, *Ann. of Math.* 126, (1987), 577–591.

[Bro5] W.D. Brownawell. Large transcendence degree revisited. I, in *Diophantine Approximation and Transcendence Theory (Bonn 1985),* ed. G. Wüstholz, Lecture Notes in Math. 1290, Springer, (1987), 149–173.

[Bro6] W.D. Brownawell. Local diophantine Nullstellen inequalities, *J. Amer. Math. Soc.* 1, (1988), 311–322.

[Bro7] W.D. Brownawell. A pure power version of the Nullstellensatz, *Mich. Math. J.,* 1998, (to appear).

[BM1]   W. D. Brownawell, D. W. Masser. Multiplicity estimates for analytic functions I, *J. Reine Angew. Math.* 314, (1980), 200–216.

[BM2]   W. D. Brownawell, D. W. Masser, Multiplicity estimates for analytic functions II, *Duke Math. J.* 47, (1980), 273–295.

[BT]   W.D. Brownawell, R. Tubbs. Large transcendence degree revisited. II, in *Diophantine Approximation and Transcendence Theory (Bonn 1985)*, ed. G. Wüstholz, Lecture Notes in Math. 1290, Springer, (1987), 175–188.

[Cav]   D. Caveny. Commutative algebraic groups and refinements of the Gel'fond-Fel'dman measure. *Sympos. Diophantine Problems (Boulder, 1994)*, Rocky Mountain J. Math. 26/3, (1996), 889–935.

[Chu1]   G.V. Chudnovsky. Algebraic independence of values of exponential and elliptic functions. *Proc ICM Helsinki*, vol 1, (1978), 339–350; See also: Transcendence methods and theta functions, *Proc. Sympos. Pure Math.*, 49, vol 2, Amer. Math. Soc., 167–232.

[Chu2]   G.V. Chudnovsky. *Contributions to the theory of transcendental numbers*, Math. Surveys and Monographs, n° 19, Amer. Math. Soc, (1984).

[Coh]   P. Cohen. On the coefficients of the transformation polynomials for the elliptic modular function, *Math. Proc. Cambridge Phil. Soc.* 95, (1984), 389–402.

[Del]   S. Delaunay. Grands degrés de transcendance pour la fonction exponentielle; Modification des hypothèses techniques dans la méthode de Brownawell, *Ann. Fac. Sci. Toulouse Math.* 5-(6), (1996), 69–104.

[Den]   L. Denis. Lemmes de multiplicités et intersection, *Comment. Math. Helv.* 70, (1995), 235–247.

[Dia1]   G. Diaz. Généralisation d'un résultat de W.D. Brownawell - M. Waldschmidt, *Bull. Soc. Math. France* 117/3, (1989), 267–283.

[Dia2]   G. Diaz. Grands degrés de transcendance pour des familles d'exponentielles, *J. Number Theory* 31, (1989), 1–23.

[Dia3]   G. Diaz. La conjecture des quatre exponentielles et les conjectures de D. Bertrand sur la fonction modulaire, *J. Théorie Nombres Bordeaux* 9, (1997), 229–245.

[DNNS1]   D. Duverney, Ke. Nishioka, Ku. Nishioka, I. Shiokawa. Transcendence of Jacobi's theta series, *Proc. Japan. Acad. Sci, Ser A* 72, (1996), 202–203.

[DNNS2]   D. Duverney, Ke. Nishioka, Ku. Nishioka, I. Shiokawa. Transcendence of Rogers-Ramanujan continued fraction and reciprocal sums of Fibonacci numbers, *Proc. Japan Acad. Sci., Ser. A* 73-Lo. 7, (1997).

[FN]   N.I. Fel'dman, Y.V. Nesterenko. *Number Theory IV. Transcendental Numbers*, Encyclopaedia Math. Sci. 44, Springer, Berlin, (1998).

[Gel1]   A.O. Gel'fond. On the algebraic independence of algebraic powers of algebraic numbers, *Dokl. Akad. Nauk SSSR (N.S.)* 64, (1949), 277–280, (Russian).

[Gel2]   A.O. Gel'fond. On the algebraic independence of transcendental numbers of certain classes, *Uspehi Matem. Nauk (N.S.)* 4, no. 5 (33), (1949), 14–48, (Russian).

[Gel3]   A.O. Gel'fond. *Transcendentnye i algebraičeskie čisla*. (Russian) Gosudarstv. Izdat. Tehn.-Teor. Lit., Moscow, (1952), 224 pp. *Transcendental and algebraic numbers*. Translated from the first Russian edition by Leo F. Boron. Dover Publications, Inc., New York, (1960).

[Jab]   E.M. Jabbouri. Sur un critère pour l'indépendance algébrique de P. Philippon. *Approximation diophantienne et nombres transcendants (Luminy, 1990)*, W. de Gruyter, Berlin, (1992), 195–202.

[Jad]   C. Jadot. Critères pour l'indépendance algébrique et linéaire. *Thèse*. Université Paris VI. 1996.

[Lau1]   M. Laurent. Hauteur de matrices d'interpolation. In: *Approximations diophantiennes et nombres transcendants, Luminy 1990* (P. Philippon ed.), W. de Gruyter, (1992), 215–238.

[Lau2]   M. Laurent. Linear forms in two logarithms and interpolation determinants, *Acta Arith.* 64/2, (1994), 181–199.

[LR1]   M. Laurent, D. Roy. Sur l'approximation algébrique en degré de transcendance un, *Ann. Inst. Fourier (Grenoble)* 49, (1999), 27–55.

[LR2]   M. Laurent, D. Roy. Criteria for algebraic independence with multiplicities and approximation by hypersurfaces, (submitted).

[Lio]   J. Liouville. Sur des classes très étendues de quantités dont la valeur n'est ni algébrique, ni même réductible à des irrationelles algébriques, *J. Math. Pures Appl.* 16, (1851), 133–142.

[Mah1]   K. Mahler. On algebraic differential equations satisfied by automorphic functions, *J.Austral. Math. Soc.* 10, (1969), 445–450.

[Mah2]   K. Mahler. Remarks on a paper by W.Schwarz, *J. Number Theory* 1, (1969), 512–521.

[Mah3]   K. Mahler. On the coefficients of tranformation polynomials for the modular function, *Bull. Austral. Math. Soc.* 10, (1974), 197–218.

[Mas1]   D.W. Masser. *Elliptic functions and transcendence*, Lecture Notes in Math. 437, Springer, Berlin, (1975).

[Mas2]   D. W. Masser. On polynomials and exponential polynomials in several variables, *Invent. Math.* 63, (1981), 81–95.

[MW1]   D. W. Masser, G. Wüstholz. Zero Estimates on Group Varieties I, *Invent. Math.* 64, (1981), 489–516.

[MW2]   D.W. Masser, G. Wüstholz. Fields of large transcendence degree generated by values of elliptic functions, *Invent. Math.* 72, (1983), 407–464.

[MW3]   D. W. Masser, G. Wüstholz. Zero Estimates on Group Varieties II, *Invent. Math.* 80, (1985), 233–267.

[Mor]   J.-C. Moreau. Démonstrations géométriques de lemmes de zéros, II, in: *Approximation diophantienne et nombres transcendants, Luminy 1982*, Progr. Math. 31, Birkhäuser, (1983), 191–197.

[Nag]   N.D. Nagaev. Division of independent variable and transcendence, *Math. Notes* 57, (1995), 292–299.

[Nak]   M. Nakamaye. Multiplicity estimates and the product theorem, *Bull. Soc. Math. France* 123, (1995), 155-188.

[Nes1]   Y.V. Nesterenko. On algebraic independence of the components of solutions of a system of linear differential equations, *Izv. Akad. Nauk SSSR, Ser. Mat.* 38/3, (1974), 492–512 (Russian).

[Nes2]   Y.V. Nesterenko. Estimates of the orders of functions of a certain class and their applications in the theory of transcendental numbers, *Izv. Akad. Nauk SSSR Ser. Mat.* 41/2, (1977), 253–284; Translated in *Math. USSR Izv.* 11, (1977), 239–270.

[Nes3]   Y.V. Nesterenko. Bounds for the characteristic function of a prime ideal, *Math. Sb.*, 123(165)(1), (1984), 11–34; Translation in *Math. USSR Sb.* 51, (1985), 9-32.

[Nes4]   Y.V. Nesterenko. On the algebraic independence of algebraic powers of numbers, *Math. Sb.* 123(165)(4), (1984), 435–459; Translation in *Math. USSR Sb.* 51, (1985), 429–454.

[Nes5]   Y.V. Nesterenko. Estimates for the number of zeros of certain functions, in *New advances in transcendence theory*, ed. A.Baker, Cambridge Univ. Press, (1988), 263–269.

[Nes6]   Y.V. Nesterenko. Estimates for the number of zeros of functions of certain classes, *Acta Arith.* 53/1, (1989), 29–46 (in Russian).

[Nes7]   Y.V. Nesterenko. Transcendence degree of some fields generated by values of the exponential function. *Mat. Zametki* 46/3, (1989), 40–49, 127; English translation in *Math. Notes* 46/3, (1989), 706–712.

[Nes8]   Y.V. Nesterenko. Modular functions and transcendence problems, *C. R. Acad. Sci. Paris, Sér.1* 322, (1996), 909–914.

[Nes9]   Y.V. Nesterenko. Modular functions and transcendence questions, *Mat. Sb.*, 187/9, (1996), 65–96 (Russian); English translation in *Sb. Math.* 187/9, 1319–1348.

[Nes10]   Y.V. Nesterenko. On the measure of algebraic independence of the values of Ramanujan functions, *Trudy Matematicheskogo Instituta imeny V.A.Steklova* 218, (1997), 299–334 (Russian); English translation in *Proc. Steklov Inst. Math.* 218, (1997), 294–331.

[Nis1]   Ke. Nishioka, A conjecture of Mahler on automorphic functions. *Arch. Math.* 53, (1989), 46–51.

[Nis2]   Ku. Nishioka. *Mahler functions and transcendence*, Lecture Notes in Math. 1631, Springer, (1996).

[Phi1]   G. Philibert. Une mesure d'indépendance algébrique, *Ann. Inst. Fourier (Grenoble)* 38/3, (1988), 85–103.

[Phi2]   G. Philibert. Un lemme de zéros modulaire, *J. Number Theory* 66, (1997), 306–313.

[PPh1]   P. Philippon. Indépendance algébrique de valeurs de fonctions exponentielles *p*-adiques, *J. Reine Angew. Math.* 329, (1981), 42–51.

[PPh2]   P. Philippon. Critères pour l'indépendance algébrique, *Inst. Hautes Études Sci. Publ. Math.* 64, (1986), 5–52.

[PPh3]   P. Philippon. Lemmes de zéros dans les groupes algébriques commutatifs, *Bull. Soc. Math. France* 114, (1986), 355–383; Errata et addenda, *ibidem* 115, (1987), 397–398.

[PPh4]   P. Philippon. Théorème des zéros effectif, d'après J. Kollár, *Publ. Univ. P. & M. Curie-Paris VI* 88, (1988), Exposé 6.

[PPh5]   P. Philippon. Théorème des zéros effectif et élimination, *Sém. Théorie Nombres Bordeaux* 1, (1989), 137–155.

[PPh6]   P. Philippon. Dénominateurs dans le théorème des zéros, *Acta Arith.* 88, (1991), 1–25.

[PPh7]   P. Philippon. Sur des hauteurs alternatives I, *Math. Ann.* 289, (1991), 255–283; — II, *Ann. Inst. Fourier* 44, (1994), 1043–1065; — III, *J. Math. Pures Appl.* 74, (1995), 345–365.

[PPh8]   P. Philippon. Deux remarques sur la géométrie des nombres. In : *Problèmes diophantiens 92–93*. See also : Approximation algébrique dans les espaces projectifs I, *J. Number Theory*, to appear.

[PPh9]   P. Philippon. Nouveaux lemmes de zéros dans les groupes algébriques commutatifs, *Rocky Mountain J. Math.* 26, (1996), 1069–1088.

[PPh10]   P. Philippon. Une approche méthodique pour la transcendance et l'indépendance algébrique de valeurs de fonctions analytiques, *J. Number Theory* 64, (1997), 291–338.

[PPh11]   P. Philippon. Indépendance algébrique et K-fonctions, *J. Reine Angew Math.* 497, (1998), 1–15.

[Rem]   G. Rémond. Sur des problèmes d'effectivité en géométrie diophantienne, *Thèse* Université Paris VI, (1997).

[Rey1]   E. Reyssat. Un critère d'indépendance algébrique, *J. Reine Angew. Math.* 329, (1981), 66–81.

[Rey2]   E. Reyssat. Propriétés d'indépendance algébrique de nombres liés aux fonctions de Weierstrass, *Acta Arith.* 41, (1982) , 291–310.

[RW]   D. Roy, M. Waldschmidt. Approximation diophantienne et indépendance algébrique de logarithmes, *Ann. Sci. École Norm. Sup.* 30, (1997) , 753–796.

[Sch1]   T. Schneider. Transzendenzunterzuchungen periodischer Funktionen 1, *J. Reine Angew. Math.* 172, (1934), 65–69.

[Sch2]   T. Schneider. Arithmetische Untersuchungen elliptischer Integrale, *Math. Ann.* 113, (1937), 1–13.

[Shi]   Shidlovskii,A., *Transcendental numbers*, Walter de Gruyter, Berlin, (1989), (Translated from Russian).

[Sie]   C.L. Siegel. Über einige Anwendungen diophantischer Approximationen, *Abh. Preuss. Akad. der Wissensch., Phys.-math. Kl., Jahrg.* 1, (1929), 1–70.

[Sou]   C. Soulé. Géométrie d'Arakelov et théorie des nombres transcendants, in *Journées arithmétiques de Luminy, 17–21 juillet 1989*, Astérisque 198–200, (1991), 355–371.

[Tij1]   R. Tijdeman. On the number of zeros of general exponential polynomials, *Indag. Math.* 33, (1971), 1–7.

[Tij2]   R. Tijdeman. On the algebraic independence of certain numbers, *Nederl. Akad. Wetensch. Proc. Ser. A* 74 = *Indag. Math.* 33, (1971), 146–162.

[Tub1]   R. Tubbs. Algebraic groups and small transcendence degree I, *J. Number Theory* 25, (1987), 279–307.

[Tub2]   R. Tubbs. Algebraic groups and small transcendence degree II, *J. Number Theory* 35, (1990), 109–127.

[Wal1]   M. Waldschmidt, *Nombres transcendants*, Springer, (1974).

[Wal2]   M. Waldschmidt. Les travaux de G.V.Chudnovskii sur les nombres transcendants, *Lecture Notes in Math.* 567, Springer, Berlin, (1977), 274–292.

[Wal3]   M. Waldschmidt. *Nombres transcendants et groupes algébriques*, Astérisque 69–70, (1979).

[Wal4]   M. Waldschmidt. Transcendance et exponentielles en plusieurs variables, *Invent. Math.* 63/1, (1981), 97–127.

[Wal5]   M. Waldschmidt. Algebraic independence of transcendental numbers. Gel'fond's method and its developments, in *Perspectives in Math.*, Birkhäuser, Basel-Boston, (1984), 551–571.

[Wal6]   M. Waldschmidt. Petits degrés de transcendance par la méthode de Schneider en une variable, *R. C. Math. Rep. Acad. Sci. Canada* 7, (1985), 143–148.

[Wal7]   M. Waldschmidt. Groupes algébriques et grands degrés de transcendance, *Acta Math.*, 156, (1986), 253–302.

[Wal8]   M. Waldschmidt. Linear independence of logarithms of algebraic numbers, *Report* 116, The Inst. Math. Sci. (IMSc.), (1992).

[Wal9]   M. Waldschmidt. Transcendance et indépendance algébrique de valeurs de fonctions modulaires, in *CNTA 5 - Carleton, Août 1996*, eds R. Gupta, K. Williams, CRM Proc. Lecture Notes, Amer. Math.Soc. 19, (1999), 353–375.

[Wal10]   M. Waldschmidt. Sur la nature arithmétique des valeurs de fonctions modulaires, *Séminaire Bourbaki, 1996-97*, n° 824.

[Wal11]   M. Waldschmidt. Algebraic independence of transcendental numbers: a survey, in *Monograph on Number Theory*, eds R.P. Bambah, V.C. Dumir, R.J. Hans Gill, Indian National Sci. Acad., Hindustan Book Agency, New-Delhi, (1999), 497–527.

[Wus1]   G. Wüstholz. Recent progress in transcendence theory, in *Number Theory, Noordwijkerhout 1983*, ed. H. Jager, Lecture Notes in Math. 1068, Springer, (1984), 280–296.

[Wus2]   G. Wüstholz. Multiplicity estimates on group varieties, *Ann. of Math.* 129, (1989), 471–500.

Index

Printing: Weihert-Druck GmbH, Darmstadt
Binding: Buchbinderei Schäffer, Grünstadt

Vol. 1706: S. Yu. Pilyugin, Shadowing in Dynamical Systems. XVII, 271 pages. 1999.

Vol. 1707: R. Pytlak, Numerical Methods for Optimal Control Problems with State Constraints. XV, 215 pages. 1999.

Vol. 1708: K. Zuo, Representations of Fundamental Groups of Algebraic Varieties. VII, 139 pages. 1999.

Vol. 1709: J. Azéma, M. Émery, M. Ledoux, M. Yor (Eds), Séminaire de Probabilités XXXIII. VIII, 418 pages. 1999.

Vol. 1710: M. Koecher, The Minnesota Notes on Jordan Algebras and Their Applications. IX, 173 pages. 1999.

Vol. 1711: W. Ricker, Operator Algebras Generated by Commuting Projections: A Vector Measure Approach. XVII, 159 pages. 1999.

Vol. 1712: N. Schwartz, J. J. Madden, Semi-algebraic Function Rings and Reflectors of Partially Ordered Rings. XI, 279 pages. 1999.

Vol. 1713: F. Bethuel, G. Huisken, S. Müller, K. Steffen, Calculus of Variations and Geometric Evolution Problems. Cetraro, 1996. Editors: S. Hildebrandt, M. Struwe. VII, 293 pages. 1999.

Vol. 1714: O. Diekmann, R. Durrett, K. P. Hadeler, P. K. Maini, H. L. Smith, Mathematics Inspired by Biology. Martina Franca, 1997. Editors: V. Capasso, O. Diekmann. VII, 268 pages. 1999.

Vol. 1715: N. V. Krylov, M. Röckner, J. Zabczyk, Stochastic PDE's and Kolmogorov Equations in Infinite Dimensions. Cetraro, 1998. Editor: G. Da Prato. VIII, 239 pages. 1999.

Vol. 1716: J. Coates, R. Greenberg, K. A. Ribet, K. Rubin, Arithmetic Theory of Elliptic Curves. Cetraro, 1997. Editor: C. Viola. VIII, 260 pages. 1999.

Vol. 1717: J. Bertoin, F. Martinelli, Y. Peres, Lectures on Probability Theory and Statistics. Saint-Flour, 1997. Editor: P. Bernard. IX, 291 pages. 1999.

Vol. 1718: A. Eberle, Uniqueness and Non-Uniqueness of Semigroups Generated by Singular Diffusion Operators. VIII, 262 pages. 1999.

Vol. 1719: K. R. Meyer, Periodic Solutions of the N-Body Problem. IX, 144 pages. 1999.

Vol. 1720: D. Elworthy, Y. Le Jan, X-M. Li, On the Geometry of Diffusion Operators and Stochastic Flows. IV, 118 pages. 1999.

Vol. 1721: A. Iarrobino, V. Kanev, Power Sums, Gorenstein Algebras, and Determinantal Loci. XXVII, 345 pages. 1999.

Vol. 1722: R. McCutcheon, Elemental Methods in Ergodic Ramsey Theory. VI, 160 pages. 1999.

Vol. 1723: J. P. Croisille, C. Lebeau, Diffraction by an Immersed Elastic Wedge. VI, 134 pages. 1999.

Vol. 1724: V. N. Kolokoltsov, Semiclassical Analysis for Diffusions and Stochastic Processes. VIII, 347 pages. 2000.

Vol. 1725: D. A. Wolf-Gladrow, Lattice-Gas Cellular Automata and Lattice Boltzmann Models. IX, 308 pages. 2000.

Vol. 1726: V. Marić, Regular Variation and Differential Equations. X, 127 pages. 2000.

Vol. 1727: P. Kravanja, M. Van Barel, Computing the Zeros of Analytic Functions. VII, 111 pages. 2000.

Vol. 1728: K. Gatermann, Computer Algebra Methods for Equivariant Dynamical Systems. XV, 153 pages. 2000.

Vol. 1729: J. Azéma, M. Émery, M. Ledoux, M. Yor, Séminaire de Probabilités XXXIV. VI, 431 pages. 2000.

Vol. 1730: S. Graf, H. Luschgy, Foundations of Quantization for Probability Distributions. X, 230 pages. 2000.

Vol. 1731: T. Hsu, Quilts: Central Extensions, Braid Actions, and Finite Groups,. XII, 185 pages. 2000.

Vol. 1732: K. Keller, Invariant Factors, Julia Equivalences and the (Abstract) Mandelbrot Set. X, 206 pages. 2000.

Vol. 1733: K. Ritter, Average-Case Analysis of NumericalProblems. IX, 254 pages. 2000.

Vol. 1734: M. Espedal, A. Fasano, A. Mikelić, Filtration in Porous Media and Industrial Applications. Cetraro 1998. Editor: A. Fasano. 2000.

Vol. 1735: D. Yafaev, Scattering Theory: Some Old and New Problems. XVI, 169 pages. 2000.

Vol. 1736: B. O. Turesson, Nonlinear Potential Theory and Weighted Sobolev Spaces. XIV, 173 pages. 2000.

Vol. 1737: S. Wakabayashi, Classical Microlocal Analysis in the Space of Hyperfunctions. VIII, 367 pages. 2000.

Vol. 1738: M. Emery, A. Nemirovski, D. Voiculescu, Lectures on Probability Theory and Statistics. XI, 356 pages. 2000.

Vol. 1739: R. Burkard, P. Deuflhard, A. Jameson, J.-L. Lions, G. Strang, Computational Mathematics Driven by Industrial Problems. Martina Franca, 1999. Editors: V. Capasso, H. Engl, J. Periaux. VII, 418 pages. 2000.

Vol. 1740: B. Kawohl, O. Pironneau, L. Tartar, J.-P. Zolesio, Optimal Shape Design. Tróia, Portugal 1999. Editors: A. Cellina, A. Ornelas. IX, 388 pages. 2000.

Vol. 1741: E. Lombardi, Oscillatory Integrals and Phenomena Beyond all Algebraic Orders. XV, 413 pages. 2000.

Vol. 1742: A. Unterberger, Quantization and Non-holomorphic Modular Forms. VIII, 253 pages. 2000.

Vol. 1743: L. Habermann, Riemannian Metrics of Constant Mass and Moduli Spaces of Conformal Structures. XII, 116 pages. 2000.

Vol. 1744: M. Kunze, Non-Smooth Dynamical Systems. X, 228 pages. 2000.

Vol. 1745: V. D. Milman, G. Schechtman, Geometric Aspects of Functional Analysis. VIII, 289 pages. 2000.

Vol. 1746: A. Degtyarev, I. Itenberg, V. Kharlamov, Real Enriques Surfaces. XVI, 259 pages. 2000.

Vol. 1747: L. W. Christensen, Gorenstein Dimensions. VIII, 204 pages. 2000.

Vol. 1748: M. Růžička, Electrorheological Fluids: Modeling and Mathematical Theory. XV, 176 pages. 2001.

Vol. 1749: M. Fuchs, G. Seregin, Variational Methods for Problems from Plasticity Theory and for Generalized Newtonian Fluids. VI, 269 pages. 2001.

Vol. 1750: B. Conrad, Grothendieck Duality and Base Change. X, 296 pages. 2001.

Vol. 1751: N. J. Cutland, Loeb Measures in Practice: Recent Advances. XI, 111 pages. 2001.

Vol. 1752: Y. V. Nesterenko, P. Philippon, Introduction to Algebraic Independence Theory. XIII, 256 pages. 2001.

Vol. 1753: A. I. Bobenko, U. Eitner, Painlevé Equations in the Differential Geometry of Surfaces. VI, 120 pages. 2001.

Vol. 1754: W. Bertram, The Geometry of Jordan and Lie Structures. XVI, 269 pages. 2001.